Intuitive Operational Amplifiers

McGraw-Hill Series in Intuitive IC Electronics

This series will help the reader gain an intuitive understanding of electronics and computers. Mathematics is kept to a minimum, as the reader gets "inside" the devices and circuits to grasp, from the electron level and up, the workings of integrated circuits, digital computers, operational amplifiers, and other electronics-related topics. The following volumes are planned for the series, and each one is written by Thomas M. Frederiksen, whose Intuitive IC Electronics (McGraw-Hill) has proved popular with engineers, managers, students, and hobbyists worldwide.

INTUITIVE DIGITAL COMPUTER BASICS: An Introduction to the Digital World (1988)

INTUITIVE ANALOG ELECTRONICS: From Electron to Op Amp (1988)

INTUITIVE OPERATIONAL AMPLIFIERS: From Basics to Useful Applications (1988)

INTUITIVE CMOS ELECTRONICS: The Revolution in VLSI, Processing, Packaging, and Design (1989)

INTUITIVE IC ELECTRONICS, Second Edition (1989)

Intuitive Operational Amplifiers

From Basics to Useful Applications

Revised Edition

Thomas M. Frederiksen

McGraw-Hill Book Company
New York St. Louis San Francisco Auckland
Bogotá Hamburg London Madrid Mexico
Milan Montreal New Delhi Panama
Paris São Paulo Singapore
Sydney Tokyo Toronto

Library of Congress Cataloging-in-Publication Data

Frederiksen, Thomas M.
Intuitive operational amplifiers.

(The McGraw-Hill series in intuitive IC electronics)
Bibliography: p.
Includes index.
1. Operational amplifiers. I. Title. II. Series.
TK7871.58.06F74 1988 621.3815'35 88-6864
ISBN 0-07-021966-4
ISBN 0-07-021967-2 (pbk.)

1234567890 DOC/DOC 8921098

ISBN 0-07-021966-4

ISBN 0-07-021967-2 {PBK.}

The editors for this book were Daniel A. Gonneau and Nancy Young and the production supervisor was Suzanne W. Babeuf. It was set in Melior by University Graphics, Inc.

Printed and bound by R. R. Donnelley & Sons Company.

This book is dedicated to the many system design engineers who, during a nationwide linear seminar in 1980, encouraged the author to write a book that used an intuitive approach to op amps.

Contents

Foreword to the First Edition

When I was a kid, I wasn't very knowledgeable about electronics, so I tried to learn something from the correspondence course my neighbor got in the mail.

Chapter 1 was easy: "The Diode."

Chapter 2 was a bit more challenging: "The Triode."

Chapter 3 was not bad: "The Tetrode."

Chapter 4 was okay: "The Pentode."

But I had problems with Chapter 5: "The 4-Band 5-Tube Superheterodyne AM Receiver." It took me several years to realize that my failure to leap the gap from Chapter 4 to full comprehension of Chapter 5 was not entirely my fault. More recently, I found that the circuits in the *Burr-Brown Handbook of Circuits for Operational Amplifiers* were just as good as the ones in the *Philbrick Nexus Applications Manual*, but I could never figure out how they got there. The former book persisted in using *MATRICES* to indicate a circuits' suitability, whereas the Philbrick guys just set out a couple equations, and if you liked the answer, well, there you were. Why is it that some people are better *explainers* than others?

So when I heard that Tom Frederiksen was going to write about an Intuitive study of op amps, I was delighted, because Tom is one of the best arm-waving engineers and explainers of this era. And as the *Philbrick Applications Manual* has been out of print for more than 10 years, the world really could use a good new book on this topic. It would only be fair to mention that the price of an op amp had fallen by a factor of 100 since the Philbrick manual was published. Thus, feasibility is enhanced by cheaper prices.

I should also go back into history and remind you that the whole idea of an *operational* amplifier, is an amplifier that is able to perform whatever *operation* you want to perform. The amplifier is intended to be the servant of the designer, and its function should be nicely determined by a few passive components connected as feedback elements around the op amp. The gain does not depend on the offset, nor vice versa, so that amplifier performance is easy to compute, first in terms of its *nominal performance* and then in terms of *deviations* (partial derivatives) away from its nominal performance. You don't need a fancy computer (analog or digital) to tell you when your op amp will let you down or will fulfill your needs. The ratio of a few passive components is usually sufficient to tell you the performance. Best of all, you may not even need equations or formulas to indicate a problem—the op amp is an exquisite tool for the analysis by gut-feel *or* arm-waving. When your friend tells you, "This circuit is in trouble because *this* capacitor is too big and *this* resistor is too small," that's a clue that fancy math is not holding the high hand over your task; but rather, the problem can be analyzed by simple analysis and *intuition*. This aspect, combined with the low cost of good op amps, has nearly revolutionized the linear circuit business.

So while Mr. Frederiksen is breaking new ground, he is doing so mainly by extrapolating linearly in the same direction pioneered by Dan Sheingold and Bruce Seddon at Philbrick, and by John I. Smith, and Jiri Dostal. He extrapolates, and then in addition he shows how his ideas fan out into broadening areas; yet he shows how these circuits are easily based on the simplest arm-waving (intuitive and very easy to analyze) arguments. Is it a breakthrough to show that no theoretical breakthrough is needed?

Then Tom has scored a breakthrough. Bravo!

Robert A. Pease
Staff Scientist
National Semiconductor Corporation
Santa Clara, CA

Preface to the Revised Edition

I am pleased with the favorable response that the first edition of this book has received, both from practicing engineers and also from the students of the many universities that recommend it as collateral reading for their linear courses.

This revised edition has provided the opportunity to include many additional helpful comments throughout the text that have resulted from my use of this book as the handout for a two-day seminar on op amps.

The last chapter, "New Developments and the Future of Op Amps," in the first edition had predicted that the linear IC design engineers would eventually solve the problems that are associated with realizing a high performance op amp with a CMOS process. This has now happened and the techniques that brought about this breakthrough in CMOS op amps are discussed as part of Chapter 1.

An improved, compatible PNP transistor was also anticipated in the last chapter of the first edition. It was expected to make use of a new, low cost dielectric isolation process. Since this writing, a high performance compatible PNP process has been put into production, but it was achieved with lower cost junction-isolation rather than a dielectric-isolation process. The significant aspects of this development are also now included as part of Chapter 1.

The ongoing search by the linear IC design engineers for a "good" PNP and the various solutions to this problem that have been used over the years, including the problems associated with each, are also discussed in Chapter 1.

I hope that those who have enjoyed the first edition will find this revised edition even more complete and useful to aid the understanding of the op amp, to learn about some interesting application circuits, and to avoid some pitfalls.

Thomas M. Frederiksen

Preface to the First Edition

Many observers note that new digital designs are being used in many systems which, in the past, were realized entirely with linear circuits. This apparent extinction of linear products is *not* showing up as a decrease in linear sales. What *is* happening is an increase in the pervasiveness of *all* semiconductor products. This increase in the total application of semiconductors has also greatly increased the demand for operational amplifiers (op amps). A few years ago, if a particular IC op amp from a major supplier sold 100 thousand per month, it was considered to be very successful. This number now approaches 1 to 3 million per month. If we consider the total number of individual op amps (duals counted as 2 and quads as 4) the total unit sales of op amps, by just one major supplier, can be approximately one-quarter billion per year!

This large consumption of op amps and the fact that op amps are basic to many of the more complex linear IC products suggests a greater need for information about op amps today than at any time in the past. Two additional factors are adding to this problem: (1) many times, system engineers have to design both the linear and digital sections on a project and (2) universities generally have had to reduce the number of linear courses to fit in the digital (microprocessor) courses. So it is not surprising that books on op amps are still in demand.

One goal of this book is to develop an understanding and appreciation for the reasons that have caused such a vast number of different IC op amp part numbers to exist. As will be seen, various applications demand that specific parameter specifications of an op amp be improved. The difficulty of simultaneously improving all the specs—especially where many times the most important one is low cost—has created a large number of op amp products.

The idea for this book came as a result of a 1980 nationwide linear seminar where the author gave a presentation entitled "Op Amp Primer." The favorable reception and the many resulting requests for a book that made use of this intuitive approach provided the encouragement to create this book.

Although many books have been written about op amps, the focus of these books has either been on the detailed design of the internal circuitry of op amps or the rigors of obtaining high precision in linear circuit design. An intuitive groundwork in the basic functioning concepts of the op amp has been missing.

Intuition involves thinking about physical systems and circuits in an almost personal way. The emphasis on only a mathematical description that is given during the formal education process tends to block this physical intuition. This is why experience must be used to acquire the *feelings* a circuit designer must have. This background is needed before the reader can fully appreciate the way application circuits are really developed by the endangered species of linear circuit designers.

Overheard conversations between op amp users reveal that most design is done with creative imaginations and discussions that produce statements like: "When this input is jerked up, this guy up here is kicked on and dumps a *gob* of current into this small cap. . . . " These comments sound like the planning of an electronic Rube Goldberg contraption—which is a valid description for most of the *really neat* application circuits.

To conceive and create circuits in this *component-personal* way is what the majority of linear circuit design is all about. This type of thinking requires an intimate understanding of and a *feeling for* the op amp and the passive and active components that are added to provide the complete application circuit.

The purpose of this book is to pass along this feeling for op amps, passive components, and op amp application circuits. Toward this end, only relatively simple mathematics will be used so as not to unnecessarily obscure the main issues involved. The major emphasis will therefore be on first order effects and the high volume, popular op amps. Information is featured that will benefit the designer who may have little or no time for research or study and is under pressure to rapidly produce functioning circuits. Interested designers will then be ready to read the extensive literature on this subject to add further details.

Chapter 1 traces the op amp to the early analog computers. The specifications of a high quality vacuum tube op amp circuit of bygone days are compared with those of a particular IC op amp that is often used as a reference point, or benchmark op amp, the 741. A simple model is then introduced for the IC op amp and this is used to explain the reasons for both op amp specs and limitations, because understanding the limitations of real-world op amps can save valuable design time. The evolution of the monolithic op amp is quickly traced to help appreciate the technical problems that were sequentially solved to arrive at the low cost, high performance op amps of today.

The Bi-FET (bipolar combined with field effect transistors) op amps are then described. Insights into the design improvements that were made possible with a new process that allowed adding JFETs (junction field effect transistors) to the bipolar op amp process, are given. Most users of Bi-FET op amps are more interested in the increased slew rate and frequency response—relatively few users want only the dc benefits.

The story of the popular quads is next. The large volumes that are shipped each month make the quads the industry's most popular linear products. How and why they came into being historically ties to the requirements of the electronic control systems for automobiles. The low prices that have resulted force many designers to use a quad: sometimes it's because they simply can't afford (or perhaps have forgotten how to bias) a transistor! Inner workings are described so the reader can appreciate some unusually good performance specs. For example, the split-collector g_m reduction trick, first used on one of these quads, is shown to solve the fabrication problems of the unsuccessful, early, dual 741. This circuit trick has been the key to the modern, small die-size, low cost op amps.

Chapter 2 opens with an intuitive approach to explain how feedback is used to control the performance of an op amp circuit. This shows that feedback can most easily be understood as *going up an attenuator*. This novel concept has been found to be an interesting and easily grasped idea that also explains why op amp circuits are less precise than a simple resistive attenuator.

Feedback control theory is then introduced to describe the op amp application circuits. The requirement for large open-loop voltage gain and the effects of changes in open-loop voltage gain on the closed-loop voltage gain of application circuits are quantitatively presented.

The major op amp error sources, from the imperfection of the feedback components to the nonideal nature of a real op amp, are discussed in Chapter 3. The benefits of the Bi-FET input stage are described and both the large-signal, high frequency, and rise-time limits of both bipolar and Bi-FET op amps are covered.

Much of this chapter is devoted to a discussion of the undesired noise sources that contaminate the output voltage of the op amp. The similarity between the analysis of the effects of the dc noise sources (that affect $V_{OUT\ dc}$) and the ac noise sources is stressed to aid understanding. The performance predictions that can be made by using these noise sources is illustrated in a number of numerical examples.

Many nonexperts shy away from considerations of ac noise because of the confusion that results from the statistical nature of noise. The purpose of the material in this chapter is to show how to predict the amount of ac noise that can be expected in the output voltage of an application circuit. Many common misconceptions are pointed out and a novel way to visually display the effects of the individual ac noise sources is given graphically. Surprising results can be obtained: for some applications the 709 may be the best op amp, for others the lowest noise voltage, high cost op amp may not help. The concepts of *noise bandwidth* and *noise gain* are introduced and some very practical examples of their usefulness are given. This chapter ends with a discussion of flicker noise (1/f) and *popcorn* noise.

Stability, or freedom from undesired oscillations, is the subject of Chapter 4. The numerical measures of how stable (or unstable) a circuit is, the stability margins (*gain margin* and *phase margin*) are defined. Ways of testing the stability of an application circuit are given and the effects of insufficient stability margins are covered.

A basic introduction to *poles, zeros,* and *root locus* is included to explain these terms and to indicate why they are of interest to the linear system designer. This chapter ends with a practical way to guide the op amp user to the cause of an undesired oscillation.

Many basic op amp application circuits are presented in Chapter 5. This is not meant to be a complete listing but will provide the operating concepts on which more complex op amp application circuits are based. Useful application circuits result from combining a few known circuit tricks in an unusual way to accomplish a desired overall function.

Practical user problems are the subject of Chapter 6. These are the generally unpublished facts that are provided by experience (and many blown-out op amps or application circuits that "hung up"). It is true that data sheets don't list everything. This can be demonstrated by having a linear system designer read a digital product data sheet or having a digital designer read a linear systems product data sheet. Today, this engineer cross-over is happening more frequently, as one designer is often doing both jobs. (If you find that this list is not complete, please send your "gotcha" to the author—in care of the publisher).

Finally, Chapter 7 discusses the role of linear MOS and the future of the op amp.

Thomas M. Frederiksen

Acknowledgments

The author acknowledges the help of many coworkers over the years and the benefits which have resulted from numerous informal discussions about op amps. The influence of Jim Solomon in the area of op amp understanding and design over many years of association has been most helpful.

Special thanks go to Robert A. Pease for his careful and helpful review of the manuscript for the first edition. Many of his comments, that result from his long association with and his basic contributions to op amps and their application circuits, have been included. Ed Harada has greatly contributed to the structure and flow of the opening chapters by providing the viewpoint of a digital designer who is learning about op amps for the first time. Much of the introductory material and basic approaches to the understanding of the op amp were suggested by him. Tim Regan carefully reviewed the final manuscript and suggested many additions and changes that have improved the content and the explanations in each chapter. The manuscript also benefitted from many helpful comments made by Steve Hobrecht, Jim Williams and Martin Giles.

Intuitive Operational Amplifiers

CHAPTER I

Background Information

This chapter provides general technical background material and appropriately starts with a look at the history of the vacuum tube analog computer. It is interesting to notice how the two early computational machines (both the digital and the analog computers) have been responsible for the most popular integrated circuits (ICs) of today: the microprocessor and the operational amplifier (op amp). Whereas the modern microprocessor (with memory on chip) is essentially replacing *all* the electronics of a digital computer of bygone days, the modern IC op amp is unfortunately replacing only one, two (the dual IC op amps), or four (the quad IC op amps) of the computational amplifiers of the analog computer. This lack of more complete linear functions on IC chips is a factor that makes many present-day system designers favor a digital approach.

Before we get into the details of the IC op amp, we will first look back in time to develop an appreciation for the contributions that were made by the early researchers.

1.1 ANALOG COMPUTERS: THE ORIGIN OF THE OP AMP

The push (and funding) for new technology during World War II stimulated the development of both digital and analog computers. During this era, Dr. C. A. Lovell of the Bell Telephone Laboratories introduced the op amp. In 1948, George A. Philbrick also independently used a single vacuum tube op amp with a controlled gain characteristic that used negative feedback, obtained by a cathode impedance, to provide a controlled plate current flow through a plate load impedance. Philbrick is also generally credited with the discovery (also in 1948) of the operational amplifier for performing integration and differentiation by electronic means.

Analog computers predate the op amp and go back to 1925 when Vannevar Bush and his associates at MIT (Massachusetts Institute of Technology) used electrical and mechanical devices to do analog computation. The basis of analog computation is to assign voltages to the variables of the physical system being investigated. Mathematical equations show the relationships that exist between these variables. The analog computer solution uses interconnected electronic and mechanical hardware to represent and follow these equations. It can be asked, "Is this a computer or rather a simulator?" Whichever way you look at it, you can obtain detailed design information about the problem that is being studied.

Early analog computers allowed significant advances in aircraft and missile design. These computers consisted of large racks of equipment containing a number of individual op amps. The name *operational amplifier* was coined because the particular type of amplifier used in these computers *performed mathematical operations.* Also, special devices (servo motors, complete with gear trains) were included to multiply by a constant or a variable. Trigonometric generators, function generators (the *memory* for tabulated data), and other specialized devices were also available. In the 1950's, a typical analog computer was 16-feet long, 8-feet high, and 4-feet deep. It needed 27,000 watts of power to function and contained 4,000 vacuum tubes, 6 miles of wire, and had 100,000 soldered joints.

Early vacuum tube amplifiers suffered from excessive drift. The technique of chopper-stabilization was used to counteract this problem. Chopper-stabilized vacuum tube op amps were the heart of general purpose analog computers.

It is interesting to compare (Table 1-1) the specs of a relatively high quality vacuum tube op amp with those of an IC op amp (the 741) that was first designed in 1967 by Dave Fullagar. The old amplifiers that used *thermionic FETs* (vacuum tubes) are seen to provide some impressive specs.

TABLE 1-1: Comparing the Specs of a Vacuum Tube Op Amp with Those of an IC Op Amp, the 741

PARAMETERS	VACUUM TUBE OP AMP	IC OP AMP (741)
Power Supplies	$\pm$ 300 V_{DC}, + 110 V_{DC}, -500 V_{DC} and 6.3 V_{AC} (filaments)	$\pm$ 15 V_{DC}
DC Voltage-Gain (open-loop)	3×8^8 V/V	5×10^4 V/V
Input Current	60 pA	60 nA
Output Current	$\pm$ 13 mA ($\pm$ 30 mA with boost R)	$\pm$ 13 mA
Output Voltage-Swing	$\pm$ 100V	$\pm$ 12V

Analog computers used the chopper-stabilized, vacuum tube op amp to realize summing amplifiers and integrators (where the variable of integration was time — although time could also be the analog of some other physical parameter of the problem, such as distance). Differentiators were avoided because of inherent noise problems. A differentiator is a high-pass filter that also *passes* broadband noise. But an integrator is a *quieter*, low-pass filter. Precise potentiometers were added to multiply by coefficients less than unity. This provided all the necessary ingredients to solve the linear, ordinary differential equations using only integrators and summing amplifiers, (as we will soon see) of many real-world problems.

The symbols that were used by analog-computer programmers in their programming diagrams are shown in Figure 1-1. (The "1"s or the "2" that are noted beside each input on the figure are the respective gain values.) These symbols are not quite the same as those used today for the IC op amp. Note that we have used a different symbol for the integrator and the "1" beside the input lead of this symbol also indicates the *gain* or transfer function of the integrator. This is shown in Figure 1-2, where the transfer functions (with the dimensions of 1/time) of both a single-input, Figure 1-2a, and a multiple input, Figure 1-2b, op amp integrator are shown. (The operation of the integrator will be discussed in Chapter 2.)

A multiple-input, preset-gain, summing amplifier results if we replace the 1 μF capacitor in Figure 1-2b, with a 1 MΩ resistor to provide similarly scaled inputs as given by

$$V_{OUT} = -[5\ (V1) + 2(V2) + (V3)]$$

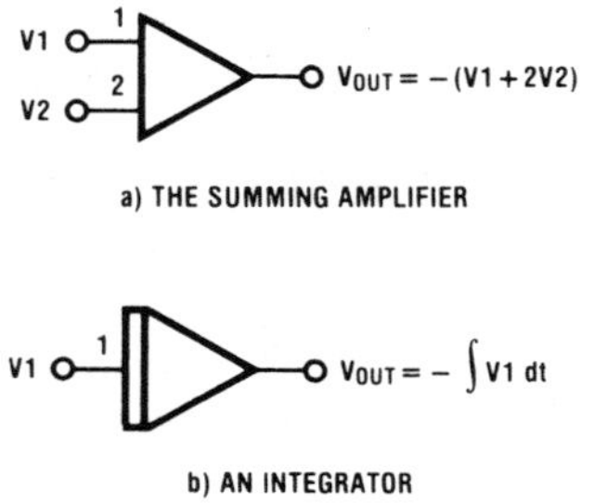

V1 1, V2 2, $V_{OUT} = -(V1 + 2V2)$

a) THE SUMMING AMPLIFIER

V1 1, $V_{OUT} = -\int V1\ dt$

b) AN INTEGRATOR

Fig. 1-1. The Basic Symbols Used by the Analog-Computer Programmers

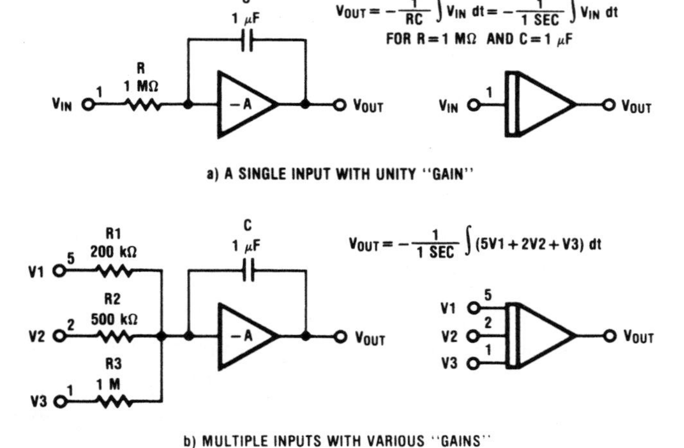

a) A SINGLE INPUT WITH UNITY "GAIN"

b) MULTIPLE INPUTS WITH VARIOUS "GAINS"

Fig. 1-2. The "Gain" of an Integrator

Programming an Analog Computer

Programming an analog computer consisted of cabling up *patch panels* that interconnected the basic op amps and other computing elements that were available in large racks of electronics. In addition, initial conditions (starting voltages for the integrators, for example), needed to be dialed in and many *coefficient potentiometers* had to be preset to predetermined values. Typical run times were 10 to 60 seconds, independent of the complexity of the problem. Special techniques were also used to get the run time to a small fraction of a second for repetitive runs. This provided a rapid indication of the effects on the system response that would result from changes made to the system. Many of these system changes could be made by simply changing a potentiometer setting. The sensitive, or most critical, parts of a system could therefore easily be found.

As today, with the digital computers, programming was one of the largest problems. The old tricks of analog-computer programming are the basis for the applications of today's IC op amps. The surprising thing is that these sophisticated analog-computer circuits of yesterday are presently so low in cost (as a result of ICs) that they now appear everywhere.

The analog programming language that was used had no relation to the modern languages of the digital computer: COBOL, LISP, or Pascal. Instead, programmers worked with a very ancient language — MATHEMATICS, which was strange to many people and added to the mystery of programming. Because we will take a quick look at analog computer programming, we will drop into this foreign tongue. Please bear with it; we only want to give you a basic idea of how programming was done, and this still can be appreciated even if MATHEMATICS is not your favorite means of expression.

As an example of how programming proceeds, assume it is necessary to study a problem that can be described by the following system equation:

$$a\frac{dy}{dt} + by - c = 0$$

where the quantities dy/dt, y, a, b, and c will be represented by either voltage analogs or potentiometer settings. The *solution* of this equation means to display voltage analogs of y and dy/dt as a function of time, starting from some initial conditions at the start of the run, at time t_0.

The first step in programming is to solve the system equation for the highest derivative. For our previous equation this is dy/dt so

$$\frac{dy}{dt} = \frac{c}{a} - \frac{b}{a}y$$

Now we notice that the two terms on the right side of this equation can be algebraically added, using a multi-input summing amplifier (this also inverts the polarity of the resulting sum, although this inversion is not always desired), to provide -dy/dt as shown schematically in Figure 1-3a. (This first step requires some faith, because we assume we can somehow create the needed unknown input quantity, y, although it is not yet clear how this will happen.)

The usefulness of the electronic integrator circuit is in producing y from dy/dt or

$$\int \frac{dy}{dt}\,dt = y$$

So we add an integrator as shown in Figure 1-3b.

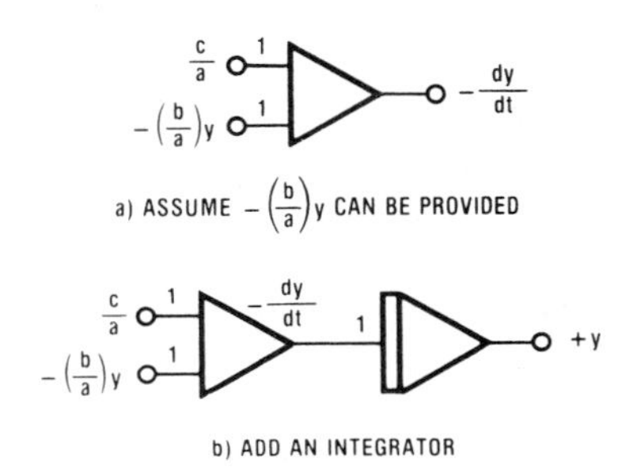

Fig. 1-3. The First Steps in Programming an Analog Computer

The variable that is required at the input to the first amplifier is -y, so a unity voltage gain sign changing amplifier is next added, Figure 1-4a.

A potentiometer is now added (indicated by the circle on the drawing, Figure 1-4b) that is preset to a value that represents the ratio of the constants b/a, and we finally have created one of the required inputs, -(b/a)y, as is also shown in Figure 1-4b. The other input, c/a, called the *forcing function*, will be provided from a reference voltage supply, V_{REF}, and a second potentiometer, that is preset to the analog of the constants c/a. Finally, an additional inverting amplifier is used to provide + dy/dt, and the output of this amplifier and the output of the integrator supply the desired solutions to this problem, + dy/dt and y, as shown in Figure 1-4c. These voltage analogs of the solution to the system equation can now be displayed on an oscilloscope or an x-y plotter.

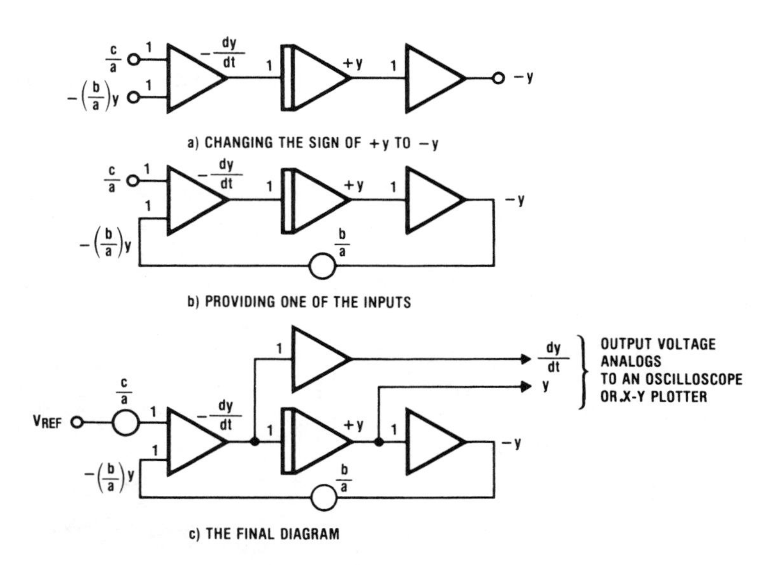

Fig. 1-4. The Final Steps in Programming

Both amplitude and time scaling were generally used to insure that the problem would *fit on the machine.* The vacuum tube op amps had a dynamic output voltage range that was limited by noise at the low end to approximately 10 mV and by the onset of large-signal saturation problems at the high end at approximately 100V. This provided an overall single-polarity voltage range of 10 mV to 100V or 10^4:1 (80 dB) for the problem variables to *wander around in.* Time scaling was used to insure that the solution to the problem would proceed at a slow enough rate for the op amps and the recording devices to keep up with all of the voltage changes that would exist.

These computers also had many additional sophistications and could handle transfer functions, time-varying coefficients, nonlinearities, discontinuities, dead zones, and even backlash (as found in gear trains, for example).

A major benefit of the analog computer is the relatively fast run time (as compared to digital computer solutions) for complex problems. A good *feeling* for the sensitivity of parameter values could also easily be obtained by simply changing various potentiometer settings during repetitive runs to immediately *see* the effect on the problem solution. This allowed an engineer to rapidly determine the relative importance of the various system parameters.

This high-speed problem solution, and other analog benefits, has created interest in *hybrid computers;* both analog and digital computational circuits simultaneously working on the same problem. "Stand-alone" analog computers are rare these days, except when used for teaching purposes.

Newer Hybrid Computers

The benefits of the analog computer have been maintained in the combination of analog computers with digital computers: the hybrid computer of today, Figure 1-5. This combines, through a sophisticated communications interface, both digital and analog computers. This type of computer is primarily used for dynamic systems simulation in the design of aircraft control systems.

— Courtesy of Electronic Associates, Inc.

Fig. 1-5. The Hybrid Computer-Facility at Grumman Aerospace Corporation.

Each processor does those tasks for which it is most efficient. The analog subsystem processes simultaneous linear or nonlinear differential equations and interfaces with other equipment that involves analog data.

The digital subsystem sets up and controls the analog processor and handles all the arithmetic, program instruction, data storage, and user communication functions. This synergistic combination of computers provides a quantum jump in computational power and ease of use over analog-only systems and offers orders of magnitude increases in the speed of a digital-only solution to a dynamic system problem.

An analog computer has the benefit that one voltage can represent 10 to 16 bits of information. (This gives resolution and even an accuracy of one part out of 1024 or 65,536 respectively.) Practical problems result when initially very accurate analog input voltages pass through electronic analog computational circuits. These circuits introduce signal distortion and noise and these added errors can greatly reduce the accuracy of analog signal processing.

A digital signal does not suffer from this gradual degradation and can pass through a large number of digital processing circuits without any distortion or noise contamination of the signal. This benefit of a digital approach has created interest in digital audio entertainment systems for the home. Combining the digital approach with the ability to increase the number of simultaneous telephone channels (by time-multiplexing) on a given existing interconnection wire link, has brought about the new digitized-voice telephone systems.

Generating Sinewaves

One of the most popular IC op amp circuits for active filters was derived from the analog computer loop used to generate low frequency sinusoids, Figure 1-6a, for use as forcing functions. As can be seen in Figure 1-6a, once started, this interconnection of amplifiers continues to *feed on itself* and will provide sustained sine and cosine waveforms of constant amplitude (at least for the duration of the computer run time) at outputs 1 and 2, respectively. Notice that the hardware realization, Figure 1-6b, eliminates the "ω" potentiometers that were indicated in Figure 1-6a to supply the "ω" coefficients. (Integrators will accumulate even very small dc errors and, therefore, will most likely end up with the output voltages of the amplifiers at maximum limiting values if left in operation too long. This special practical problem requires either periodic resetting or the use of an auxilliary dc feedback-loop to keep the integrators operating within their linear range.)

This overall negative feedback loop, two inverting-integrators and one sign changer, has been slightly modified to provide the *bi-quad* and the *state-variable* RC active filters that we will discuss in Section 5.5

Before we leave this section, we will analyze (although somewhat superficially) the operation of one of these early chopper-stabilized vacuum tube op amps. This is included for the old timers who may remember hearing of (or working with) these amplifiers. Newcomers to the analog art should temporarily skip over this discussion and return, if interested, after completing Chapters 1 and 2. (We don't want to cause unnecessary confusion by introducing vacuum tube circuits; but, after all, these were the first op amps.)

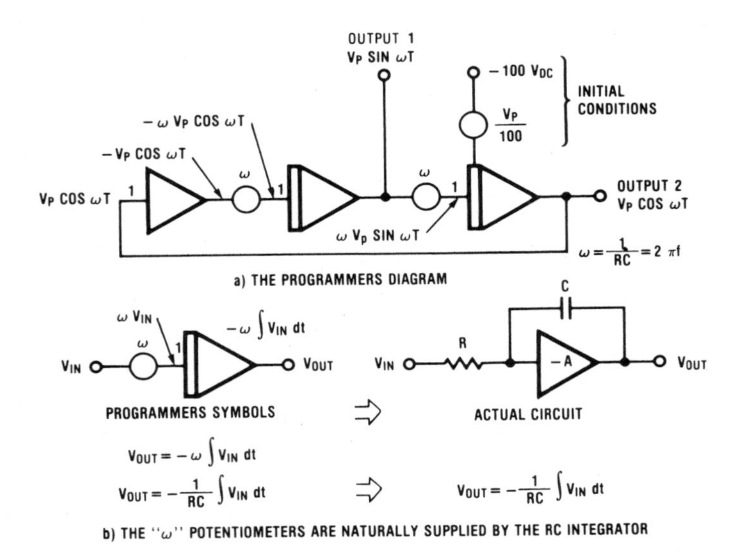

Fig. 1-6. A Low-Frequency Sinewave Generator

The Chopper-Stabilized Vacuum Tube Op Amp

If we take a closer look at one of these chopper-stabilized, vacuum tube op amps, Figure 1-7, we notice that it is made up of two amplifiers: *the four-tube inverting dc amplifier* across the top of the figure (this has a dc gain of approximately -100,000) and *the two-tube ac amplifier* shown across the bottom of the figure (this has a gain — dc output to ac peak-to-peak input — of -3000).

An electro-mechanical chopper was used to convert the dc voltage that exists at the inverting input of the op amp, shown as V_ϵ on Figure 1-8, to a squarewave ac signal at the input to the ac amplifier, shown as V2 on the figure (all of the voltage waveforms on this figure are shown exaggerated for clarity, actual voltages are in the mV range). This squarewave voltage is amplified by the 2-stage, non-inverting, ac amplifier (with an ac gain of 6000), and is then *synchronously detected* at the output (V5) by the extra contacts (shown as SW2 on the figure) on the chopper. These contacts are *synchronized* to be closed when the input contacts (SW1) are open, and vice versa. This provides a ground reference for the amplified output squarewave, V5, and causes it to swing in only one direction (this is the rectification or *detection* function), the reverse of the polarity of the dc input voltage, V_ϵ.

This *detected* output waveform is then filtered by an RC network (22 MΩ and 1 μF) to provide a dc output voltage ($V1_b$) that is one-half the peak-to-peak value of the single-polarity output squarewave. The negative dc output voltage that results across the 1 μF filter capacitor is applied to the control grid of the differential input stage, $V1_b$, of the upper dc amplifier. For example, a negative change of 1 mV at this input would also cause a negative change of 1 mV in V_ϵ (the $V1_a$ input). This input differential amplifier can, therefore, be used to cancel or remove the dc voltage, V_ϵ, that otherwise would exist at the input of the op amp. *This is the key*

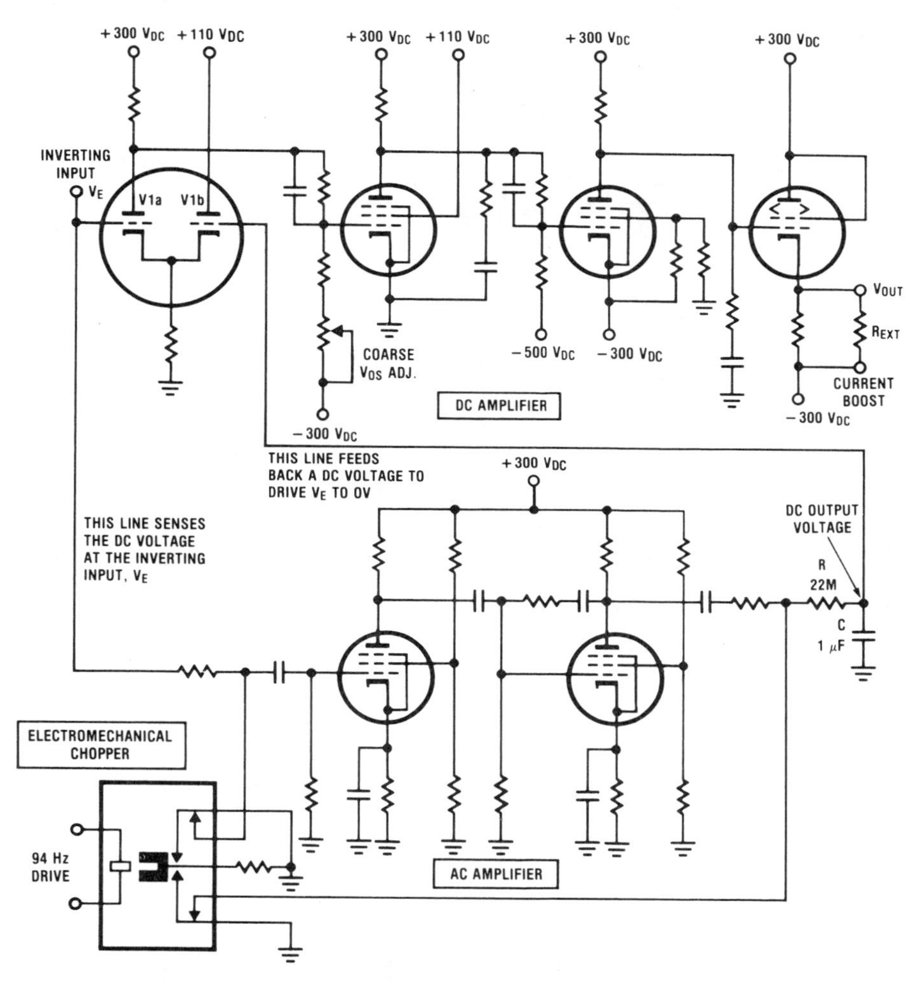

Fig. 1-7. A Vacuum-Tube Chopper-Stabilized Op Amp

idea of the offset reduction that takes place with a chopper-stabilized op amp. The action of this offset voltage correcting loop therefore forces the dc voltage at the inverting input, V_ϵ, of the op amp to essentially 0 V_{DC}.

Notice that a chopper-stabilized op amp has only a single input. (The non-inverting input is not available for an application circuit because it has been used to cancel the offset voltage.) *This inverting input must also always be operated at ground potential.* Therefore, *only inverting applications are possible with a chopper-stabilized op amp.* (This is worth spending some time on because the new MOS amplifiers, being similar to vacuum tube amplifiers, generally use some form of chopper-stabilization to reduce drift.)

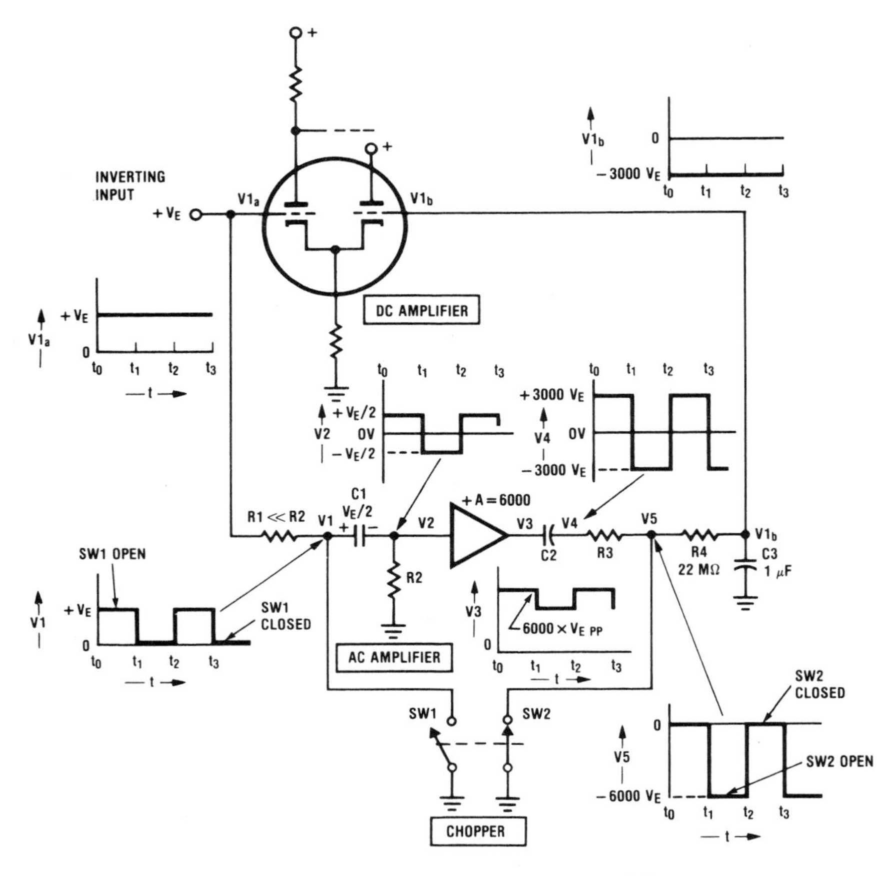

Fig. 1-8. Waveforms of the AC Amplifier

It is not so obvious that the overall dc gain of this op amp is the product of the gains of the two amplifiers, or 3×10^8 (100,000 × 3000). This can be more easily comprehended if we imagine momentarily disconnecting the offset correcting ac amplifier. For example, the dc gain of −100,000 of the main amplifier now requires only 1 mV dc at the input to support a −100V dc output voltage. (This assumes that we have no initial offset voltage; that is, precisely 0V IN gives 0V OUT and there is no drift in this offset voltage during the short time it takes to gather the data.) Now, if we reconnect the offset correcting chopper-stabilized amplifier, this 1 mV (that appears at the inverting input, V_ϵ) will be reduced by an amount that is equal to the gain of the chopper-stabilized amplifier (3000).

To appreciate this action, notice that a dc correction voltage of −1 mV must be supplied by the chopper-stabilized amplifier to cancel the +1 mV that existed at the input to the main amplifier *before this chopper-stabilized amplifier was connected. When the chopper-stabilized amplifier is connected,* to create a correcting output voltage of −1 mV requires a relatively small input voltage, because this chopper-stabilized amplifier has a gain of 3000. The overall resulting input voltage, V_ϵ', is therefore now

$$V_\epsilon' = \frac{V_{OUT}}{A_V} = \frac{1 \times 10^{-3}}{3 \times 10^{3}} = 0.33\ \mu V$$

and the output of the main dc amplifier is still at -100V.

Looking now at the overall dc gain of the op amp, we find

$$A_V = \frac{V_{OUT}}{V_\epsilon'} = \frac{100V}{0.33 \times 10^{-6}} = 3 \times 10^{8}$$

which is the product of the gains of both amplifiers, as previously stated. We will show the IC equivalent of this circuit in Chapter 5 (Section 5.2 — Getting the Best of Two Op Amps) where we couple a high frequency Bi-FET op amp with an op amp that has more precise dc performance to end up with a composite amplifier with very good overall specs.

A lower-cost, "8-pin version" for plugging into a standard octal tube socket was also available: the K2-W op amp. This operational amplifier, Figure 1-9, was an invention shared between Loebe Julie (of Julie Research Labs, Inc.) and George A. Philbrick that the latter developed, produced, and popularized.

This op amp operated from *standard* split supplies ($\pm$ 300 V_{DC}) and the performance specs are also listed on the figure. The low cost of this amplifier (approximately $22.00) and the useful performance specs made it one of the first of the high-volume op amps: the "741" of

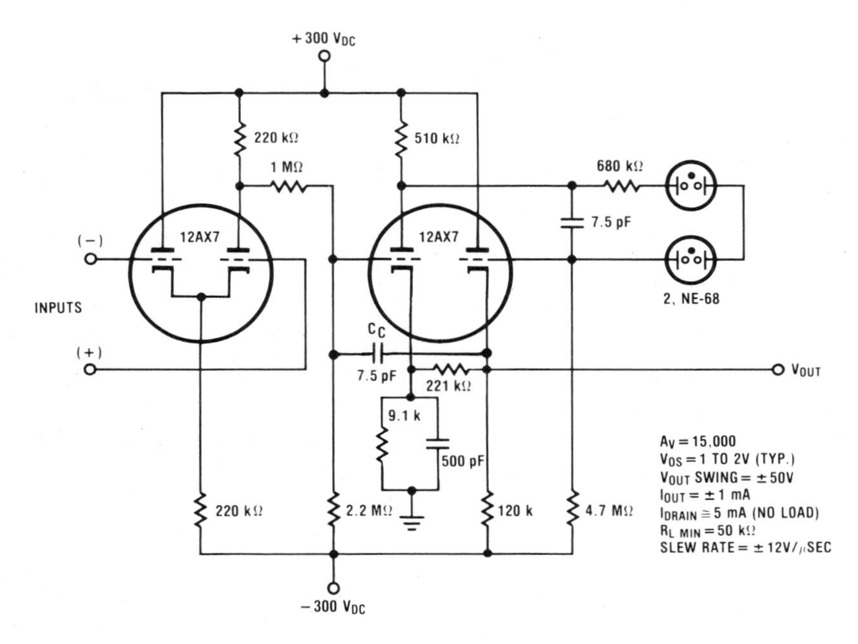

Fig. 1-9. The Popular K2-W Op Amp

its era. For an increase in cost, a chopper-stabilized version was offered with a *very much* lower offset voltage spec.

Armed with these historical perspectives, let's now start our investigations of the modern IC op amp; the uses and the performance specs of which have been accelerating since the mid-1960's. Perhaps now you can understand why the world has bypassed a logical evolutionary step — the analog home computer.

1.2 GETTING INSIDE THE OP AMP

An IC op amp is usually made up of the basic blocks shown in Figure 1-10. The external input to the op amp is a differential voltage [the difference that exists between the individual voltages applied to the input pins $V_{IN}(+)$ and $V_{IN}(-)$]. The overall output is the voltage that is available at the output pin, V_{OUT} (called a single-ended output voltage because only one output voltage exists). This output voltage is usually referenced to ground.

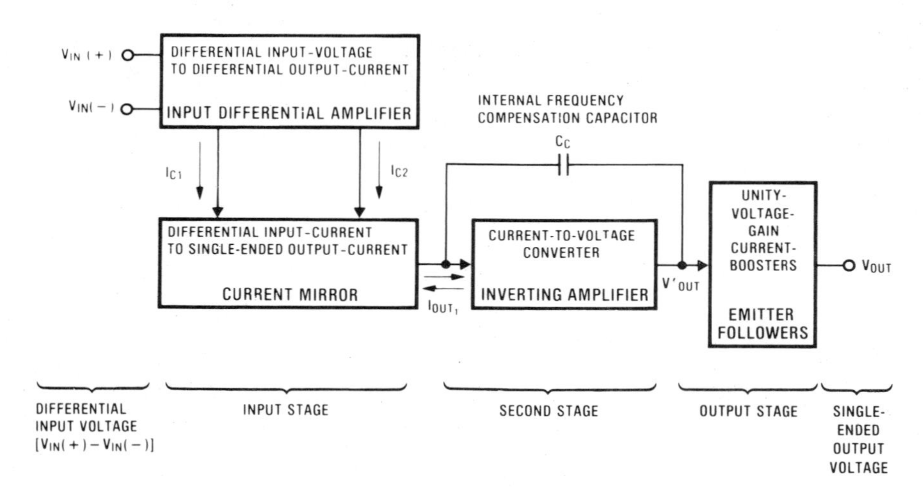

Fig. 1-10. Block Diagram of a Basic Op Amp

Although this is a voltage in, voltage out amplifier, notice there is an immediate conversion to differential currents in the Input Differential Amplifier block. This design approach has emerged as the most popular for IC op amps. These differential currents, I_{C1} and I_{C2}, that are provided by this first amplifier are then converted to a single-ended output current, I_{OUT1}, in the Current Mirror block. This is often called a *differential-to-single-ended converter.* This single ended output current, I_{OUT1} (the *output* of the *Input Stage*), can be of either polarity and therefore can both be supplied (sourced) or accepted (called a current sink) by the Current Mirror block.

The Second Stage of the basic op amp converts this current back into a voltage V_{OUT}', that will become the output voltage of the op amp. An internal capacitor, C_c (the frequency-compensation capacitor), is the component that does this conversion of current to voltage. As we will see later, C_c is also used to keep the op amp *stable* so undesired oscillations will not result in the op amp application circuits.

Finally, this second-stage output voltage, V_{OUT}', goes to the Output Stage where unity voltage gain Emitter Followers boost the output current that the op amp can supply to the external circuitry.

With this overall diagram in mind, we will now consider the basic circuits that accomplish these major functions.

The Input Differential Amplifier

The Input Differential Amplifier responds to an input differential voltage [$V_{IN}(+)$ minus $V_{IN}(-)$]. Ideally, any input voltages that are the same at both inputs (called a common-mode input voltage) will be rejected and will not affect the output currents of this amplifier. A differential input voltage causes the current that is supplied to the emitters to be split between its collectors, I_{C1} and I_{C2}, as shown in Figure 1-11. The total available emitter dc biasing current, $2I_E$, is usually a relatively small current, 20 μA, to keep the dc input current of the op amp small. (In a bipolar transistor, the emitter current is the sum of the collector and the base currents. For simplicity, we will neglect base current, so we will than have: $I_{C1} + I_{C2} = 2I_E$.) For example in Figure 1-11a, a differential input of +1V (the + input is one volt more positive than the − input) will cause the collector of Q1, I_{C1}, to carry essentially all of the $2I_E$ current and Q2 will be OFF ($I_{C2} = 0$). Similarly, as shown in Figure 1-11b, a differential input of −1V will cause the collector of Q2, I_{C2}, to carry the full $2I_E$ current. With a differential input voltage of 0V, Figure 1-11c, the biasing current will be equally split in our idealized amplifier, or in this case $I_{C1} = I_{C2}$.

Many of the performance specifications, and especially the input specifications, of an op amp are directly determined by the characteristics of the input differential-amplifier circuit that is used. (Most IC op amps use NPN transistors rather than the PNP transistors that have

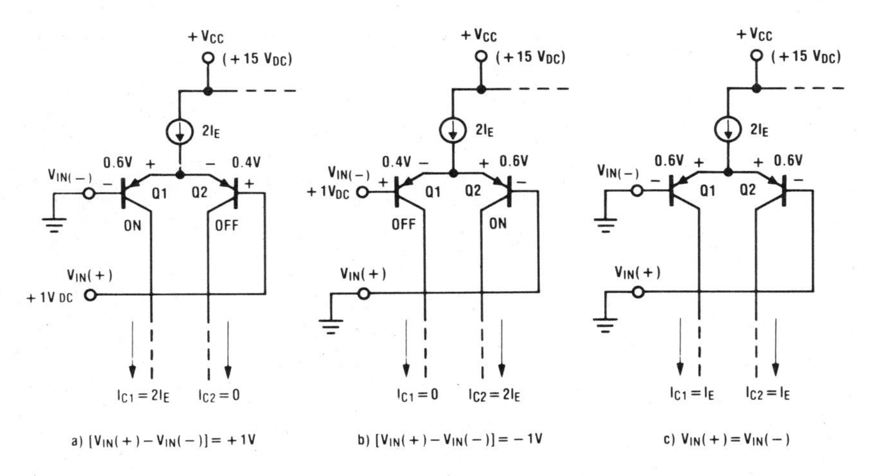

Fig. 1-11. The Current-Splitting Action of the Input Differential Amplifier

been used here for illustration.) It has always been necessary to keep the transistors of the input circuit well matched in their electrical characteristics, and also they should be kept at the same junction temperature. For example, a difference of only 1°C between the junction temperatures of these two input transistors will upset the symmetry and cause a thermally induced difference in the V_{BE} voltages of these two transistors of 2 mV (as a result of the approximately −2 mV/degree centrigrade temperature change in the V_{BE} voltage of a bipolar transistor). If this 1°C temperature difference should occur, an externally applied voltage of 2 mV (and not 0 V as we previously had with our ideal amplifier) would be needed to "offset" this undesired thermal imbalance, to cause the two collector currents to be equal. This required external voltage that is needed to restore symmetry within an op amp is called the "offset voltage."

An additional problem exists in an IC op amp because the power that is dissipated by the relatively high-current output transistors creates temperature differences across the surface of the chip. Special layout techniques must be used to locate the transistors of the input differential amplifier along the anticipated isothermal lines on the surface of the chip. This chip layout for thermal symmetry is a large concern for an IC op amp, and therefore these input transistors are always located as close together as possible, along isothermal lines, and as far away from the output transistors of the op amp as possible. Rectangular chips are generally favored where the heat-producing output transistors are placed along one of the narrow sides of the chip and the sensitive input transistors are located as far away as possible: at the opposite narrow edge of the chip, as shown in Figure 1-12.

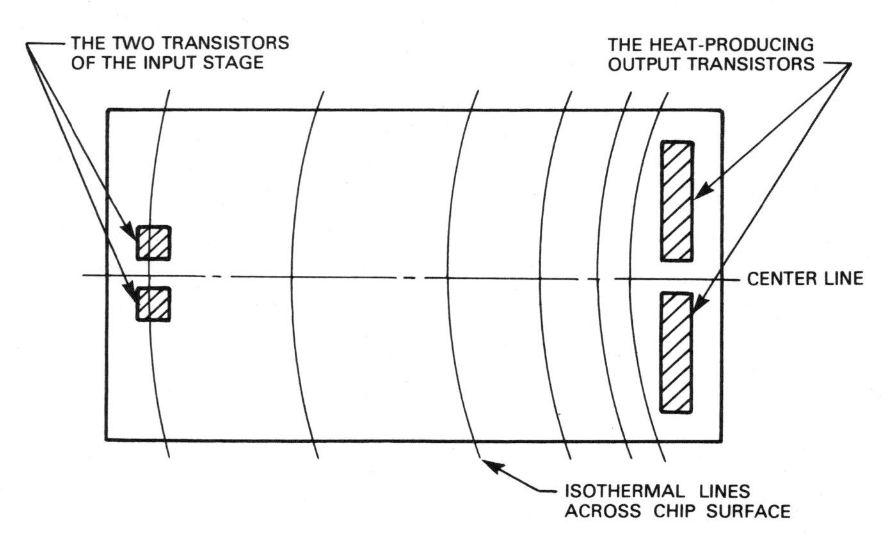

Fig. 1-12. An Isothermal Chip Layout

Further improvements in both thermal and electrical matching of these two-input transistors can be obtained by making use of a quad grouping of four separate transistors, where each of the input transistors is the parallel interconnection of two of these quad transistors (the ones that are diagonally opposite to each other), as shown in Figure 1-13. This is called a "cross-coupled quad" and is extensively used in IC op amp layouts.

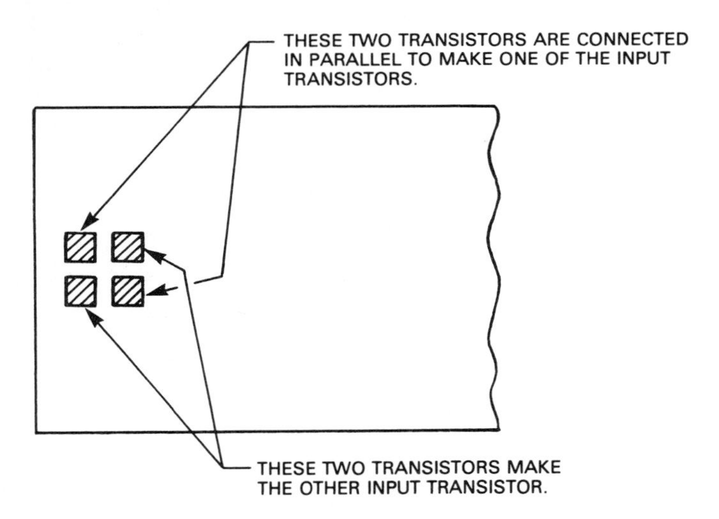

Fig. 1-13. A Cross-Coupled Quad of Bipolar Input Transistors

High values of offset voltage in CMOS op amps have generally been accepted because MOS transistors are very difficult to match. A solution to this problem is the extension of the cross-coupled quad to a checkerboard, parallel-interconnection of 8 MOS transistors out of a square, 4 by 4 grouping of 16 total transistors, a "checkerboard-coupled hexadecimal," as shown in Figure 1-14.

The input current required by the input differential amplifier becomes the input current for the complete op amp. Many schemes have therefore been used to reduce this input current, such as Darlington-connected input transistors, special chip-compatible high current gain NPN transistors, and the use of both JFET and MOSFET transistors. We will consider some of these schemes later in this chapter.

The Current Mirror

Notice that the collectors of the Input Differential Amplifier, Q1 and Q2, as shown in Figure 1-15, are connected to a *Current Mirror* (sometimes also called a current reflector) that is composed of Q3 and Q4.

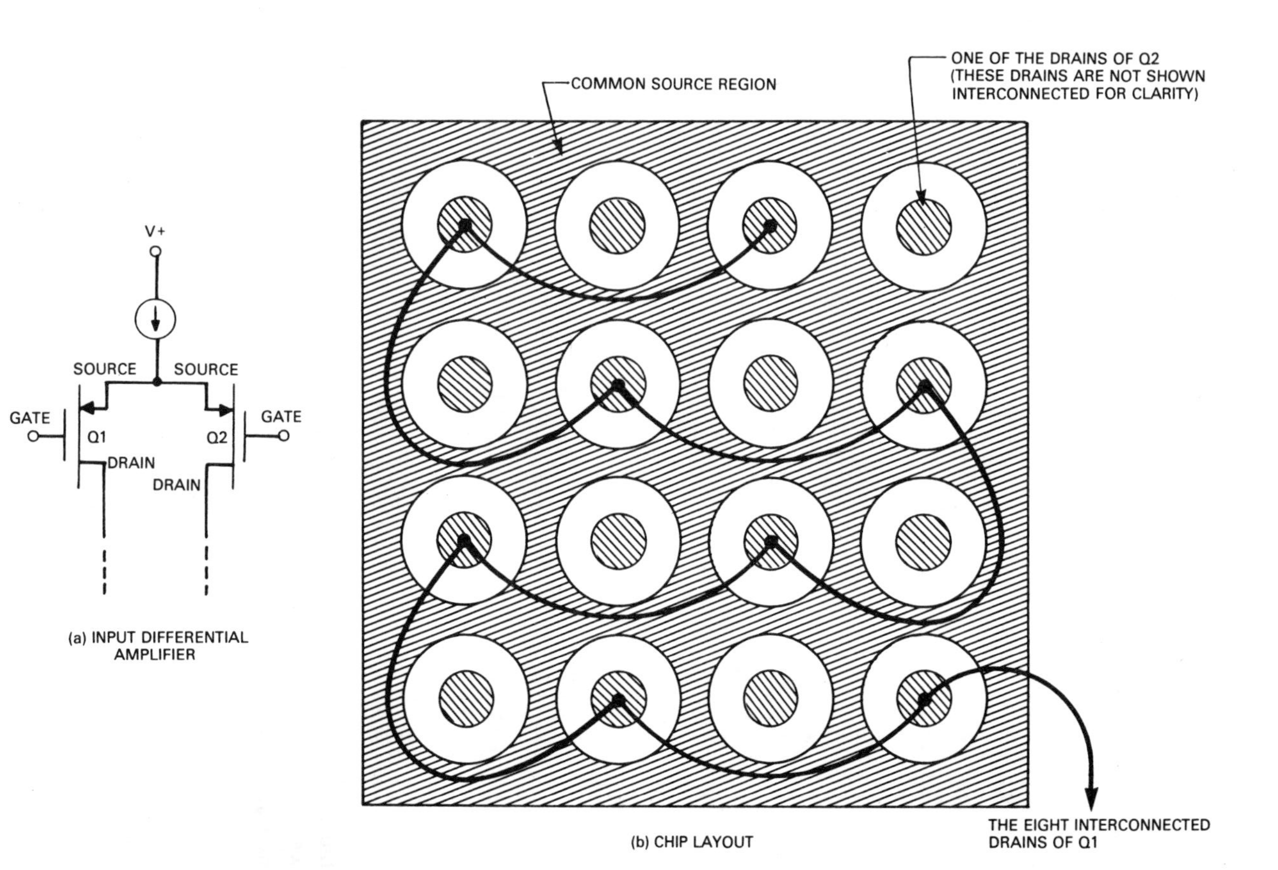

Fig. 1-14. A Checkerboard-Coupled Hexadecimal of MOS Input Transistors

The collector-base short shown for Q3 is a way to force the collector of Q3 to carry or *sink* a current that is equal in magnitude to the input current, I_{C1}. The current I_{C1} causes the base-emitter voltage of Q3 to increase until the collector of this transistor can carry I_{C1}. For example, if the collector of Q3 were to attempt to carry less current than I_{C1}, a difference current

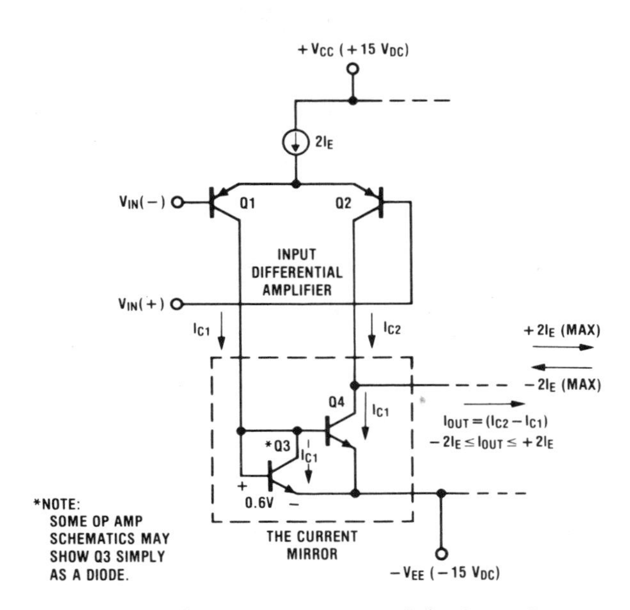

Fig. 1-15. The Current Mirror of the Input Stage

would exist at the point where the collectors of Q1 and Q3 join. This difference current would then cause the base-emitter voltage of Q3 to increase. This increased drive voltage to Q3 would increase the conduction of this transistor (or we could say that the difference current increases the current-drive to Q3) and therefore would force the collector current of Q3 to equal I_{C1} (we are still neglecting base currents).

The exact matching (identical geometries and matched electrical characteristics) of the transistors Q3 and Q4 (that we are assuming is achieved on an IC chip) will cause Q4 to also carry a current equal in magnitude to I_{C1}. This will result because the base-emitter junction of transistor Q4 is wired in parallel with the base-emitter junction of transistor Q3 and two identical transistors with the same V_{BE} voltage will conduct the same amount of collector current (if we neglect the complicating effects of unequal collector voltages). Thus, the current I_{C1} is said to be *mirrored* and *appears* as collector current for Q4. A *single-ended* output current, I_{OUT1}, is provided that is equal to I_{C2} minus I_{C1}. Therefore, this current mirror connection is also known as a *differential-to-single-ended converter.* This is believed to have been first used in a linear IC in 1966 by James E. Thompson. From this point on, in the rest of the op amp circuit, we no longer have a differential signal.

When the input differential voltage is 0V, the single-ended output current from this first stage is 0 mA. Notice that I_{OUT1} will range between the maximum values of $+2I_E$ to $-2I_E$, in response to differential input voltages to the op amp.

This Input Differential Amplifier Stage *will (ideally) respond only to the difference voltage* that exists between $V_{IN}(+)$ and $V_{IN}(-)$. *Any voltages that are common to both inputs will be rejected* and will not cause any effects at the output. It is important to remember that signals that appear simultaneously at both of the inputs, called *common-mode* signals, do not (ideally) result in a change in the dc biasing current magnitude ($2I_E$) or the split of this biasing current between the two collectors I_{C1} and I_{C2}. Therefore, *an ideal op amp will not respond to the common-mode input signal and, thus, has excellent common-mode rejection.*

A Basic Op Amp Circuit

Much of the mystery of an op amp can be easily dispelled if we consider the simplest circuit that still performs an op amp function. After this is understood, the reasons for many of the specifications and also the limitations of real op amps will become very clear.

A simplified op amp circuit is shown in Figure 1-16. If this were actually constructed, it would operate, although a more complex IC op amp would provide performance improvements.

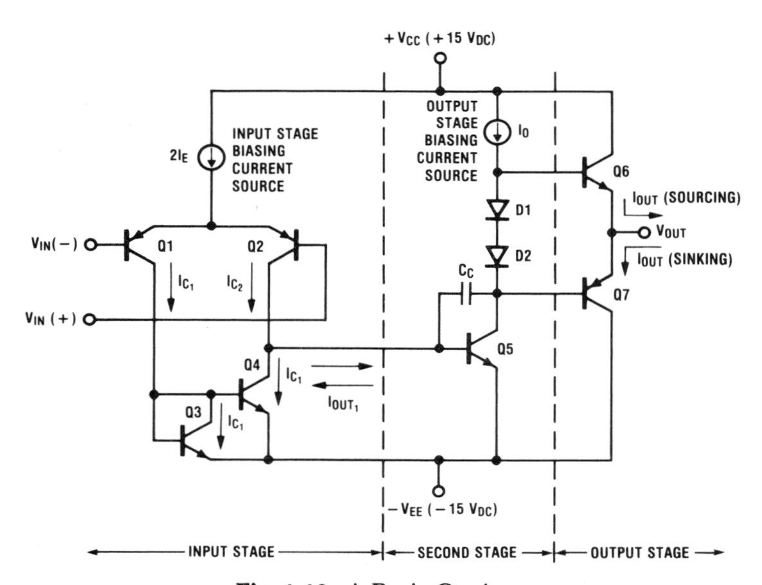

Fig. 1-16. A Basic Op Amp

Here we have added the Second Stage, the transistor Q5, and the frequency compensation capacitor, C_C, that is connected from the output to the input of this Second Stage. This common-emitter stage accepts I_{OUT1} from the Input Stage and converts this to V_{OUT}' (this will become the output voltage for the op amp, V_{OUT}, after passing through the unity gain Output Stage).

The Output Stage consists of two emitter-follower transistors, Q6 and Q7, that are used to provide both a low output impedance and a large output current capability. Notice that the upper dc biasing current source, I_0, is shared as collector current for the Second Stage (Q5) and base-drive current for the Output Stage emitter-follower, Q6.

The overall voltage gain of this basic op amp is supplied by the Input Stage (for example, by the I_{OUT1} that is supplied for a given differential input voltage) and the Second Stage (by the ability of this stage to convert this I_{OUT1} into the output voltage for the op amp, V_{OUT}). [The overall voltage gain (also called *the open-loop voltage gain*) of a typical IC op amp is 50,000.]

As we will see later, the usefulness of the op amp results from the ability of the input leads to accept a wide range of common voltage levels and to still respond to the slight difference voltage that may exist between these input voltages. Further, it is desired that these inputs to the op amp should not require large values of input current (base current for the input transistors Q1 and Q2) so that they will not disturb (load) the circuit nodes that they are measuring. (This is the same advantage that a high input-impedance voltmeter provides.) Finally, a large value of open-loop voltage gain is desirable for an op amp so that any required output voltage can be supplied by the op amp from only a small difference voltage at the inputs. For example, if an output voltage of + 10V is needed from the op amp, the differential input voltage required (using the previous 50,000 value as the voltage gain) is only

$$V_{IN} = \frac{10V}{50{,}000} = 200\ \mu V$$

so we can usually make the assumption that the voltages at the inputs to an op amp are identical. This, as we will later see, is one of the keys to the usefulness of an op amp in many of the application circuits.

Basic Amplifier Applications of the Op Amp

We will now consider the three basic amplifier application circuits that can be obtained with an op amp. These relatively simple amplifier circuits account for a very high percentage of the actual applications of op amps.

The unity-gain amplifier, or the voltage follower, is an application circuit that seems very wasteful of voltage gain. This circuit exchanges the high open-loop voltage gain (50,000 typical) of the op amp for a predictable voltage gain of very nearly 1. No resistors are needed in this circuit, shown in Figure 1-17a. We see that the output of the op amp is simply tied back or "fed back" to the inverting input and the input signal is connected to the non-inverting input of the op amp. (We will see, in Chapter 2, that this application is using 100% feedback).

This circuit is also called a "voltage follower" because *the output voltage* almost exactly *follows the input voltage* because the high open-loop gain of the op amp doesn't allow a large differential input voltage to exist. This voltage-follower is similar in function to a single-transistor emitter-follower (Figure 1-17b) and both of these circuits are used to isolate a load from a signal source. For this reason, these circuits are sometimes also called "buffers".

The op amp circuit has many operational advantages over the emitter-follower circuit. There is no V_{BE} dc voltage drop between the input voltage and the output voltage. There is less loading of the input signal source (because the input current of the op amp is usually smaller

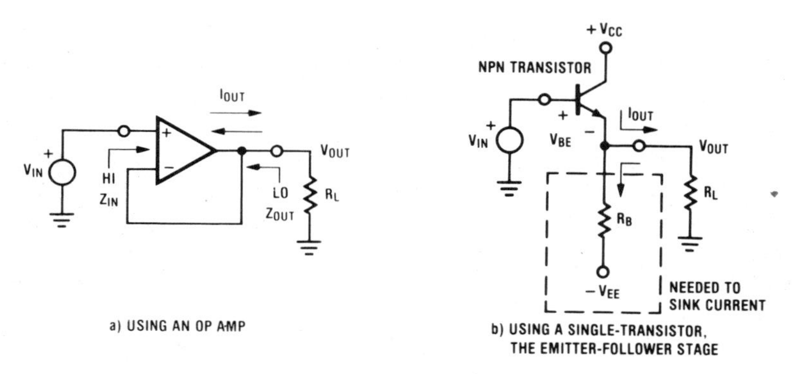

Fig. 1-17. Noninverting Unity-Gain Amplifiers

than the base current of the emitter-follower). Plus, the op amp can both source current to the load, R_L, and accept current (sink current) from the load. The NPN emitter-follower of Figure 1-17b only sources current—the additional biasing resistor shown in dotted lines, tied from the emitter to a negative power supply voltage, is needed to sink current.

The Noninverting Amplifier Application. Resistors are often used with an op amp to provide predictable noninverting voltage gain as shown in Figure 1-18. The equation for the resulting voltage gain, A_V, is also shown on this figure. This equation is an approximation and the reasons for errors in the voltage gain of this application circuit will be considered in Chapter 3.

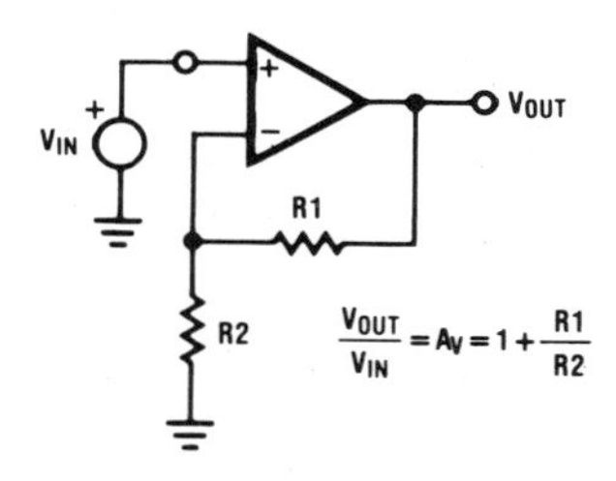

Fig. 1-18. The Noninverting Amplifier Application

The Inverting Amplifier Application. An inverting amplifier application of an op amp is shown in Figure 1-19. Notice that the equation for the approximate voltage gain of this circuit is simply the ratio of the values of the two resistors that are used in the circuit. The reasons for this difference in voltage gain between the inverting and the noninverting voltage amplifiers, even though the same feedback resistors are used, will be discussed in Chapter 2.

In these amplifier applications, we usually simply consider the resulting voltage gain to be as given by the approximate equations. In more precise applications, we may wonder what the actual voltage gains really are and how the voltage gain of a particular circuit is expected to

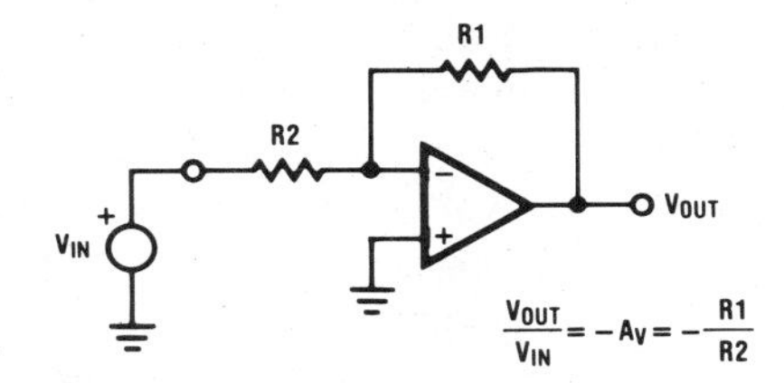

Fig. 1-19. The Inverting-Gain Application

change if we insert different op amps into the same circuit board or change the ambient temperature. We will consider these effects in Chapter 3. The practical problem of finding a non-zero dc output voltage to exist with any of these amplifiers even though the input voltage pin is tied to ground ($V_{IN} = 0\ V_{DC}$) is also considered in Chapter 3.

Finally, the reasons that any of these amplifiers can be unstable and supply their own input signals — the unintentional oscillation problem — are discussed in Chapter 4.

We will now consider some of the limitations of the op amp and we will use our basic op amp circuit to show why these limitations exist.

Limitations of the Op Amp

The voltage swing limitations (at the inputs or the outputs) of op amps result because of the magnitudes of the power supply voltages that are used ($\pm 15\ V_{DC}$ is most common). These limitations also exist for a discrete circuit op amp that is operating from these same power supplies. (We will consider the limitations that are unique to the IC op amps in Section 6.2, Being Unkind to an IC Op Amp.)

Differential Input Voltage. Large values of differential input voltage will cause one of the base-emitter junctions of the Input Differential Amplifier (Q1 or Q2) to be reverse biased, as shown in Figure 1-20. If this junction actually enters breakdown, then large values of differen-

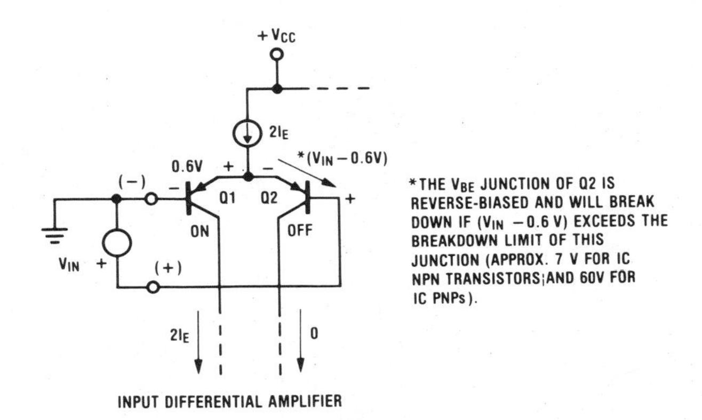

Fig. 1-20. Large Differential Inputs Can Cause Base-Emitter Voltage Breakdown

tial input current can flow. Operating a transistor with this emitter-base junction in breakdown can degrade the current gain, β (where $\beta = I_c/I_B$), of the transistor and for this reason, should always be avoided. The magnitude of this degradation is not controlled. It depends upon the chip temperature and the time-duration of the breakdown, but significant β loss can result.

Input Common-Mode Voltage. In our basic op amp circuit, the common-mode voltage that is allowed at the inputs is limited to both a maximum-positive and a maximum-negative voltage value. If an application circuit is designed to keep the common-mode voltage within this restricted dynamic range, the op amp will properly respond to any small differential signal that may exist.

For this basic op amp circuit, the positive common-mode voltage limit, shown in detail in Figure 1-21 is established by the base-emitter ON voltages of Q1 and Q2 (0.6 V_{DC} each) and a small voltage drop (say, 0.3 V_{DC}) that must be provided across the $2I_E$ biasing current source to keep it operational (and still provide a large value of output impedance) for a total of 0.9V. Therefore, the positive input common-mode limit is approximately +14.1V for a +15 V_{DC} power supply voltage. It is useful to keep in mind that the different input circuitry of the LM301 and the Bi-FET op amps allow a positive common-mode voltage range that extends to the $+V_{CC}$ supply level.

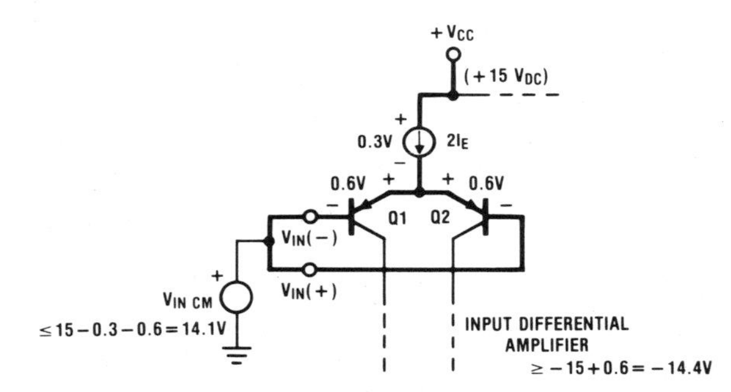

Fig. 1-21. The Positive Limit on Input Common-Mode Voltage Range

The negative input common-mode voltage limit is reached when there is insufficient biasing voltage remaining to keep the transistors of either the input differential amplifier (Q1 and Q2) or the current mirror (Q3 and Q4) from saturating. (A transistor is saturated when the base-collector junction is forward biased.) If, somewhat arbitrarily, we were to restrict the allowed minimum biasing voltage for the collector-base junctions of any of these input stage transistors to 0V, then the negative input common-mode voltage can come to within one V_{BE} (0.6 V_{DC}) of the negative power supply voltage value, or −14.4V with a −15 V_{DC} power supply, as shown in Figure 1-22. (A different input circuit is used with the LM324, LM358, and the LM339 comparators that allows the input common-mode voltage to include ground even when operated with a single positive power supply voltage.)

As long as the common-mode voltage at the inputs is kept between −14.4V to +14.1V, the basic op amp that we have been considering will properly respond to a differential input sig-

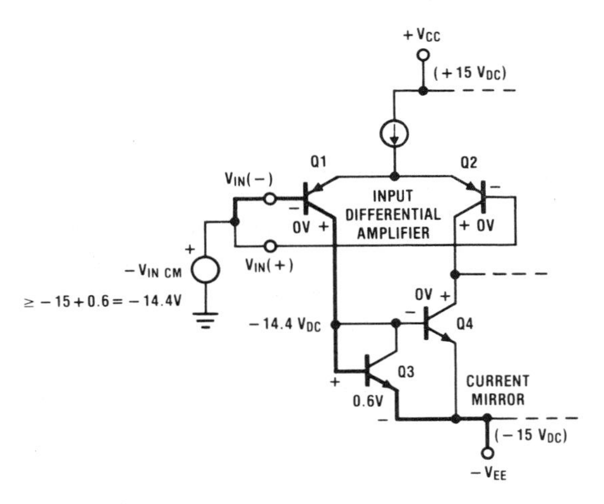

Fig. 1-22. The Negative Limit on Input Common-Mode Voltage Range

nal. Actual IC op amps usually have more complex input circuitry (with more than one V_{BE}) so the input common-mode voltage range may be approximately -13V to + 13V. Some ICs allow the common-mode voltage to range clear to + V_{CC} or to -V_{EE} (as we have previously noted) but rarely to both for the same IC.

When the common-mode input voltage is at the extreme limit established by dc biasing requirements, some of the performance specs of the op amp may suffer degradation; and, for this reason, less input common-mode voltage range is typically allowed. Because these common-mode limits involve V_{BE} voltages, they are temperature dependent (at approximately -2 mV/°C for each V_{BE}) and they are always referenced to the actual supply voltages that are used.

Output Voltage Swing. The output voltage swing of our basic op amp is limited by the V_{BE} drop of Q6 and Q7 and circuit drive considerations, as shown in Figure 1-23.

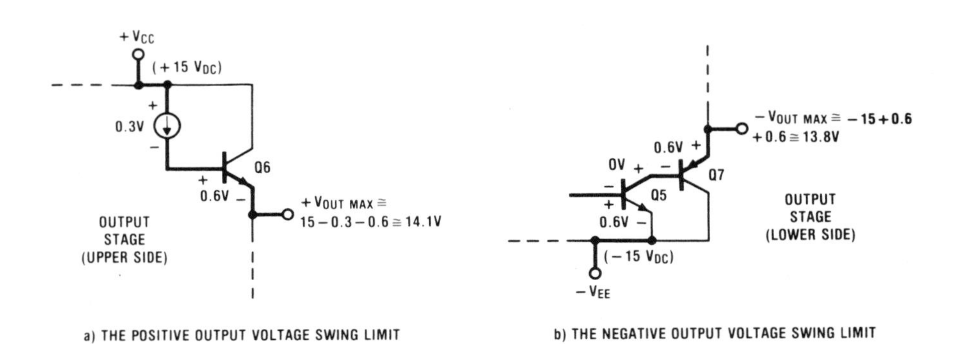

Fig. 1-23. Output Voltage-Swing Limits

When many op amps are used in a system, it is important to insure that the maximum output voltage swing of one will not exceed the input common-mode voltage range of the next. This is handled in the design of the IC op amps by keeping the output voltage swing slightly less than the allowed maximum input common-mode voltage range. (This should be verified in a system that uses many different op amp types or different power supply voltage levels.)

This oversight in the design of some early IC op amps (the 709) has caused application circuit *latch-ups* because the *sign* or phase of the inverting input pin could abruptly change owing to saturation of the input differential amplifier. This input-stage saturation resulted from the use of resistors as collector-loads for the input NPN differential amplifier as shown in Figure 1-24. When an input voltage (at the base of the input transistor) became too positive, the collector of that input transistor could enter saturation because of the large voltage that is dropped across the large-valued collector load resistor. An input signal at the base of the inverting input transistor, Q2, would now directly couple to the collector, without the normal signal inversion of this common-emitter connection. This phase change at the $V_{IN}(-)$ input would cause *negative feedback* to become *positive feedback* and application circuits would therefore latch-up.

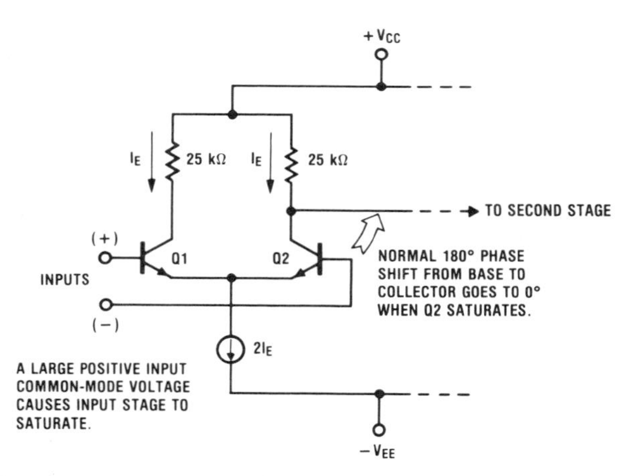

Fig. 1-24. Resistive Loads for Input Stage Allow Saturation

Output Current. The maximum output current that can be obtained from most IC op amps is approximately ± 10 mA. For some op amps, the spec on output current is given as a particular output voltage swing when driving a stated value of load resistor (such as ± 10 volts across a 2 kΩ or 10 kΩ load resistor). Output short-circuit protection circuitry is usually also provided to prevent op amp destruction for momentary shorts to ground or to either of the power supply voltages. (The time duration of an allowed short-circuit condition depends upon the power that is being dissipated in this fault-mode, the package used, the heat sinking, and the ambient temperature.)

Response Time. The capacitor C_C, that was shown on our basic op amp, Figure 1-16, is called the *comp cap*. This *frequency compensation capacitor* (of value 5 to 30 pF) is usually provided on the IC chip, but ICs are available (called *uncompensated op amps*) that require the user to add this capacitor as an external component. Now high frequency performance can be improved and response time reduced by tailoring the value of C_c to suit a specific application.

A minimum value for C_c is usually required to frequency-stabilize the closed-loop op amp application circuit (that is, to prevent undesired oscillation; we will have more to say about how this happens in Chapter 4). When more gain is taken in an application circuit, C_c can be made smaller in value and the circuit response time will therefore improve.

Notice (Figure 1-16) that one side of this capacitor is connected to a point that is at a relatively fixed voltage (the V_{BE} of Q5) above the negative power supply voltage ($-V_{EE}$), and the other side is essentially at the V_{OUT} potential (actually a diode or V_{BE} less in dc voltage, but the output voltage will follow any changes in this voltage). Therefore, to change the magnitude of V_{OUT}, we must change the magnitude of the voltage across C_c.

The voltage across a capacitor can only be changed relatively slowly, unless lots of current is available to charge or discharge the capacitor. This is why the V_{OUT} of an op amp is generally slow to respond to a rapidly changing input signal. The compensation capacitor, C_c, is charged or discharged by the I_{OUT} of the input stage. (Digital circuit designers avoid placing capacitors in shunt with the signal path because capacitors slow up circuit response time, but this is the price the op amp designer must pay for freedom from undesired oscillations.)

Notice that hum or any other noise that may exist on the $-V_{EE}$ power supply can couple to the output of the op amp because of the placement of this capacitor. This makes most op amps more susceptible to noise on the $-V_{EE}$ power supply than on the $+V_{CC}$ supply.

Where's the Ground Pin?

Another thing you may have noticed in this schematic of the basic op amp is that *there is no ground pin!* This discovery usually bothers anyone who is exposed to an op amp for the first time, because all of the basic *transistor gain stages* and even all of the *logic circuits always have a ground pin.* (The emotional desire by some designers to actually have a ground pin on an op amp product was, at one time, almost realized. Many years ago, a hybrid product that had a spare pin available on the package, was being discussed at a semiconductor factory. Rather than call this "NC," *no connection;* an astute designer, Barry Siegel, who was keenly aware of this emotional desire for a ground pin, suggested that it, instead, be called "optional ground." Unfortunately, this was not allowed, so *the conceptual problem of no ground pin on op amps remains to this day.)*

In many ways, this lack of a ground pin is the key to the op amp. The output voltage of an op amp responds only to the differential voltage that exists between the two inputs. As will be seen in Chapter 5, *ground can be introduced to the op amp in many different ways by the external components of the application circuit. This adds to the flexibility of op amp uses.*

A Simple Model for the IC Op Amp

An equivalent circuit for an ideal op amp is shown in Figure 1-25. Notice it has an infinite input impedance, no dc input currents are needed, there is no offset voltage (the very small differential dc input voltage that is actually required to make the output voltage go to zero for a zero input voltage condition), and neither input lead is ground referenced. There is also no restraint on the magnitudes of input voltages that can be applied. The sign of the zero impedance active voltage source (shown with a voltage gain, A_V, of infinity, ∞), that provides the ground-referenced output voltage, V_{OUT}, depends only on the difference voltage, V_D, that exists between the two inputs. Additionally, if $V_{IN}(+)$ is only slightly greater than $V_{IN}(-)$, V_{OUT} will be at the most positive voltage limit. But, if $V_{IN}(-)$ is only slightly greater than $V_{IN}(+)$, V_{OUT} will be at the most negative voltage limit. (This control of the sign of the output voltage is the reason for the "+," *non-inverting;* and "–," *inverting,* distinction between the input terminals.)

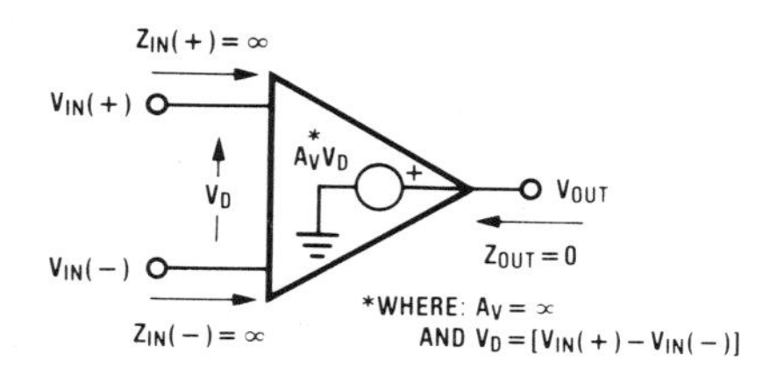

Fig. 1-25. The Idealized Op Amp Equivalent-Circuit

This drastic output voltage response to small input voltage differences results from the large value of voltage gain of an op amp. Even if we assume that this voltage gain is only 50,000 (a typical value), a signal of only 1 mV would *attempt* to drive the output voltage to 50V ($1 \times 10^{-3} \times 5 \times 10^4$). If we have only $\pm 15\ V_{DC}$ power supplies for the circuit ($V_{CC} = +15\ V_{DC}$ and $V_{EE} = -15\ V_{DC}$) the output voltage can, at most, only range from + 15V to –15V (and will typically be less, owing to voltage swing loss in the output stage as we discussed earlier in this chapter).

When a *large* differential input voltage (as this 1 mV example) is applied directly to the input pins of an op amp, the op amp is said to be *overdriven* or driven into *saturation* and the resulting output voltage is said to be *against the rails.* (This term comes from the use of the word *rails* or *power supply rails* to replace *power supply lines.)* If we had an op amp that had a maximum output voltage swing of ± 12V and a gain of 50,000, it would produce the idealized transfer function (V_{OUT}/V_{IN}) shown in Figure 1-26. This seems to indicate that the op amp is almost always overdriven — *it is only linear over a 500 μV input voltage range (–250 μV to + 250 μV).*

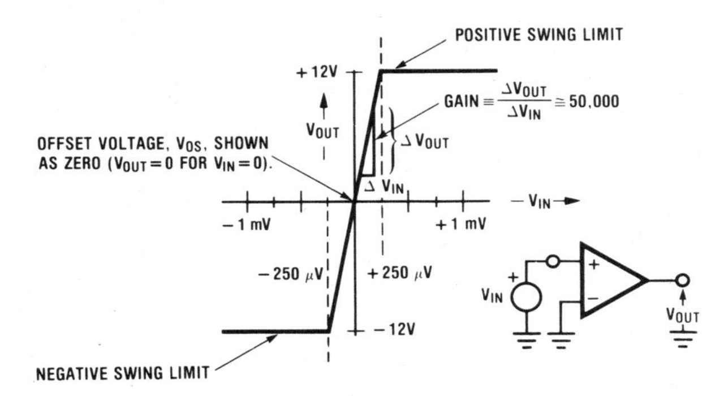

Fig. 1-26. Open-Loop Transfer Function of an Op Amp

High-gain op amps (A_V > 10,000) cannot easily be held within their linear range when operated without feedback, because offset voltage drift and noise (including ac signal pickup at the input) have magnitudes that are comparable to the small linear input voltage range of only ± 250 μV.

Any desired output voltage over the complete range from −12V to + 12V can be provided with such a small change in the input voltage that the actual value of the input voltage is usually neglected and assumed to be equal to zero. This is the basic reason for the usefulness of the op amp in feedback applications — *we can count on $V_{IN}(-)$ being equal to $V_{IN}(+)$* as long as the output voltage is under the control of the input voltage, or an equivalent statement is: *providing the amplifier is operating within its linear range* (and has not been driven into saturation causing V_{OUT} to be against the rails).

If an op amp has no feedback loop (that is, it is not operating in the usual closed-loop condition: there are no external components that *loop* or tie from the output back to either input), the op amp is operating *open loop*. This is one of the many uses for an op amp: a voltage comparator.

A Model for AC Gain. A modern IC op amp consists of a very complex interconnection of a large number of transistors, resistors, and capacitors (sometimes there is more than one). This large number of components obscures the understanding of the basic functioning of the overall circuit. Even the IC op amp designers benefited when Jim Solomon presented a simple ac model for the typical IC op amp, Figure 1-27. (Not *all* op amps fit this two-stage model, but the majority do.) The surprising thing is that this relatively simple ac model is very useful, and was even responsible for indicating the ways to improve the performance of the IC op amp.

The input differential amplifier stage is modeled as a *transconductance, g_m, block.* As we noticed earlier, this stage accepts a differential input voltage, v_{IN}, and produces a single-ended output current, i_{OUT1}. Therefore, the transfer function (or *gain*) of this stage is i_{OUT1}/v_{IN} and has the dimensions of 1/resistance or conductance. Because we are transferring from the input to the output, this is called *transconductance* and has the symbol g_m. The magnitude of g_m indicates how much output current is produced from a given, small-signal, differential input voltage, or

$$i_{OUT1} = g_m v_{IN} \tag{1-1}$$

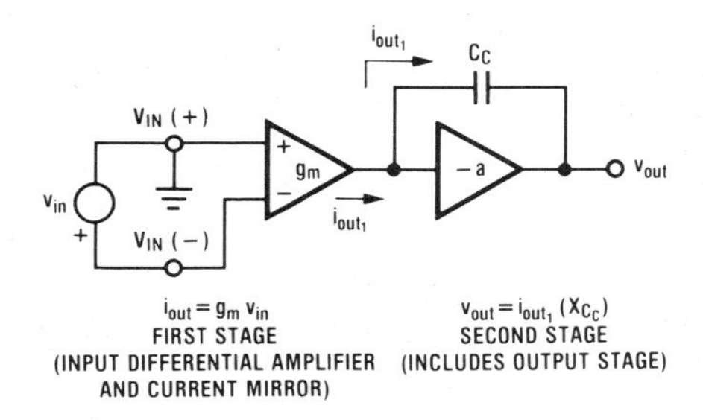

Fig. 1-27. A Simple Model for the AC Gain of an IC Op Amp

The second stage of the op amp is modeled as an inverting amplifier that has the frequency compensating capacitor, C_C, connected as a feedback element from the output to the input. This is called a *transimpedance amplifier* because the transfer function of the circuit (the output voltage to input current relation) is held equal to the *impedance* of the feedback element (the output voltage results from the flow of I_{OUT1} through C_C). In this case, this impedance is the capacitive reactance of C_C, or

$$\frac{v_{OUT}}{i_{OUT}} = X_{C_C} = \frac{1}{j\omega C_C} \tag{1-2}$$

where $\omega = 2\ \pi f$

Now, to determine the overall transfer function for the op amp, we can solve (1-2) for v_{OUT}, as

$$v_{OUT} = \frac{i_{OUT1}}{j\omega C_C}$$

and substituting for i_{OUT1} from (1-1),

$$v_{OUT} = \frac{g_m\ v_{IN}}{j\omega C_C}$$

or the open-loop ac gain, A_V, of the op amp becomes

$$A\ (\omega) = \frac{v_{OUT}}{v_{IN}} = \frac{g_m}{j\omega C_C}$$

which provides a surprisingly simple result: the first stage can be completely specified by g_m and the second stage by C_C, and *this is all there is to the ac gain of an op amp.* [This is restricted to ac because the capacitive feedback provided by C_C does not control the dc gain of the second stage. In fact, this equation (1-3) implies an infinite gain at dc ($\omega = 0$), but the overall dc gain of the op amp is limited by the finite dc gain of the second stage.]

This simple gain equation is very useful. Notice that the ac gain of an IC op amp at a particular medium frequency (say, 1 kHz) is under better control (owing to the stabilizing effects of the local negative feedback that is provided by C_C) than the dc gain because of the uncertainty of the dc open-loop gain of the second stage. Therefore, large variations in the dc gain among op amps of the same type can exist, yet all units can have approximately the same unity gain frequency.

This equation has also pointed the way to the solution of the fabrication problems of the early dual 741 (the 747) that resulted because of the two, large, die-area consuming, 30 pF, comp caps. Equation (1-3) shows that *if C_c is reduced by the same ratio as g_m — the ac gain will remain unchanged. This was the key to reducing comp caps from 30 pF to 5 pF.* Low-cost quad op amps then could be built and small die-size 741s could then be introduced. (We will consider this again when we discuss the LM324 later in this chapter.)

Finally, this equation also indicates how the ac open-loop gain can be made essentially independent of temperature. To achieve this, g_m is made independent of temperature in the circuit design, because the on-chip capacitor, C_c, is essentially independent of temperature. (This was done in the design of the LM324 quad op amp.)

In general, the voltage gain, A, of an op amp can be expressed as a non-dimensional numeric value (the ratio of volts OUT to volts IN), V/mV (volts OUT to mV IN, to reduce the magnitude of the number), or it can also be expressed in decibels (dB), as linear system designers like to do, where

$$A\ (\text{dB}) = 20 \log \frac{V_{OUT}}{V_{IN}}$$

So, if $V_{OUT} = V_{IN}$ at some high frequency, we have a gain of 1, or a gain of zero dB. [This frequently appears in dealing with op amps and can cause some initial confusion because *"a gain of zero (dB) equals one"* seems in error even if you say it slowly.]

To show some more uses for this simple equation for the ac gain, we will now determine the high frequency gain limits of an op amp.

Predicting the Unity-Gain Frequency. A measure of the high-frequency usefulness of an op amp is provided by the specification called the *unity-gain frequency,* f_u, [this is also sometimes called *gain-bandwidth product* (GBW) or *unity-gain cross frequency* because the open-loop gain characteristic *passes through or crosses unity gain* at this frequency, as shown in Figure 1-28]. Input signals at this frequency will therefore experience an open-loop gain of only unity, so the amplifying benefits of the op amp are limited to frequencies less than f_u.

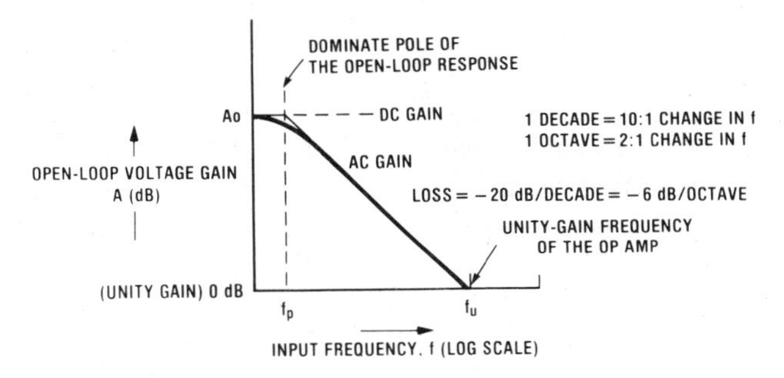

Fig. 1-28. The Open-Loop Response of an Op Amp

Once f_u is known for an op amp, the open-loop gain at any frequency (except dc and very low frequencies) can be easily determined because

$$A_V(f) = \frac{f_u}{f}$$

If we operate at a frequency which is 0.1 f_u, for example, this equation indicates that the open-loop voltage gain will be 10 (20 dB). The slope or rate of the gain loss is −20 dB/decade (10 to 1 change in frequency) or −6 dB/octave (2 to 1 change in frequency). This results from the steadily decreasing impedance of C_c as frequency is increased; capacitive reactance varies inversely with frequency.

We can determine what controls f_u by solving equation (1-3) for the frequency that makes $A(\omega)$ equal to unity, or looking again at this equation

$$A_V(\omega) = \frac{g_m}{j\omega C_c}$$

an expression for the magnitude of gain becomes

$$\left|A_V(\omega)\right| = \frac{g_m}{\omega C_c}$$

and an expression for f_u is obtained as

$$\left|A(f_u)\right| = \frac{g_m}{2\pi f_u C_c} = 1$$

or

$$f_u = \frac{g_m}{2\pi C_c} \tag{1-4}$$

This indicates that to obtain a large value for f_u requires a large value for g_m and a small value for C_c. (This ability of C_c to control or limit f_u is the main function of this capacitor.) Large values of f_u are desired to increase the frequency performance of the op amp, but the upper limit that can be allowed for f_u is dictated by the requirements for closed-loop stability (we will consider this in Chapter 4).

An Op Amp Versus a Low Pass Filter. The open-loop gain of an op amp, from (1-3), indicates that the gain falls as the frequency is increased, and a maximum or limiting value exists at zero frequency (dc). This is the same general transfer function as is obtained with an RC low pass filter. Both are shown in Figure 1-29.

Here, we can see that the *corner frequency* or *first pole* (we will have more to say about *poles* and *zeros* in Chapter 4) of this function occurs at a frequency represented by f_p. For the IC op amp, this is usually in the range of 3 Hz (741) to 50 Hz (for high frequency op amps).

Programmable op amps (such as the LM346) allow the user to control *(to program)* the biasing currents and, therefore, the magnitude of the first-stage g_m can be varied. This also controls the magnitude of f_p and allows a controllable low pass filter response characteristic. (This has been used as a means of obtaining a class of RC active filters that do not require external capacitors.)

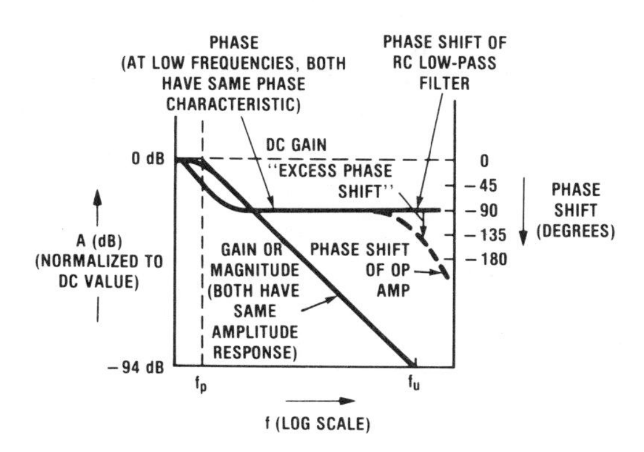

Fig. 1-29. Gain and Phase of Both an RC Low-Pass Filter and an Op Amp

Notice that the shape of the gain or magnitude responses of the RC low pass filter and the op amp are the same (over the frequency range shown) — the only difference is in the dc gain that is provided: 1 (0 dB) for the passive filter; 50,000 (94 dB) for the typical op amp.

The phase responses of these two circuits depart as the frequency approaches f_u, the unity-gain frequency of the op amp. The low pass filter has a single capacitor, so the limiting or maximum phase lag is -90°. The comp cap of the op amp provides the dominant pole, f_p, that is similar to the single pole of the RC filter. Additional phase shift occurs at high frequencies for the op amp because the signal must pass through all the complex active and passive circuitry of the op amp. This gives rise to what can be called *excess phase*. (This term has been used for the excess phase shift that occurs in a single transistor.) *It is this rapid accumulation of phase lag at high frequencies that can cause instability in a closed-loop op amp application circuit* and forces the op amp designer to provide a large enough value for the comp cap to keep f_u low enough to prevent operating where this excess phase shift comes too close to -180°.

Larger values of C_c will reduce f_p (move f_p to the left), so the transfer function will shift to the left to a new line that is parallel to the one shown in the figure. This new magnitude response will therefore intersect both the upper dc gain line and the lower -94 dB gain line at lower values of frequency (both f_p and f_u will reduce). Stability is improved because less excess phase is picked up at the highest frequency where the op amp still has at least unity gain, f_u. (There will be more on this in Chapter 4.)

We will now consider the slew rate limits of an op amp and indicate how the design of an IC op amp balances or trades off these various specs.

Predicting Slew Rate Limits. The presence of the comp cap limits the slew rate, *the maximum rate-of-change of the output voltage of the op amp.* (This name was derived from the terminology of servomechanisms where "slew rate" implies how fast an electronically-positioned mechanism, such as a radar antenna, can be moved.) The reason for the slow response of an op amp can be understood if we return to the simple models of the IC op amp and redraw them as shown in Figure 1-30. The maximum current out of the first stage is $\pm 2I_E$ ($\pm 20\ \mu A$ in this example of the 741 input stage). This is the only current that is available to change the voltage at V_{OUT}, and C_c limits just how rapidly this can be done, because, for any capacitor,

$$\left.\frac{dV_{cap}}{dt}\right|_{max} = \frac{I_{max}}{C}$$

So, for the 741 op amp,

$$\text{Slew rate} = \left.\frac{dV_{OUT}}{dt}\right|_{max} = \frac{\pm 2I_E}{C_C} = \frac{\pm 20\ \mu A}{30\ pF} = \pm 0.67\ V/\mu sec$$

This is the reason for the typical slew rate limit of the common IC op amps.

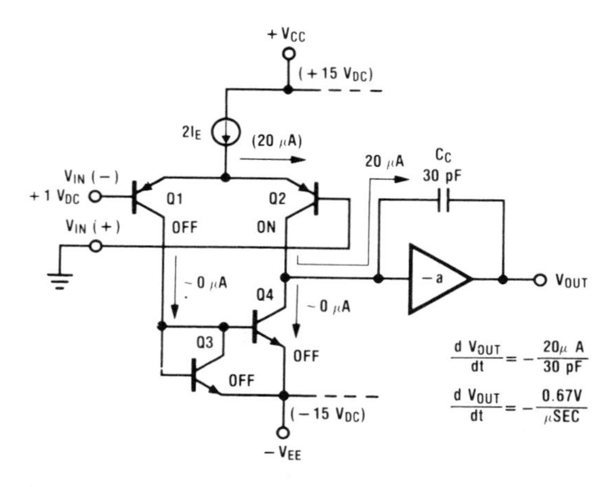

Fig. 1-30. The Reason for the Slew-Rate Limit of the 741 Op Amp

This requirement for current out of the input differential stage is also the reason why an input error voltage is necessary anytime the output voltage of an op amp is changing or ramping. To ramp the voltage across C_C, an output current is required from the first stage and, to produce this output current, an input dc error voltage must exist. This is the *rate, or ramp, error.* (We will consider this in the next chapter.)

It is important to notice that *to realize the slew rate spec requires a relatively large differential input voltage* (greater than ± 120 mV for a bipolar op amp and greater than ± 1 to ± 3V for FET input op amps) to insure that the input differential amplifier is *fully switched* (with one side essentially OFF). This will guarantee that the full $2I_E$ current is available to change the voltage across C_C at the maximum rate.

To obtain a large slew rate, we need lots of first-stage biasing current, $2I_E$, and a small value for C_C. Unfortunately, if the internal biasing current, I_E, of the input transistors is increased, the g_m (and, therefore, the ac gain) also increases; because, for a bipolar transistor,

$$g_m = \frac{qI_E}{kT} = \frac{I_E}{26\ mV}$$

But, if we could increase the first-stage biasing current ($2I_E$), and/or decrease C_c, we would increase the slew rate. Both of these changes, unfortunately, also undesirably increase f_u and thereby cause frequency stability problems. *To improve only slew rate, we need to increase biasing current without also increasing g_m. This naturally results with the relatively poor g_m-per-milliampere of bias current performance of the JFET.* The unusual thing is that this *poor performance is just what is needed for the input stage of an op amp.*

For a JFET, operating at I_{DSS}, the relationship between source current, I_S, and g_m is

$$g_m = \frac{2I_S}{V_{PO}}$$

where: V_{PO} = the pinch-off voltage of the JFET (2V typical for IC JFETs). The ratio of transconductance to source current for this JFET, a figure of merit for an amplifying device, becomes

$$\frac{g_m}{I_S} = \frac{1}{1V}$$

This clearly shows the small value of g_m/I_S that is obtained with the JFET (1/1V) when compared with the g_m/I_E of a bipolar transistor (1/26 mV). The biasing current of the FET can, therefore, be increased in the ratio of 1V/26 mV or approximately 40:1 (and this will increase the slew rate by this same factor) for the same value of g_m. *This simple fact, that higher biasing currents can be used for a given g_m in an FET transistor, is the basis for the high slew rate of the Bi-FET op amps where JFETs are used for the input differential amplifier.* (We will have more to say about Bi-FETs later in this chapter.)

Large-Signal Frequency Limits

The slew rate limit in the time domain appears as a large-signal limit in the frequency domain. For a sinusoidal output voltage, the maximum rate-of-change of this voltage can be determined by looking at the derivative of a sinewave with respect to time, or if

$$V_{OUT} = V_p \sin \omega t$$

where V_p is the peak value of the sinewave, then the time derivative becomes

$$\frac{dV_{OUT}}{dt} = \omega V_p \cos\omega t$$

and because the maximum value of the cos is + 1,

$$\left.\frac{dV_{OUT}}{dt}\right|_{max} = \omega_{max} V_p$$

This explains why op amps can handle low amplitude high frequency signals. Because slew rate depends on both the operating frequency and the peak voltage, a large-amplitude output voltage swing is difficult to achieve at high-frequency. (This effect can be easily noticed, because reducing the amplitude of a relatively high frequency sinewave will remove the slew rate distortion that is caused by the slow response time of the op amp.)

We have found that slew rate limits the dV_{OUT}/dt of an op amp, so we can equate slew rate to the maximum rate-of-change of a sinewave to determine the large signal frequency capabilities of an op amp, as

$$\omega_{max} V_p \leq \text{slew rate}$$

or

$$\omega_{max} \leq \frac{\text{slew rate}}{V_p}$$

The inequality in this equation for the maximum frequency (also called *power bandwidth*) is indicating that this frequency should be less because *slew distortion* will occur at a somewhat lower frequency. For the 741 op amp with a slew rate of 0.67 V/μsec, the maximum frequency that can be provided with small distortion at a 20V peak-to-peak output amplitude (V_p = 10V) is

$$\omega_{max} \leq \frac{0.67\ \text{V}/\mu\text{sec}}{10\text{V}} = 6.7 \times 10^4 \text{ radians/sec}$$

or

$$f_{max} = \frac{\omega_{max}}{2\pi} = \frac{6.7 \times 10^4 \text{ radians/sec}}{2\pi \text{ radians/cycle}} = 10.7 \text{ kHz}$$

When an op amp is operating at the slew rate limit, there is no longer any feedback; the op amp is operating open-loop (the input differential amplifier is fully switched). This loss of the linearizing benefits of feedback causes increased distortion of the output waveform. Slew distortion eventually reduces the output voltage response to high-frequency sinusoidal input voltages. Triangular waveforms are produced because the op amp can't follow even the slower portions of the input voltage waveform. Therefore, the op amp will eventually, at high frequencies, simply switch between the positive and negative slew rate limits as the input sinewave alternately changes the sign of the differential input voltage. (This will cause a dc shift in V_{OUT} if the positive and negative slew rates are unsymmetrical.) If the input frequency continues to be increased, then the peak-to-peak swing will steadily reduce in magnitude until the triangular output voltage waveform of the op amp eventually disappears.

If we now consider the time domain or step response of an op amp (the output voltage response to an abrupt or step change in input voltage), we find that there is a difference in the response that depends on the magnitude of the change in output voltage. We will first look at the small-signal case.

Small-Signal Rise-Time Limits

The closed-loop, small-signal bandwidth of an op amp, [f(−3 dB)], limits the rise time, t_r, that can be obtained in the output voltage waveform following a small-signal step change in the input voltage. (The rise time is defined as the time it takes a voltage to change from 10% to 90% of the final voltage change.)

The one key equation (sometimes printed on oscilloscopes) that links the time domain to the frequency domain is that the rise time of a system with a single high frequency roll-off depends on the closed-loop small signal bandwidth as

$$t_r = \frac{0.35}{f\,(-3\text{ dB})}$$

This equation can be derived by considering the small-signal step response of a single-pole, low-pass amplifier (where we are assuming there is no significant overshoot or ringing in the response) which is given by

$$\Delta V_{OUT}(t) = \Delta V_{OUT,final}\,[1 - \exp\,(-t/\tau_c)]$$

where

$$\tau_c = \frac{1}{2\pi f_c}$$

and

$$f_c = -3-\text{dB bandwidth of the amplifier}$$

The previously used concept of "rise time" is often used because it provides a description which is based on easily identified points on this constantly changing exponential waveform.

We can calculate the rise time of this exponential waveform by making use of the definition, or

$$t_r = t_{90\%} - t_{10\%}$$

where $t_{10\%}$ can be found from

$$0.1\ V_{OUT,final} = \Delta V_{OUT,final}\,[1 - \exp\,(t_{10\%}/\tau_c)]$$

$$0.9 = \exp\,(-t_{10\%}/\tau_c)$$

$$\ln\,(0.9) = -t_{10\%}/\tau_c$$

$$t_{10\%} = -[\ln\,(0.9)]\tau_c$$

$$t_{10\%} = 0.11\tau_c$$

and $T_{90\%}$ is found from

$$0.9V_{OUT,final} = \Delta V_{OUT,final}\,[1 - \exp\,(-t_{90\%}/\tau_c)]$$

or

$$t_{90\%} = -(\ln\,0.1)\tau_c$$

$$t_{90\%} = 2.3\tau_c$$

now

$$t_r = (2.3\tau_c - 0.11\tau_c) = 2.19\tau_c$$

but

$$\tau_c = \frac{1}{2\pi f_c}$$

so

$$t_r = 2.19\left(\frac{1}{2\pi f_c}\right) = \frac{0.35}{f_c}$$

which is the previous equation that will now be used to simplify the calculations when there are only small changes in V_{OUT}.

To determine the small-signal bandwidth, we look again at the unity-gain frequency, f_u, often called the *gain bandwidth* (or *gain-bandwidth product)* of an op amp because f_u limits the product of the noninverting closed-loop gain and the resulting small-signal bandwidth. (This will be discussed in more detail in the next chapter.) This unity-gain frequency, f_u, spec indicates how the closed-loop bandwidth, f (−3 dB) or f_c, falls as we take more noninverting, closed-loop gain, A_{cl}, as shown in Figure 1-31. This ration is given by

$$f\,(-3\text{ dB}) = \frac{f_u}{A_{cl}} = f_c$$

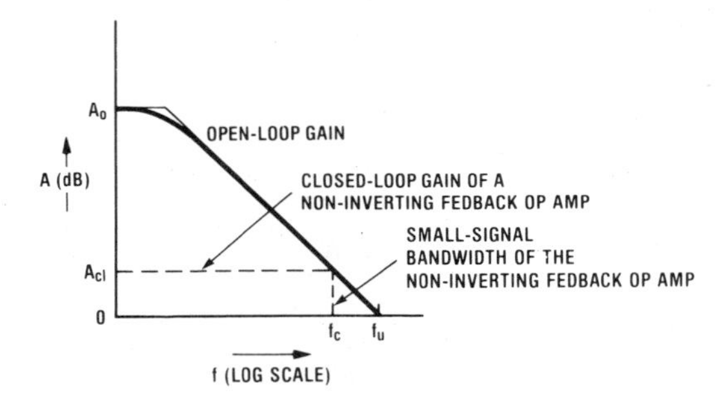

Fig. 1-31. Closed-Loop Bandwidth Depends on Closed-Loop Gain

This equation can be substituted for f (−3 dB) in the previous equation to provide

$$t_r = \frac{0.35}{f_u}\,(A_{cl})$$

For an op amp with f_u = 1 MHz (like the 741 and most of the popular op amps), the rise time can be related to the value of noninverting, closed-loop gain as

$$t_r = 0.35\,(A_{cl})\ \mu\text{sec}$$

This equation now shows that at unity gain (closed-loop gain equals + 1), the rise time to expect from a 1 MHz op amp is 350 nsec. But if a closed-loop gain of + 100 were taken, then

$$t_r = 0.35\,(100)\ \mu\text{sec} = 35\ \mu\text{sec}$$

A higher frequency op amp will reduce the rise time because the above equations will be divided by the higher value of unity-gain frequency (expressed in units of MHz). So a Bi-FET op amp with $f_u = 4$ MHz, for example would provide

$$t_r = 0.09\ (A_{cl})\ \mu sec$$

In a typical application, either this rise-time limit or limits that are associated with slew rate will apply. *The slew rate is the ultimate limit on the time rate-of-change of the output voltage. It is independent of the closed-loop gain. For small changes in V_{OUT}, the slew rate may not be the limiting factor. It may be the small-signal, closed-loop bandwidth.* We will now consider this additional limit on the output voltage response time.

Small-Signal Settling-Time Limits

The small-signal settling time of an op amp application circuit relates to the upper corner-frequency, f_c. For a small-signal input voltage step, the resulting exponential output waveform asymptotically approaches the steady-state value, V_F. This exponential output voltage response can be represented as

$$V_{OUT}\ (t) = V_F\ [1 - \exp\ (-t/\tau_C)]$$

where the time constant, τ_C, is given by

$$\tau_C = \frac{1}{2\pi f_c}$$

Responses to small-signal input voltage steps will be within 1% of the final value in 4.6 τ_C and within 0.01% in 9.2 τ_C. For example, an op amp application circuit with an $f_c = 100$ kHz would provide a τ_C of

$$\tau_C = \frac{1}{2\pi \times 10^5} = 1.59\ \mu sec$$

Therefore, in 7.3 μs the output voltage will be within 1% of the final value (and in 14.6 μs within 0.01%) following a small-signal input voltage step.

When a large step change in input voltage is made, the output voltage can exhibit an initial time delay and then it will move at the slew rate limit; the op amp is initially operating open-loop. As the output voltage approaches the steady-state value, the differential input voltage to the op amp reduces in magnitude and ultimately the feedback loop again starts to operate. Additional time is then needed for the amplifier to settle to the final value of the output voltage. This combination of slewing and then settling to within a predetermined error band of the final output voltage change is usually simply called *settling time* and will now be considered.

Large-Signal Settling-Time Limits

Op amps that are used in high speed pulse application or that are associated with digital computer interfacing (such as digital-to-analog converters) often operate with abrupt input voltage changes. Special op amps have been designed to operate with these rapidly changing signals.

The key feature of these op amps is a good settling-time characteristic. Op amps that were not specially designed for good setting-time response can exhibit overshoot and an oscillatory convergence to the final output voltage value that greatly extends the settling time.

The settling-time specification of an op amp has been standardized to be associated with an output voltage change of 10V in a unity-gain inverting amplifier configuration. (Inverting amplifiers are capable of faster response because there is essentially no common-mode voltage change.) We are now concerned with slew rate, the previously discussed rise-time effect and the dynamics of the fed-back amplifier.

In the simplified settling-time response, shown in Figure 1-32, (where we are neglecting the small initial delay in the output response), a final exponential waveform starts as soon as the op amp stops slewing and is again operating closed loop. (We assume that the slopes of the output voltage responses match at the changeover point and further assume that there are no internal circuit design deficiencies that cause problems in this transition interval.) The asymptotic approach that this exponential waveform makes to the final output voltage, V_F, causes the settling time to increase beyond the shorter time that would result if the op amp could slew all the way to V_F. For example, if we demand that the output voltage of the op amp be within a very small error band of the final value, we must wait for a time that represents many time constants of the final exponential waveform. For this reason, the settling-time spec must also state how much error is allowed. We may want to be within 1% (100 mV error), 0.1% (10 mV error) or; more typically, for the settling-time spec of the faster op amps, 0.01% (1 mV error) of the final 10V value.

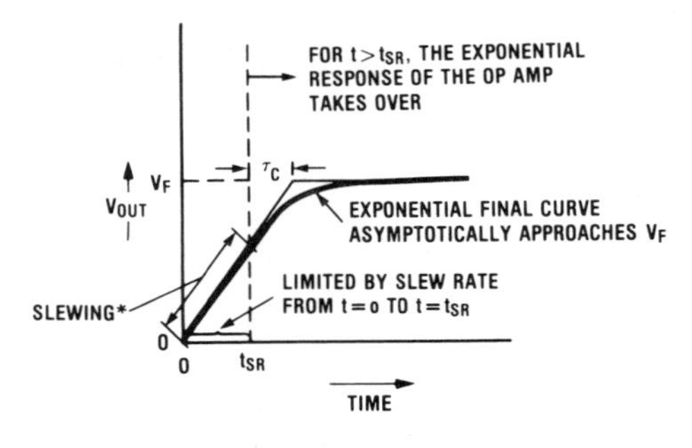

Fig. 1-32. Settling Time of an Op Amp

Op amps that are designed for good settling time typically use a single comp cap as in our simple model; feedforward and other complex compensation schemes (using multiple caps) usually don't settle as rapidly. The op amp design must also insure a well-behaved closed-loop response that is adequately compensated to prevent overshoot and ringing. Even for fast settling op amps, a major part of the settling-time spec is consumed by slew rate. (In an application that does not require a full 10V output voltage change, the time to settle is, of course, shortened.)

We can use an op amp with a good settling spec, the LF356, as an example to show the relative components of a 0.01% settling time. Typical specifications for this amplifier are: slew rate = 12V/μsec and f_u = 5 MHz.

The test circuit for the settling-time spec is a unity-gain inverting amplifier, so the small signal bandwidth, f_c, is reduced to 2.5 MHz. (The reason for this factor of 2 loss in bandwidth will be discussed in Chapter 2.)

If we assume the amplifier immediately starts slewing, the output voltage could change the required 10V in a time, t_f, given by

$$t_f = \frac{\Delta V_{OUT}}{SR} = \frac{10V}{12V/\mu sec} = 0.83\ \mu sec$$

The exponential waveform that results near the end of the settling-time interval (when the small-signal response takes over) has a time constant, τ_c, that is given by

$$\tau_c = \frac{1}{2\pi f_c} = \frac{1}{2\pi\ (2.5 \times 10^6)}\ sec$$

or

$$\tau_c = 0.064\ \mu sec$$

so the amplifier will slew for 0.77 μsec ($t_f - \tau_c$ as shown in Figure 1-26), and then the exponential waveform will smoothly take over for the last ΔV_{OUT} that is given by

$$\Delta V_{OUT} = \frac{0.064\ \mu sec}{0.83\ \mu sec} \times 10V = 770\ mV$$

This is not a large enough differential input voltage to develop full slew rate in a Bi-FET op amp.

We have to wait until this final exponential waveform gets to within 0.01% of 10V (within 1 mV of 10V) of an exponential waveform that will only change a total of 770 mV. The time to accomplish this, t_e, can be found from the final exponential equation as

$$V_{OUT}\ (t) = V_{OUT}\ [1 - \exp\ (-t_e/\tau_c)]$$

where:

$$V_{OUT}\ (t) = 769\ mV \text{ and } V_{OUT} = 770\ mV$$

therefore

$$\left[1 - \frac{769}{770}\right] = \exp\ (-t_e/\tau_c)$$

and

$$t_e = -\tau_c\ \ell n \left(1 - \frac{769}{770}\right)$$

$$t_e = (0.064\ \mu sec)\ (6.65)$$

or

$$t_e = 0.43\ \mu sec$$

The settling time can be approximated as the sum of the 0.77 μsec slewing time and this final 0.43 μsec exponential delay time or $t_s \cong 0.77 + 0.43\ \mu sec = 1.2\ \mu sec$. As a result of neglecting the initial time delay and the loop dynamics, this is 20% less than the 1.5 μsec settling-time spec, but it does provide some insight on settling time. The design problems of achieving a settling-time spec of 400 nsec in a relatively new Bi-FET op amp, the LF400, can now be appreciated.

1.3 THE EVOLUTION OF THE MONOLITHIC OP AMPS

Op amps became popular because system designers did not have to be concerned about the *details* of their design work. The use of op amps allowed them to concentrate more on the requirements of the system they were building. Systems with improved performance could be built at low cost. The actual number of transistors used in both linear and digital systems today has increased many orders of magnitude. The surprising thing is that there is usually less total power consumed.

The performance of the monolithic op amp has been limited over the years by the fabrication technologies and the active devices that could be used in a monolithic design. In 1963, the first IC op amps appeared and these made use of only NPN transistors and diffused resistors.

The standard NPN transistor that was developed for the early digital IC circuits is shown in Figure 1-33. These very high-performance NPN transistors were the starting point for the linear IC designers. The first change that was necessary was to raise the breakdown voltage capability of these transistors. This was accomplished by using lighter doping regions and deeper base diffusions.

The diffusions that are used to fabricate this transistor structure are also used to provide many other useful circuit elements. For example, it was well known by the early digital IC designers that the diffusion that was used to form the base of this transistor could also be used to make P resistors and the diffusion that is used for the emitter of this transistor could also be used to make N^+ resistors, as shown in Figure 1-34.

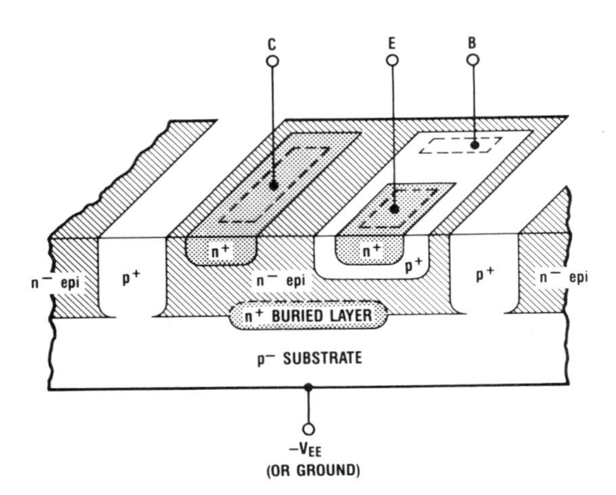

Fig. 1.33. The Standard IC/NPN Transistor

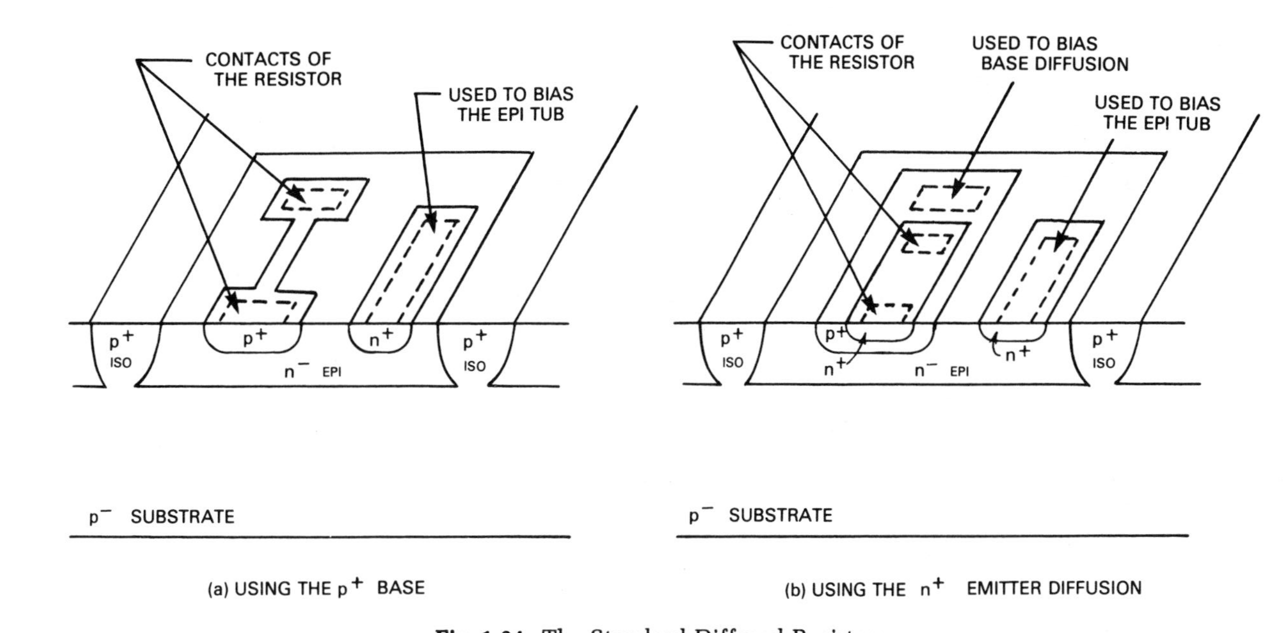

Fig. 1-34. The Standard Diffused Resistors

The linear IC designers added the pinch resistor and the epi FET resistor shown in Figure 1-35. Notice that in both of these resistor structures a diffusion of the opposite type is used to constrict the resistance region and thereby raise the resistance value. Applications of the pinch resistors must limit the voltage across these resistors to less than the zener breakdown voltage (typically 7 V).

A few other useful circuit elements, which are compatible with the standard process that is used to make the NPN transistor, are shown in Figure 1-36. The MOS capacitor shown in Figure 1-36a uses a large sheet of the N^+ emitter diffusion to form the lower or "silicon" plate of this capacitor, and a thin layer of silicon dioxide over this diffused region is the dielectric layer of this capacitor. This oxide layer is then covered with the aluminum interconnect metal to form the upper plate, the "metal" plate, of the metal-oxide-silicon (MOS) capacitor.

Zener voltage-reference diodes are formed by making use of the voltage breakdown (the BV_{EBO}) of the emitter and the base diffusions of the standard NPN transistors. Operation at low current flow can be enhanced by making use of the "finger" structure shown in Figure 1-36b. Because the area that is in breakdown is made very small, it is only along the perimeter of the N^+ diffusion that overlays the P-type base diffusion.

The Search for a "Good" PNP

The linear IC designers have always been plagued because the standard IC process does not provide a good PNP transistor. Many ingenious substitutes have been used, but this problem remains today (although more complex processes are now solving this problem). It is interesting that the bipolar digital ICs did not require a PNP and it is only the more recent shift to a CMOS process that has provided the benefits of a good PNP for the design of low current-drain logic circuits.

Lateral PNPs Are Discovered. The low dc voltage gains of the early IC op amps were greatly improved by the use of the lateral-PNP transistor shown in Figure 1-37. Notice that the transistor action is no longer in a path that is vertical to the die surface (as with the NPN transistor); the path is now horizontal, or lateral, to the surface. This lateral device was first presented in a paper in the Proceedings of the IEEE in December 1964, by H. C. Lin, T. B. Tan, G. Y. Chang, B. Van Der Least, and N. Formigoni. These PNP transistors also solved the dc voltage level shifting problems in the design of op amps and also made good constant-current biasing sources.

The early lateral PNP transistors could only handle relatively small values of collector current. This limitation was solved by incorporating a couple of NPN transistors with a lateral PNP to make an overall "composite PNP" transistor as shown in Figure 1-38. The collector current of the lateral PNP is very low and this composite PNP has been used in the output stages of many linear ICs. A major problem with this composite PNP is that the relatively low frequency response of the lateral PNP tends to cause this circuit to exhibit a local oscillation. Much design effort has been expended to reduce this problem, but this composite PNP circuit has not provided a very satisfactory solution to the problem of providing a good PNP.

With the later appearance of high β lateral PNPs came the μA709 in 1965, the μA741 op amp in 1967 and the LM101A in 1968 (with an order-of-magnitude reduction in input current). The performance of these op amps, although not as good as the higher cost modular op amps,

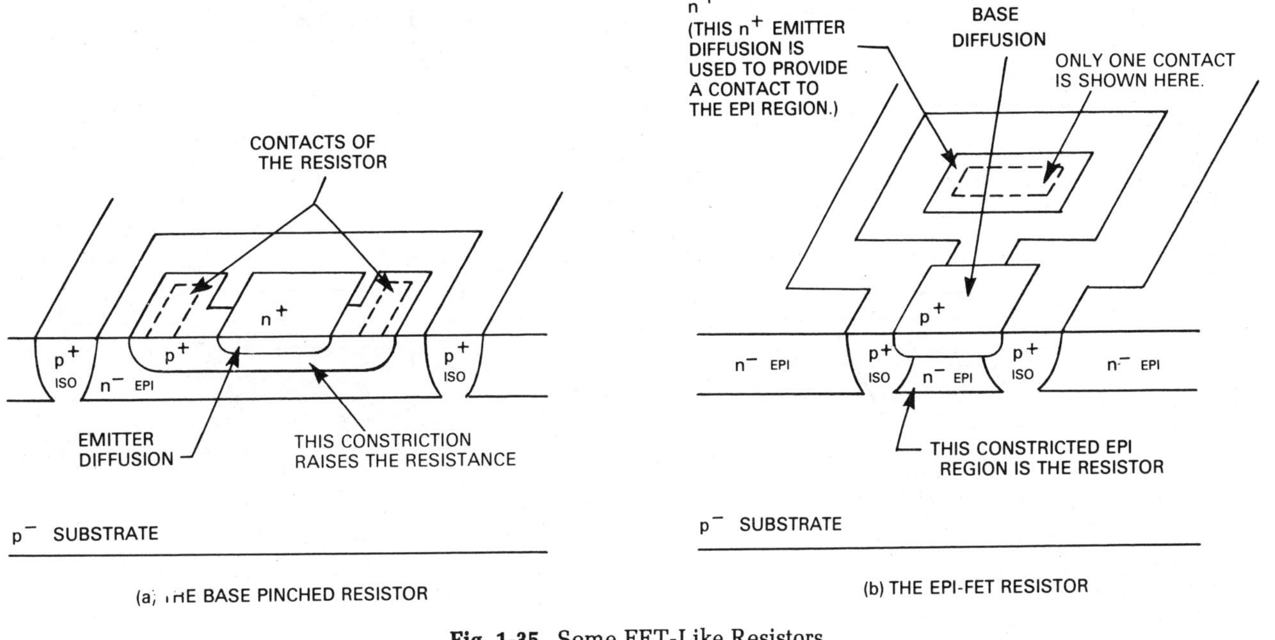

Fig. 1-35. Some FET-Like Resistors

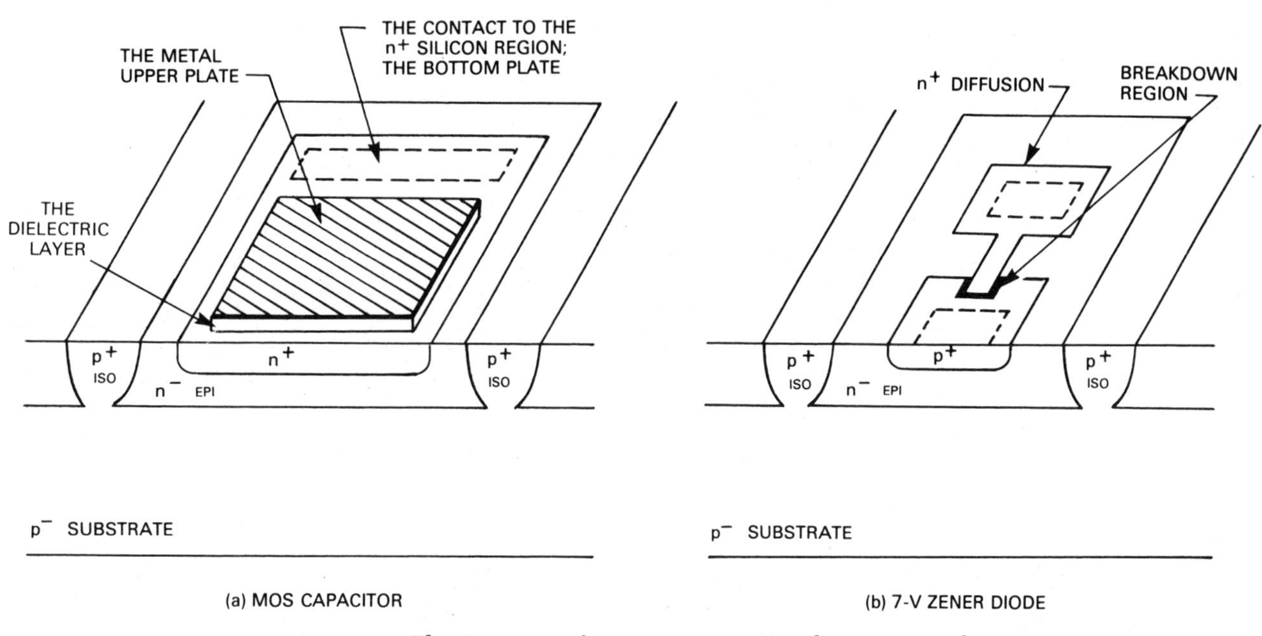

Fig. 1-36. The Structure of a MOS Capacitor and a Zener Diode

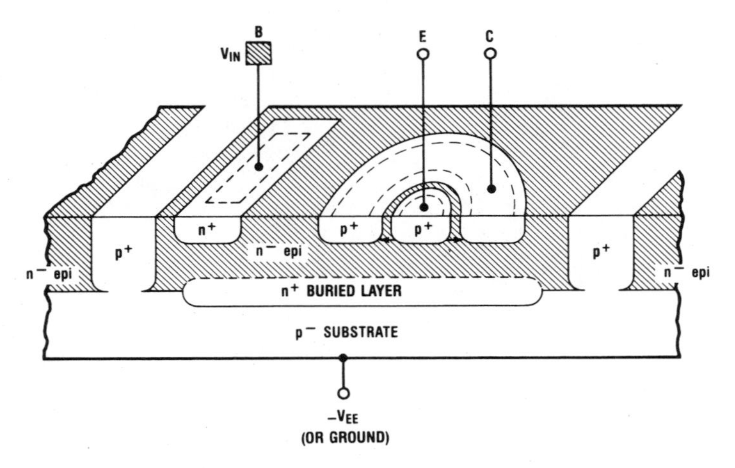

Fig. 1-37. The Lateral-PNP Transistor

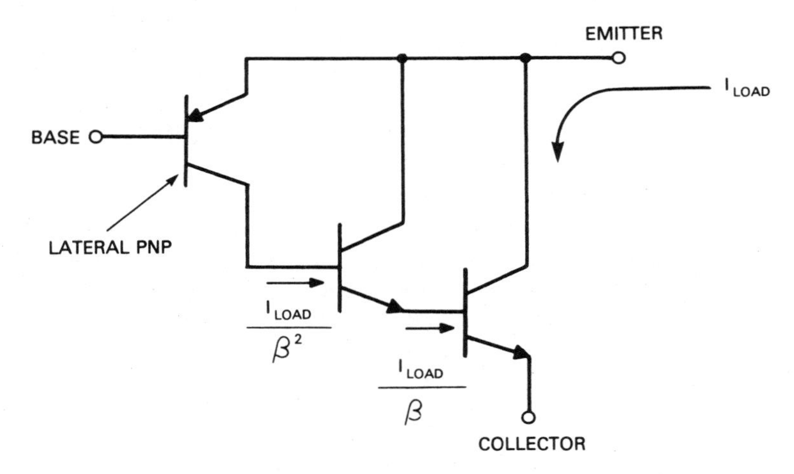

Fig. 1-38. A Composite PNP Circuit

was now adequate for many systems. The overwhelming advantage of the IC op amp was low cost, even when the small size of a monolithic circuit was unimportant to the system designer.

The lateral-PNP transistor has a relatively poor frequency performance and this limited the high-frequency capabilities of these early IC op amps. Additionally, these bipolar op amps; when biased for good frequency performance, had relatively high input currents.

The Vertical PNP. The discovery of a vertical PNP occurred relatively quickly, but this transistor, shown in Figure 1-39, didn't allow access to the collector. This limited its useful-

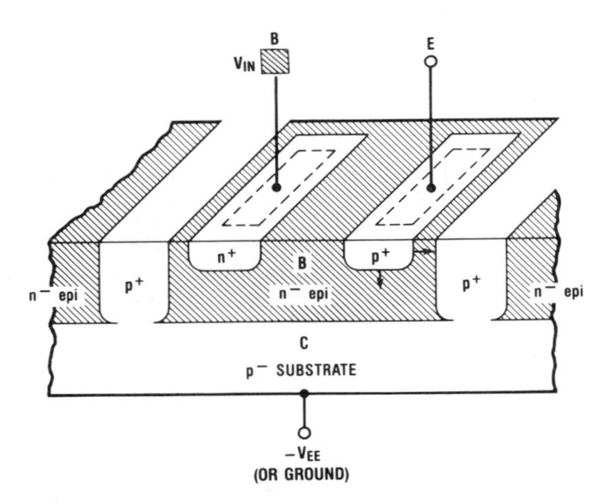

Fig. 1-39. The Vertical PNP Transistor

ness; it could only be used as an emitter follower. The current gain and frequency response, although better than the lateral PNP, were still not as good as that of the NPN transistor.

Use a Separate PNP Chip. A good solution, from the circuit designer's viewpoint, to the problem of providing a good PNP was to simply go buy one. This approach was used in an early high-frequency unity-gain current booster that was designed to boost the output current of any op amp to ± 300 mA. This was really a simple hybrid product that contained two chips, the IC main chip and the separate power PNP transistor chip, and was therefore called a "monobrid" product.

A greatly simplified circuit schematic of this current booster is shown in Figure 1-40. The die attach of the PNP chip to the header of the IC provided the heat sinking for the PNP (a special multipin TO-66 package was used for this medium power product) and also simultaneously made the proper electrical contact to the collector of the PNP. The emitter of the PNP was simply wire-bonded to the common output pin of the package. To complete the hookup, a spare pin on the package was used to tie the base of the PNP back into the IC control chip.

This product suffered because two chips were necessary for its construction, and one of these chips was not manufactured by the IC group and therefore had to be kept in stock. Unfortunately, this production problem restricted this solution to the PNP problem to this single product.

Use a P-Channel JFET. When the idea came to develop a process-compatible JFET transistor, this transistor was made a P-channel device and was used, in addition to its FET benefits, to provide a substitute for a good PNP transistor.

The dc benefits of a FET transistor have been known for a long time. We have also seen how the relatively poor g_m-per-milliampere of biasing current of a JFET front-end can also greatly increase the slew rate of an op amp. The big problem was how to provide FET devices in a standard bipolar monolithic process.

The first op amp with a combination of bipolar and JFET transistors was announced in a paper by George Wilson at the 1968 Solid State Circuits Conference. This used a compatible

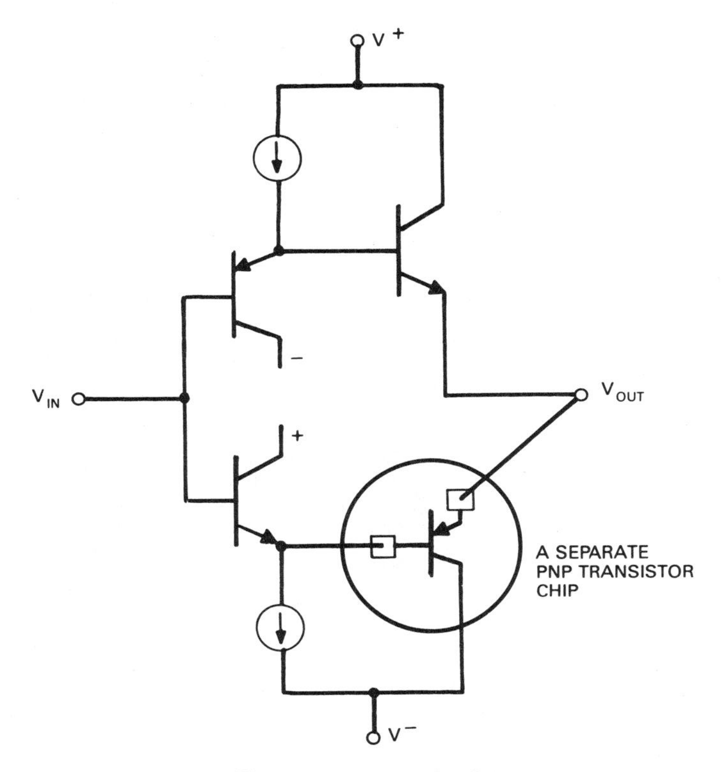

Fig. 1-40. A Monobrid IC

monolithic JFET process that was never adopted by the large-volume semiconductor suppliers.

To provide matched JFETs requires careful control of the impurity doping of the silicon wafers. It wasn't until the ion implant machines were introduced into the linear IC fabrication lines that this combination of high performance JFET devices with a bipolar process commercially appeared. The idea for the Bi-FET was nucleated when Ronald (Rod) W. Russell was observing the *bad* resistors that were made with the new ion implant machine. These thin, lightly doped resistors were easily voltage modulated; a frustrating observation when you want a *good* resistor. It then occurred to Rod that these looked very much like JFETs—and so the Bi-FET program was launched to see what could be done to convert these poorly performing resistors into a more useful active device. With the processing and device design help of James L. Dunkley, Rod W. Russell designed the first Bi-FET op amp, the LF356, and it was announced in 1973.

This compatible P-channel JFET required adding two ion implants to the standard linear wafer fabrication process. These implants put in the channel of the JFET (the P^- implant) and also a top gate (the N^+ implant), as shown in Figure 1-41. The gate lead of this JFET attached to the epi tub that was used to accommodate this transistor and the leakage current of this relatively large area, reversed-biased junction provides the input current for the Bi-FET op amps and the other ICs that use this transistor as the input device.

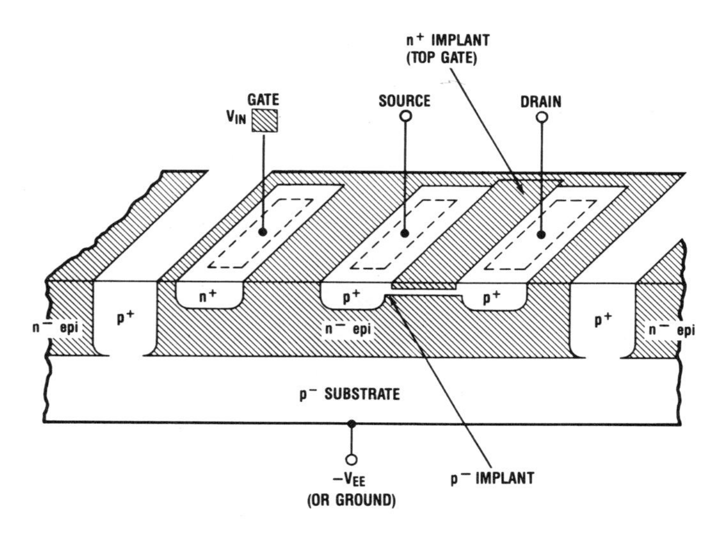

Fig. 1-41. The P-Channel JFET

The main motivation to develop a compatible JFET was to use these FET transistors at the input to an op amp. As we will soon see, the P-channel FET was also used to form a composite PNP transistor, similar to the composite PNP that earlier made use of the lateral PNP transistor. This composite no longer suffered from the poor frequency response of the earlier circuit and was therefore a very useful substitute for a good PNP in the design of the output stage of an op amp.

CMOS Solves the Problem. A good PNP is inherent in the advanced CMOS processes. The existence of complementary transistors, where both are high-frequency devices, provides a major simplification to the circuit designer for both linear and digital products. While CMOS solves the problem of obtaining a good PNP, it presents a few new problems of its own. We will consider some of the solutions to these problems later in this chapter.

A Good, Compatible Bipolar PNP. The continuing need for a good, compatible bipolar PNP transistor has finally been solved by adding to the processing complexity of linear ICs. Compatible PNP transistors that make use of dielectric isolation have been available for a number of years, but these costly processes have not been popular with the large-volume suppliers of linear ICs.

A way of providing a vertically integrated PNP (called, by some manufacturers, a "VIP" process) is shown in Figure 1-42. Many changes have been made to create this PNP. A rather deep N^- "well" is located under the PNP transistors, and a P^+ buried layer is located within

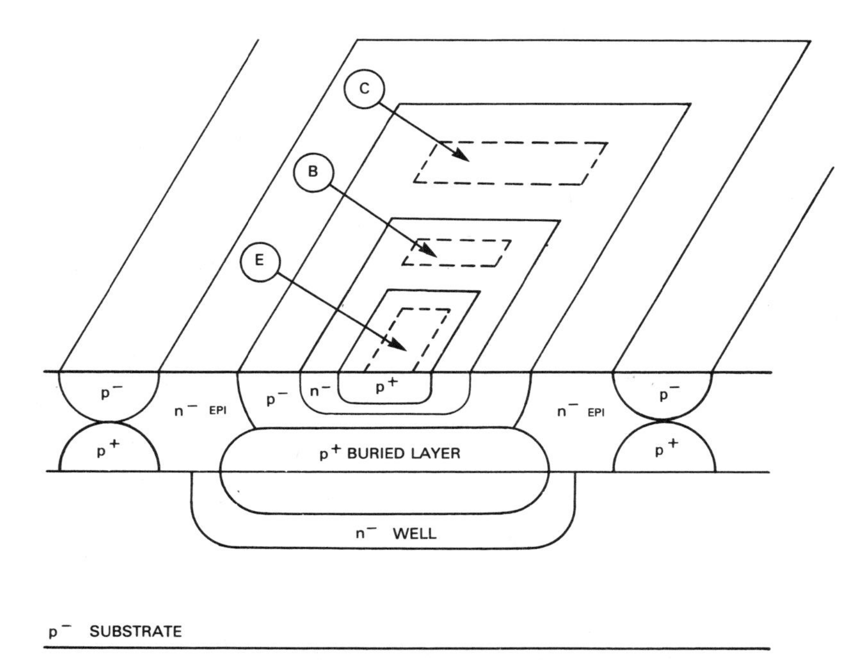

Fig. 1-42. A Structure Used to Form a Compatible PNP Transistor

this well. This buried layer is used to reduce the resistance in the collector bulk regions. It provides the same benefits as the N^+ buried layer that is used for the NPN transistors.

The epi isolation has been changed to include diffusions that move both down from the surface and up from the bottom through the epi layer (this lower diffusion also provides the buried layer), the "iso up" process. The upper isolation diffusion is also used to form the high resistivity P-type collector region for the PNP transistor. This collector diffusion is driven into the buried layer.

A special N-type base region is then formed within the collector, and a P^+ region, located within this base region, forms the P-type emitter. A contact (for simplicity, not shown on the figure) to the epi layer is used to provide a positive reverse-biasing voltage to insure that the collector of the PNP transistor never forward biases into the epi region. In addition, a P-type guard ring (also not shown on this figure) is added at the surface, around the perimeter of the P^- collector region.

The addition of this high-performance PNP transistor to the linear designer's bag of tricks has provided, and will continue to provide, many new IC op amps and other products with greatly improved-high frequency capabilities.

Super-β NPNs Reduce Input Current

This input current problem was attacked by making use of an additional NPN transistor: the Super-β structure. When transistors are being fabricated, if they are left in the emitter diffusion furnace too long, a narrow basewidth results, Figure 1-43, and they are rejected because of low values of breakdown voltage (because of reach-through or punch-through limitations). Bob Widlar liked the high β (3,000 to 10,000 as opposed to 50 to 300) that resulted and by making use of a careful circuit design that kept high voltage off of the specially processed Super-β input transistors, he came up with the LM108A in 1969 with a maximum input current spec of 2 nA. This was a large step forward for the monolithic op amp, but the use of lateral-PNP transistors still limited the frequency performance.

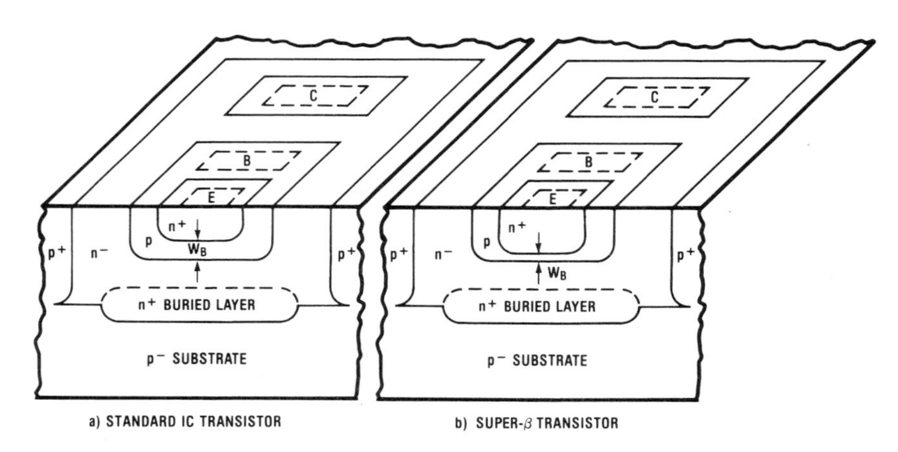

Fig. 1-43. Narrow Basewidth, W_B Raises β

Slew Rate Improvements

Various attempts were made to improve slew rate, but larger offset voltage and offset voltage drift resulted because of the increased circuit complexity that was usually associated with the fast slewing input stages. Additionally, a larger differential input voltage overdrive was also generally needed to obtain the increased slew rate.

Bandwidth Improvements

The small-signal bandwidth of the IC op amp was improved by changing the frequency compensation technique. The first use of *feedforward compensation* in an IC op amp, Figure 1-44, was accomplished by Bob Dobkin with his LM118 in 1971. The basic idea of this scheme is to use a capacitor to shunt or bypass the signal around the slow-responding PNP stage. The ac signal is thus *fed forward* and is not attenuated by the slow-responding stage. This significantly improved the small-signal bandwidth (15 MHz), but the input current was relatively unaffected.

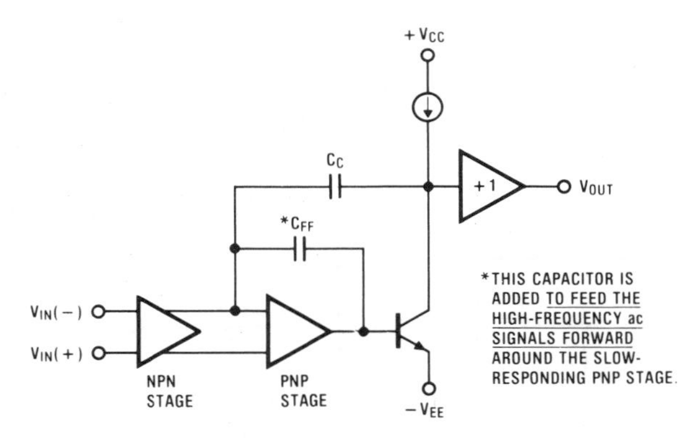

Fig. 1-44. Feedforward Bypasses the Slow-Responding PNP Stage

Op amps designed and made with the VIP process are more recent additions in the area of wide-bandwidth IC products. The major benefit of a good PNP is to provide a high-frequency op amp. A simplified schematic of one of the first design approaches for enhanced high-frequency response is shown in Figure 1-45. The differential signal currents out of the collectors of the input NPN differential amplifier stage are fed to the emitters of a second differential amplifier, a PNP stage and the only one with voltage gain. A current mirror in the output of this second stage accomplishes the differential-to-single-ended conversion. The output of this second stage then drives a pair of complementary emitter followers that handle the output current requirements of the op amp. The three PNP transistors shown on this schematic (Q3, Q4, and Q6) are the improved VIP PNPs.

There are several unique things about this op amp design. For example, a frequency compensation capacitor is not used across an inverting amplifier, so there is no Miller multiplication provided for this capacitor. The dominate pole of the open-loop response therefore occurs at a much higher frequency than with most op amps. It is located in the vicinity of 10 kHz rather than the more familiar 10-Hz range. The frequency compensation capacitor is provided by the unintentional stray capacitance that exists at the only high-impedance node of the circuit: the output of the second stage.

The amplifier, as shown in this figure, is stable only for closed-loop gains greater than or equal to 25. The gain-bandwidth product therefore has the very large value of 725 MHz (as opposed to the more typical values of 1 or 5 MHz for unity-gain compensated op amps). Further, the dc open-loop voltage gain is relatively small; typically 10,000. In addition, the slew rate has the very respectible value of 300 V/μs.

To tame this amplifier, emitter degenerating resistors are added to the input stage to reduce the first stage transconductance. This is used to provide an additional op amp product which is stable for closed-loop gains greater than or equal to 5. This reduces the gain-bandwidth product to 125 MHz but retains the same slew rate spec. The dc open-loop voltage gain also similarily decreases to a value of 2,500. The unusual thing is that both the dc and the ac open-loop voltage gain decrease by approximately the same factor. Usually, the value of the intentional frequency compensating capacitor is altered and this is why the term "decompen-

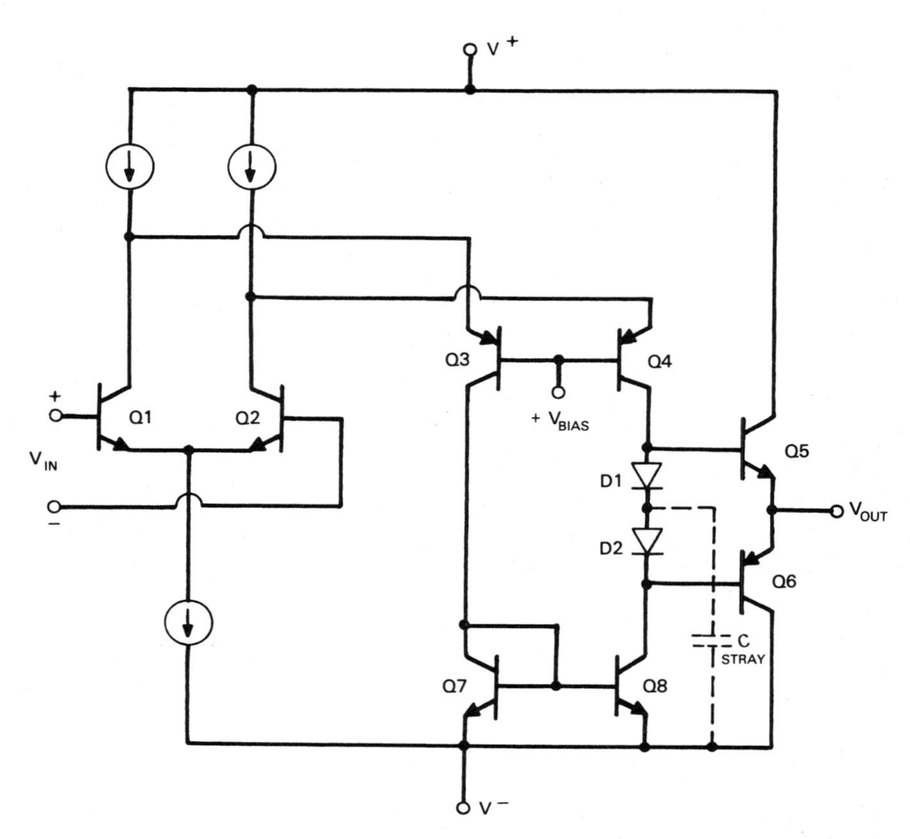

Fig. 1-45. A Simplified Schematic of a Fast VIP Op Amp

sated" op amp has been used. If only the value of the comp cap is altered, the dc gain would remain unchanged. With this "transconductance boost" design, the dc gain improves along with the ac gain for the "decompensated" versions of this op amp design.

To round out this product family, an op amp that is stable with unity closed-loop gain is also provided. This reduces the gain-bandwidth product to 50 MHz and also retains the same slew rate spec. Again, the dc open-loop voltage gain also reduces to a value of approximately 750.

Reducing the Size of the Comp Cap

A new transconductance-reduction trick (J. E. Solomon and R. W. Russell, "Transconductance reduction using multiple-collector PNP transistors in an operational amplifier," U.S. Patent No. 3,801,923, Mar. 1974), Figure 1-46, was responsible for small die size in IC op amps because

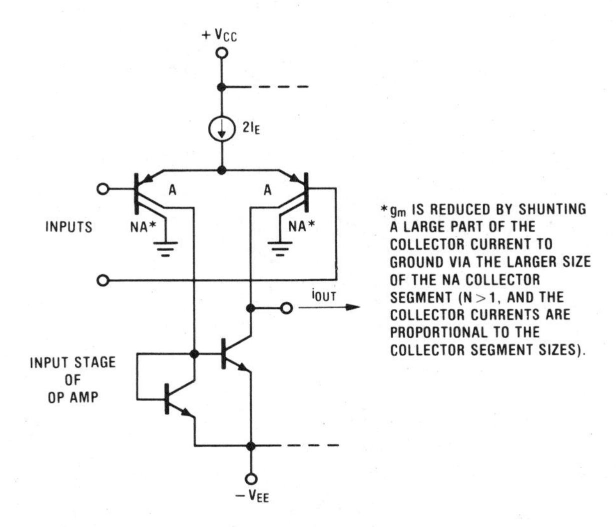

Fig. 1-46. Transconductance Reduction Using Multi-Collector Lateral PNPs

the compensation capacitor could now be reduced in area. This transconductance-reduction circuit has the advantage that a relatively large biasing current can be used for the emitters of the input differential amplifier. Phase lag that normally results from the stray capacitance at this common-emitter node is therefore greatly reduced because of the lower values of the dynamic emitter resistance of these transistors when operated at large currents. It is this detrimental phase lag, and the reduced positive voltage slew rate at the common-emitter node, that prevents simply reducing the magnitude of the input differential amplifier biasing current as a way to reduce g_m.

This new transconductance-reduction trick, plus both circuit and layout optimizations, has resulted in a modern version of the 741 that has essentially the same performance but is achieved with a factor-of-four reduction in die size. (This is important because small die cost less to manufacture.) This transconductance-reduction trick has been used in many bipolar op amps.

The Move to CMOS Op Amps

The shift by the digital designers to CMOS processes has caused a corresponding interest in CMOS for linear ICs. A major problem in the design of an MOS op amp is that the MOS transistor has a relatively poor g_m-per-milliampere of bias current, Figure 1-47. In looking at this figure it may appear that a larger geometry device would eventually allow better g_m performance than that obtained with a bipolar transistor. The thing that is missing on this figure is that such a large-geometry device, operating at a low level of current, would put the MOS transistor in the subthreshold region of operation and the g_m that is provided in this region is actually worse than is shown.

This lower g_m and also the lower output impedance of the MOS device reduces the gain

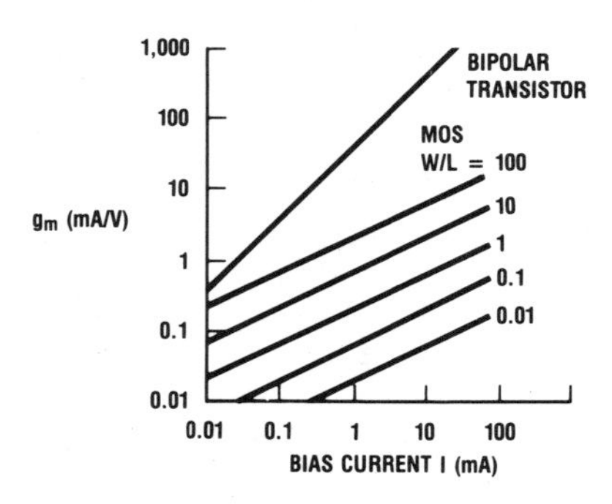

Fig. 1-47. Transconductance of an MOS Versus a Bipolar Transistor

that can be obtained with a single stage of amplification. Two-stage op amps therefore provide open-loop voltage gains of only a few thousand for NMOS op amps and a few tens-of-thousand for CMOS op amps. (The frequency compensation difficulties of a three-stage op amp favor design approaches that raise the gain that can be obtained with a two-stage op amp.)

Linear designs have always been greatly aided by the availability of complementary devices. It is only recently that attention is being given to the benefits of CMOS: the switching load device, the power dissipation savings, and the elimination of dc current flow through the interconnect metal. What appears to be emerging as *the optimum technology for any type of IC is*, therefore, *a CMOS process.*

Using New Circuit Approaches. The existing linear ICs evolved from the early bipolar digital process. Circuit designs have resulted that made use of the large number of matched bipolar transistors that were available for relatively low cost. Newer CMOS op amp designs are now making use of the parasitic bipolar transistor and analog designers are finding ways to depend on well-controlled biasing currents rather than biasing voltages.

The limited transconductance of the MOS transistors has forced the use of a three-stage op amp rather than the conventional two-stages that have become common for bipolar IC op amps. An example of a three-stage CMOS op amp is shown in the block diagram of Figure 1-48. In addition to the frequency compensation capacitance, C_c, there are two additional feed-forward capacitors that are used. The output of this op amp is the same point as the output of the integrator; a separate unity-gain buffer stage is not used.

The previously mentioned checkerboard-coupled hexidecimal of 16 total MOS transistors was used to solve the usual problem of high offset voltage in an MOS op amp. Native (unimplanted) P-channel transistors were used for the input stage to provide the lowest noise performance.

The circuit diagram for this input stage looks very familiar. It is similar to an emitter-coupled pair of PNP bipolar transistors (with FETs, the Darlington connection is not needed). The tail current is provided out of the drain of a P-channel transistor and a 2-transistor N-channel current mirror serves both as the load for this input differential stage and the differential-to-single-ended converter.

Both of the inputs to this CMOS op amp have on-chip protection circuitry that is similar to what is used on CMOS digital circuits: an input protection resistor and a diode from the

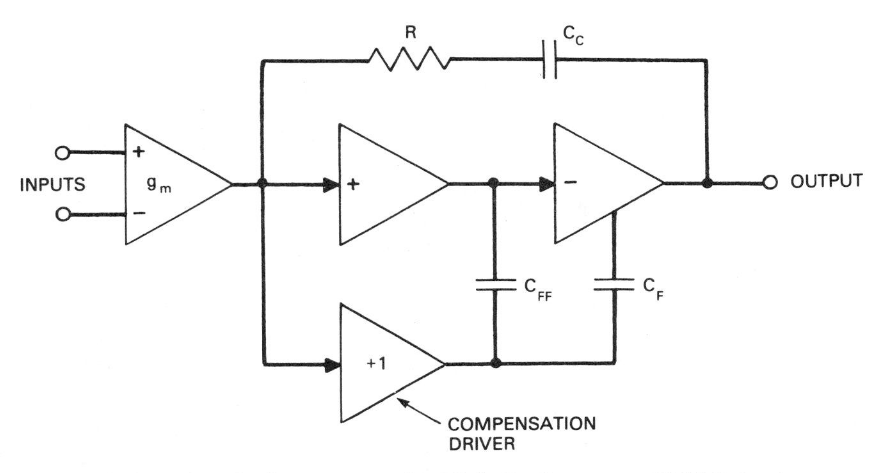

Fig. 1-48. The Block Diagram of a High-Performance CMOS Op Amp

internal side of this resistor to each supply. It is the reverse-biased leakage currents of these input protection diodes that cause this op amp to have a dc input current. In practice, the upper diode (to $+V_{CC}$) provides more leakage current than the lower diode (tied to $-V_{EE}$) and therefore the input current of this op amp comes out of the package (the direction you would expect for a PNP differential input stage). This current is typically only 50 fA at room temperature (less in a plastic package), varies with the input voltage at the rate of about 10 fA/V (an equivalent input resistance of 100 TΩ), and increases to only 5 pA at +125°C.

To assist the design of this CMOS op amp, use has been made of the parasitic lateral NPN transistor that is available with the process as shown in Figure 1-49. Parasitic vertical NPN transistors have been used for many years in metal-gate CMOS digital products. These bipolar vertical transistors, although only emitter followers, are useful to carry the heavy currents that are needed for LED drivers. The new thing with this op amp design is to make use of a lateral NPN transistor so that a collector is available to the circuit designer.

This lateral NPN transistor has been used to provide controlled-magnitude biasing current references by making use of the circuit shown in Figure 1-50. The vertical collector of the NPN still exists and collects approximately two-thirds of the total emitter current. This bias circuit makes use of a 4 to 1 emitter-area scaling between the differentially connected NPN transistors Q7 and Q6. This provides a difference in the V_{BE} voltages of these transistors (a ΔV_{BE}) of 40 mV. In operation, with the NPN differential amplifier at balance, the 4-k resistor in the emitter of Q7 has this 40 mV developed across it and therefore provides a 10-μA reference current flow.

In designing with MOS transistors, the equivalent of emitter-area scaling is provided by scaling the channel width to channel length ratios (W/Ls). Knowing that the total tail current of the NPN differential amplifier is 20 μA allows W/L scaling to be used between Q9 and Q8

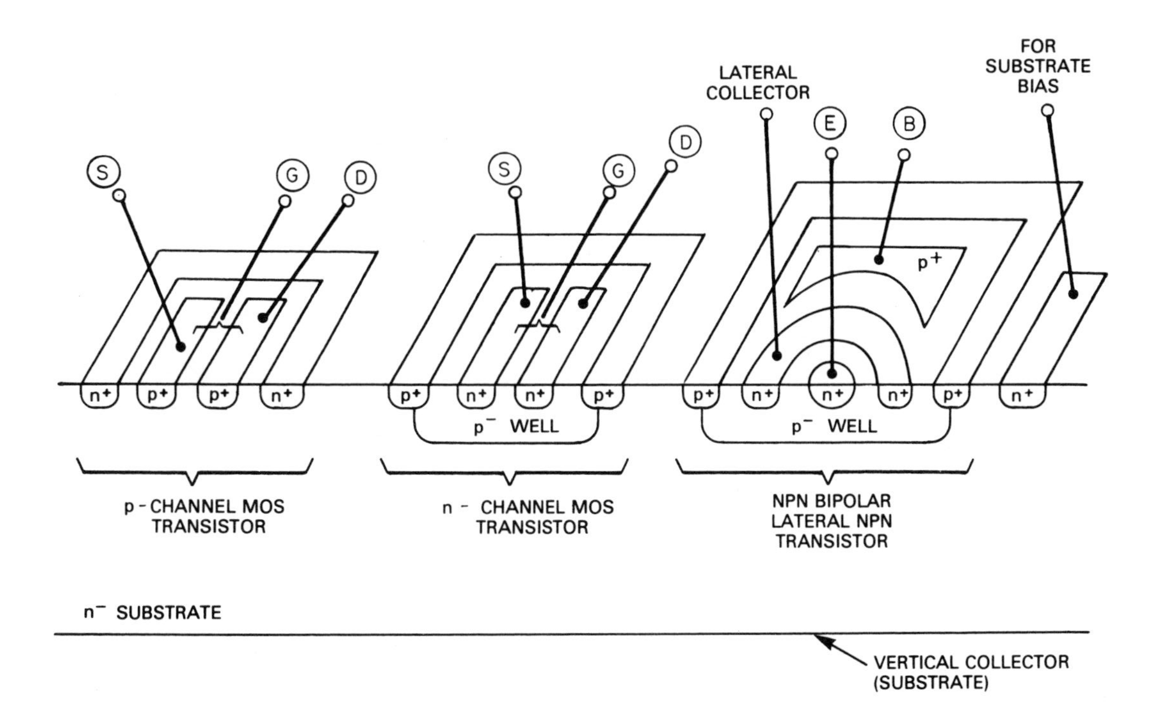

Fig. 1-49. The CMOS Transistors and the Parasitic NPN Bipolar Transistor

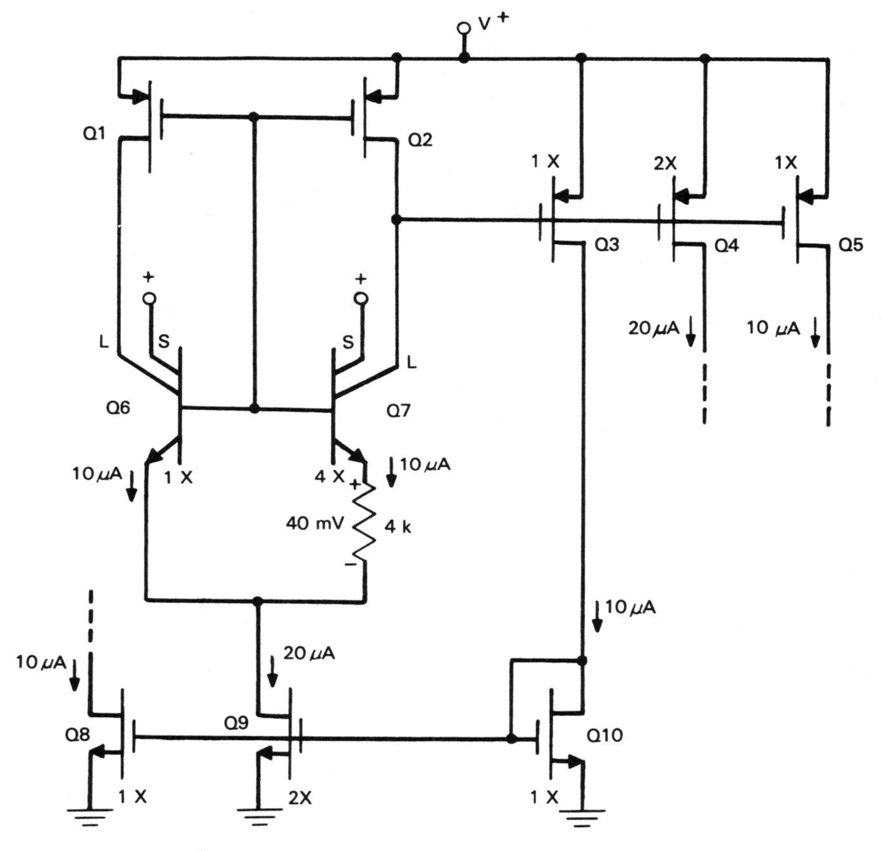

Fig. 1-50. The Δ V_{BE} Biasing Current which Uses Lateral NPN Bipolar Transistors

to provide, for example, the proper gate-source voltage biasing such that Q8 will carry 10 μA of drain current. Similarily, Q10 is scaled to carry 10 μA, and the drain current of this transistor is used as a reference for transistors Q3, Q4, and Q5. In this figure, Q4 will supply 20 μA of drain current and Q5 will supply 10 μA of drain current. The use of a familiar bipolar circuit (and a parasitic lateral NPN bipolar transistor) has allowed the design of a relatively predictable biasing circuit (not easily done with only CMOS devices) that provides biasing currents for the rest of the CMOS op amp circuitry (via transistors such as Q4, Q5, and Q8—and any additional transistors and scalings that may be needed).

Another interesting aspect of this design is the way that the output short-circuit current is limited. In bipolar op amps, the output current is usually passed through a sense resistor just prior to coming out of the output lead. The voltage that is dropped across this sense resistor will eventually be of a large-enough magnitude (when the maximum amount of output current is reached) that special current limiting circuitry that is normally dormant will come to life.

This current limiting circuitry then removes or steals base current from the output transistor and thereby provides an output current limiting function.

In this CMOS op amp, the complementary output MOS transistors are connected as common-source amplifiers. The output of the op amp is the common connection of the drains of these output transistors. This connection (collectors out) is rarely used in bipolar op amps but does provide the advantage of a large output voltage swing. The unusual thing with this op amp is that output current limiting is provided by simply limiting the magnitude of the gate-source drive voltage that is applied to these output FETs.

The overall performance, on a 5-V supply, that has been achieved with these CMOS op amps is equal to or better than that of most commercially available bipolar, Bi-FET, or CMOS quad amplifiers. This performance has been achieved by a combination of both circuit design and layout techniques, and therefore these op amps are suitable for incorporation on system chips that are built with the latest digital CMOS processing technology. (For more information on this op amp see: "A Quad CMOS Single-Supply Op Amp with Rail-to-Rail Output Swing," by Dennis Monticelli in the December 1986 issue of the *IEEE Journal of Solid-State Circuits.*)

1.4 A LOOK AT SOME OF THE POPULAR IC OP AMPS

The 741 is still the lowest-cost single op amp available and is supplied by a number of semiconductor companies. Two things happened that affected the market for high-volume op amps; both Bi-FET op amps and low cost quad op amps appeared. We will now look more closely at these two developments.

LF356, The First Bi-FET

The earlier, all-diffused JFETs that were reported by Wilson had the large offset voltage problem (± 50 mV) and the increased noise voltage that have always been associated with an FET input op amp. The new thing with the Bi-FETs was that the ion implanters could provide doped regions in silicon with more precision than could be achieved with the older diffusion processes. This control of the total doping within the channel region allowed the fabrication of matched JFETs with a low V_{OS}: a significant improvement when compared with the products that were then available from discrete JFET suppliers. Noise voltage was competitive (a factor of two less than a 741), a low 1/f-noise corner frequency (that we will consider in Chapter 3) was provided, the frequency response was improved, the slew rate was *greatly* increased, and the low input current and the nearly infinite input resistance were all benefits of this new Bi-FET op amp: a large step toward the *ideal op amp.*

The higher-frequency P-channel JFETs of the Bi-FET process were also used to replace the lower frequency lateral-PNPs in other parts of the op amp circuit. The first Bi-FET op amp, the LF356, therefore simultaneously solved many of the performance problems of the previous bipolar op amps. Bi-FET processes are used today by a number of suppliers to provide a broad line of op amps.

The basic design of the LF356, shown in Figure 1-51, also makes use of JFETs as the load devices for the input differential stage. A second differential bipolar stage is then used and a common-mode feedback loop (using Q1) insures the proper dc biasing of the input stage.

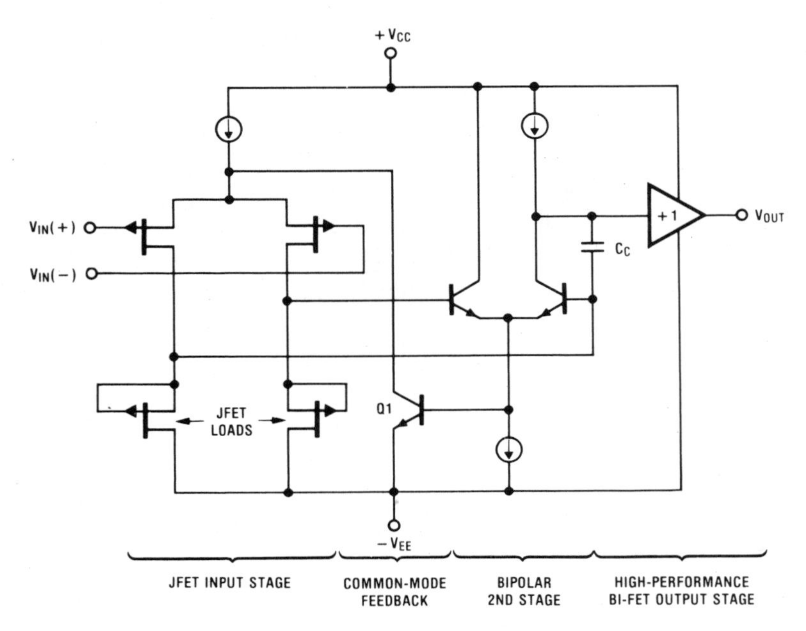

Fig 1-51. The Basic Design of the LF356 Bi-FET Op Amp

The JFET was also used to design an improved output stage for the LF356, Figure 1-52. The key to this design was to keep the upper transistor, Q2, ON during the complete output voltage swing to improve the frequency response of the output stage. The design of the lower half of the output stage (that accepts current from an external load) of an IC op amp is always more difficult, because of the lack of a high-frequency PNP transistor, and the phase characteristics of this lower half are poorer than the upper NPN half. Therefore, if the upper NPN transistor can always be kept ON, an improved output stage results.

To understand this output stage, notice that under a static output voltage condition, the base-emitter voltage drop of the upper NPN output transistor, Q2, is matched by the forward-voltage drop, ϕ, of the diode D1. This provides a 0V gate-source voltage for the P-channel JFET, Q3, and therefore this depletion-mode JFET is conducting the current that keeps Q2 biased ON. The lower-side Darlington output transistors, Q5 and Q6, are both OFF because the matched JFET, Q4, is conducting the same current as Q3 and the current mirror (Q7 and Q8) causes Q8 to absorb all of the current that is supplied by Q3.

The upper NPN, Q2, supplies current to the external load for an increase in the input voltage, V_{IN}, to this output stage. When this input voltage, V_{IN}, is decreased, the gate voltage of Q3 also falls and this increases the current flow in Q3. This current increase [out of the drain (D) of this JFET (Q3)] therefore enters the base of the lower-side Darlington pair, Q5 and Q6, and is multiplied by a large current-gain factor. This causes the collector of Q6 to accept large currents from an external load. (It is this improved output stage design that allows the LF356, when used as a unity-gain inverting amplifier, to drive a 0.01 μF capacitive load without frequency stability problems.)

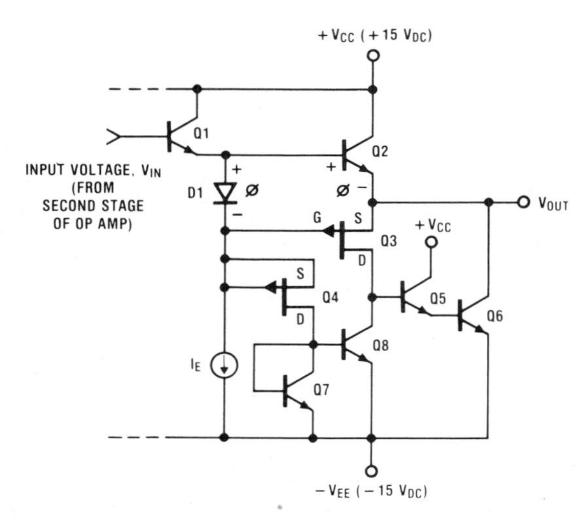

Fig. 1-52. The Basic Output Stage of the LF356 Op Amp

The appearance of this first Bi-FET op amp breathed new life into the monolithic op amp designers because it gave them a new device in their bag of design tricks. Also, new linear monolithic products became possible: sample-and-hold circuits, FET-input instrumentation amplifiers, and analog switches.

It was soon realized that lower-cost, smaller die-size, Bi-FET op amps would also be valuable additions to the product line, and so the next move was made in this direction to produce the LF351, LF353, and LF347.

Reducing the Power Drain. Special applications of op amps, such as those that operate with battery power, require a minimum power drain. To meet this need, the LF441, 442, and 444 (single, dual, and quad) Bi-FET op amps have been introduced. The power dissipation for the LF441 (200 μA max supply current) is one-tenth that of the 741, yet the same ac specs and improved input dc specs have been achieved.

Improving Spec Guarantees. Users of low-cost Bi-FET op amps can obtain products with maximum guarantees on V_{OS} and V_{OS} drift with the LF411 and LF412 (single and dual). The gain-bandwidth product is also guaranteed greater than 3 MHz and the minimum slew rate is 10V/μsec.

A Special Design for Fast Settling. The problem of designing an op amp with a fast 0.01% settling-time spec (400 ns) has finally been solved with the LF400. This op amp has a bandwidth of 20 MHz and a slew rate of 57V/μsec. It is useful for the combined digital and linear systems and other high-speed applications.

The Popular Quads

A quad op amp provides the lowest unit op amp cost and was necessary to solve the economic problems of the emerging automotive control electronics. The difficulty of placing four op amps on the same die was due to the large size (and, therefore, large die-area) of the frequency compensation capacitor. For this reason, a simple op amp stage with only a 3 pF comp cap was the first of the quads.

LM3900, The First Quad. The LM3900, the Norton Amplifier or the Current-Differencing Amplifier, consists of only a single gain-stage; a moderately high voltage gain (3000 V/V or 70 dB) common-emitter amplifier. Because application design approaches are very different from those used with standard op amps, the LM3900 is usually not called an op amp to avoid confusion with the standard op amps.

This current-differencing amplifier, Figure 1-53a, makes use of a new symbol, Figure 1-53b. It uses only a few transistors and provides an output voltage swing of 4 volts when operated with a single 5-volt power supply. A current mirror (Q7 and Q8) at the noninverting input is the novel current-differencing element of this design. The collector of Q8 subtracts (or differences) current from the current that is supplied to the inverting input. (A wideband current-differencing amplifier is also available, the LM359.)

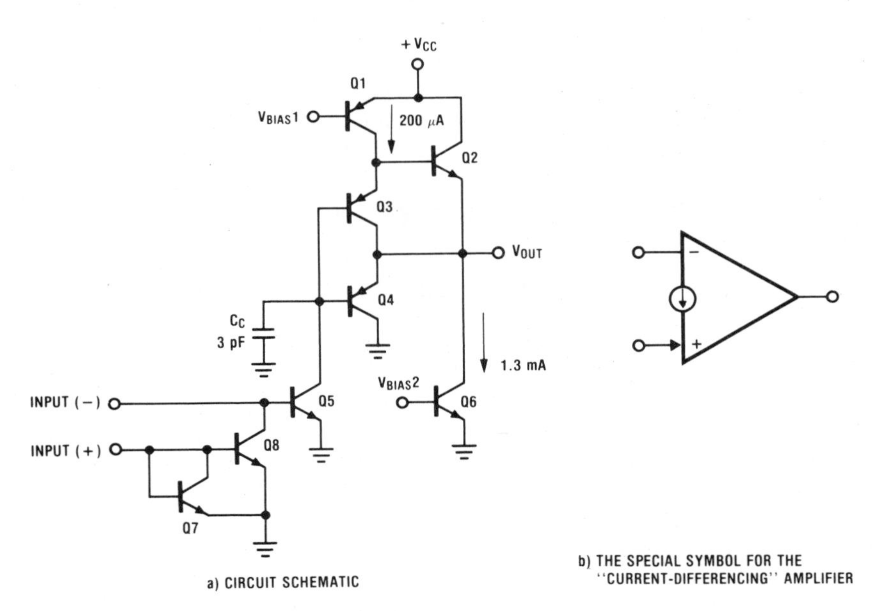

Fig. 1-53. The LM3900 Current Differencing Amplifer

All of the voltage gain is supplied by the one common-emitter amplifier, Q5. The emitter-follower transistors; Q2, Q3, and Q4 reduce the loading on this gain stage. Many users consider this as simply a "super transistor" (Q5) that has low base-current and has a low-impedance voltage source available from its collector (via Q2).

LM324, Quad Op Amp with g_m Reduction. The first op amp commercially available that made use of the split-collector g_m reduction technique was the LM324. The main feature of the LM324 was *an input stage that had a common-mode voltage range that included ground,* *even though operated with a single power supply voltage.* This allowed almost all the application design flexibility of dual supplies for low-cost, low-voltage (5 V_{DC}), single-supply linear systems.

Another feature of the LM324 is the relative immunity to the effects of high power-density rf fields. This is a result of the relatively slow PNPs that are used directly at the input of this op amp. This RFI (Radio Frequency Interference) immunity is a big advantage in many high noise applications.

The LM324 op amp design has emerged as the best value (lowest cost per op amp function) and the total sales of this op amp have made it *the highest-volume op amp circuit ever produced.* For example, the LM324 has four op amps per package, the LM358 is the same op amp design with two op amps per package, and the LM392 has one of these op amps — and one comparator — per package.

LM339, A Quad Voltage Comparator. A quad voltage comparator, with essentially all of the features of the LM324, was the next development: the LM339. A comp cap is not used (because linear operation is very rarely required with a comparator) and the output is the uncommitted collector of a grounded-emitter NPN output transistor. This comparator, therefore, is faster than the popular op amps and the free collector is useful as a voltage clamp, or switch, to ground. The use of a comparator improves the speed of those op amp application circuits that use positive feedback to produce digital functions.

Special bipolar op amps are also now available that have improved dc characteristics or special low noise performance. The current catalog offerings of IC op amps are becoming very complete; this product is in an advanced stage of its evolution.

1.5 THE INSTRUMENTATION AMP VERSUS THE OP AMP

Special problems are associated with amplifying the analog signals that are provided by transducers. These voltages indicate the magnitude of a physical parameter that is being measured. Signal levels can be extremely low (measured in microvolts) and are often provided from a differential signal source (to eliminate grounding problems and to allow the rejection of interfering signals that can be many volts in magnitude that exist as common-mode voltages).

The requirements on amplifiers for this type of service are very stringent: low initial offset voltage, low drift, low input current, accurate and stable voltage gain, high common-mode rejection, and high input impedance (on both of the differential inputs, with no external circuit loading owing to the presence of feedback resistors).

Although two- and three-op amp application circuits (with critical resistor ratios) are often used, specially designed *instrumentation amplifiers* are recommended for this service. These high performance instrumentation amplifiers make use of a basic circuit design that guarantees high common-mode rejection without requiring matched resistors. The closed-loop gain is established by the ratio of two resistors, but neither resistor is directly connected to an input lead of the amplifier. These gain-setting resistors are now usually precision thin-film resistors that are deposited on the IC chip.

Additionally, a separate *output sense* pin is usually provided to eliminate voltage drops in the output path from the amplifier to the load and a *reference* pin allows large dc shifts to be introduced in the output voltage. These features increase accuracy and provide application advantages, although instrumentation amplifiers are not intended for the wide variety of applications that op amps cover.

To add to this confusion, op amps with excellent dc performance specs (comparable to those required for instrumentation amplifiers) are called *instrumentation op amps* (as the LM725, for example). We will have little more to say about either of these relatively high-cost precision amplifiers because our emphasis is on the large-volume, low cost op amps.

Now that we have covered some of the basic background of the op amp, let's see what the magic of feedback is all about.

CHAPTER **II**

Feedback Control Theory Is for Op Amps, Too

It is *feedback* that *allows accurate linear systems to exist.* Without this discovery, linear approaches to the solutions of real-world problems would be severely limited. The problems in amplifier performance that feedback solves are due to the wide initial tolerances, the poor linearity, and also the drifts in critical amplifying characteristics that exist with the basic amplifying devices, whether they are vacuum tubes, bipolar transistors, or FETs.

This was perceived as a major problem with the early telephone-line amplifiers because a signal must pass through many of these amplifiers to complete a long distance call. When Bell Laboratories researcher, Harold S. Black, considered the overall tolerance and linearity problems of a long distance multi-amplifier telephone link, he realized that the amplifiers needed improved linearity. The overall amplifier performance specs should have very little to do with the uncontrollable characteristics of the individual amplifying devices. The solution came to him in a flash on his way to work one morning and resulted in U.S. Patent No. 2,102,671, filed in 1927, and was the basis for Black's induction into the National Inventors' Hall of Fame in Arlington, Virginia. His solution depended *on the more accurate and stable, passive components: this is the key idea of feedback.* (Many earlier examples of other uses of feedback also exist in marine equipment, speed governors, to prevent oscillations in multi-stage amplifiers, and to stabilize radio frequency amplifiers.)

The price paid for this predictable performance is a loss in the amount of gain that is achieved in the final amplifier from a given number of amplifying devices. Gain is therefore exchanged for higher accuracy in an amplifier with negative feedback.

The details of how this tradeoff in gain is best made is the subject of Feedback Control Theory. For example, if feedback is so beneficial, we are certainly tempted to use lots of it. We generally like to use all the feedback that is possible, so we can achieve our ideal, predictable amplifier. But, *there is a limit* which is reached when the closed-loop or fed back system (whether it is an op amp circuit or a radar antenna controller) *breaks into oscillations.* That is, it is out of control, supplying its own input signal and producing the exasperating waveforms often seen on oscilloscopes or even breaking up and destroying the mechanical components of servomechanisms. (This lack of mechanical parts that could become dislodged is one of the benefits of being a circuit designer as opposed to a servo designer, should your system break into oscillation.) Certainly, this tragedy must be brought under control during the design phase; thus, the importance of Feedback Control Theory.

It is sometimes thought that this theory is only useful for the study of servo-mechanisms: the electro-mechanical systems. The surprise is that this same theory applies to the tiny, all-electronic, IC op amp and all of its closed-loop (or feedback) application circuits.

Let's try now to develop some intuition for this *very useful, but also very dangerous,* thing — *feedback* — so we can avoid the *plague of the linear designers: undesired oscillations.*

2.1 CONSIDERING FEEDBACK AS: GOING UP AN ATTENUATOR

Before we get into all the complications of feedback, let's first take a close look at a less complex, resistive attenuator circuit, Figure 2-1. This circuit is used whenever we have too large a signal, V_{IN}, and we want only some fraction of it as an output signal, V_{OUT}. Notice that neither an amplifier nor an additional energy source is needed in this situation.

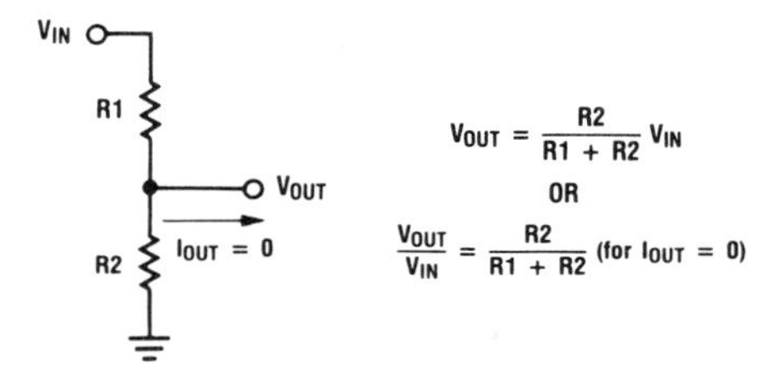

Fig. 2-1. The Resistive Attenuator

The transfer function (V_{OUT}/V_{IN}) for this circuit contains only the resistor values R1 and R2. This is a large advantage because resistors are the most accurate passive components available to a circuit designer. For this reason, the transfer function or *gain* of an attenuator can be made with high accuracy and it is also possible to insure only a relatively small change in this transfer function as the ambient temperature changes. (For example, if the temperature effects on each resistor still provide the same ratio of the resistor values, there will be no change at all.)

A major problem with this attenuator has to do with *loading errors*. If any current, shown as I_{OUT} on the figure, flows into or out of the V_{OUT} tap point, the transfer function given is in error, because in the derivation, I_{OUT} was assumed equal to zero. As long as this no-loading condition is met, there are no errors. The performance will be limited only by the exact values of the two resistors that are used.

Now, if we had the opposite system problem, a voltage V_{IN}; which, instead of being too large, was too small, we would need an amplifier. So, let's use an op amp.

We would like to find a way to connect this op amp so as to cause this small input voltage, V_{IN}, to be created at the tap point of a resistive divider. If we can accomplish this, then a larger voltage, V_{OUT}, must, of necessity, exist across both of these resistors in series. This larger voltage will then represent the precise amplification of V_{IN} that we are looking for.

If we draw an op amp and the associated feedback resistors (both made equal to 10 kΩ, for clarity) in an unconventional way, Figure 2-2, we can obtain an intuitive understanding of feedback. Notice that the resistors are in the same relative positions that were used in the attenuator example, but now we are trying to force the tap point voltage to equal V_{IN}, our smaller input voltage. So we are attempting to *go UP the attenuator*, which requires an energy source or *pump:* the op amp and its associated power supplies.

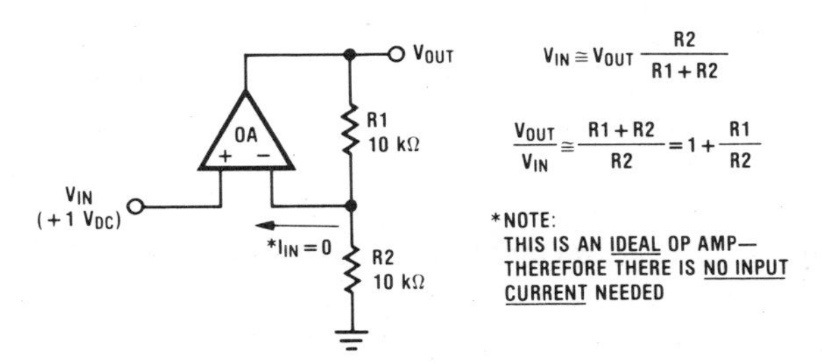

Fig. 2-2. Going Up an Attenuator

Let's see what happens to this circuit if we arbitrarily make V_{IN} equal to $+1\ V_{DC}$. We can start by assuming that the output voltage of the op amp (V_{OUT}, at the top of the resistors) is initially at 0V. This also provides 0V at the (−) or inverting input of the op amp, $V_{IN}(-)$. So the differential input voltage to the op amp is 1V and we notice that $V_{IN}(+)$ is larger than $V_{IN}(-)$ by this 1V.

With an open-loop gain spec of 50,000 (again, a typical value to assume), the output of the op amp is sent on its way to 50,000V! This large 1V differential input signal also insures that V_{OUT} is moving at its maximum rate — the slew rate limit of the op amp.

Notice, from the figure, that the resistor R1 is tied directly to the output of the op amp, so this resistor is *feeling* or responding to this rapidly increasing output voltage. In fact, exactly one-half of V_{OUT} is continuously *fed back* to the *negative* input of the op amp (this is where the expression *negative feedback* comes from).

Now op amps are somewhat like people, they can't strangle themselves. For example, if $V_{IN}(-)$ of the op amp were to try to become larger in magnitude than $V_{IN}(+)$, then the output voltage of the op amp must turn around and go negative. This contradicts the original assumption that the output voltage was large and positive, so this can't be allowed. What really happens is that as the voltage at $V_{IN}(-)$ rises, it continuously reduces the size of the differential input signal that is being applied to the op amp. This $V_{IN}(-)$ voltage will stop increasing and will reach an equilibrium value that allows just enough differential input voltage to exist for the op amp to continue to supply the required dc output voltage, V_{OUT}.

Applying some insight, let's guess that this is all going to work out properly and that we will be able to force the tap voltage of these two resistors to be equal to the input voltage, 1V. The voltage V_{OUT} would therefore stop increasing when it reached a value of $2\ V_{DC}$. To supply this amount of output voltage from the op amp requires an input differential voltage that is given by

$$V_{IN}(+) - V_{IN}(-) = \frac{V_{OUT}}{A_V} = \frac{2\ V_{DC}}{50{,}000} = 40\ \mu V_{DC}$$

Because $V_{IN}(+)$ is held at $+1.000000\ V_{DC}$ by the input voltage V_{IN}, $V_{IN}(-)$ must have a value of

$$V_{IN}(-) = 1 - 40\ \mu V = 0.999960\ V_{DC}$$

(We are assuming we have an ideal op amp so there is no dc offset voltage.)

You can see that our first guess, that the output would stop at 2 V_{DC}, is in error because the output voltage is really going to be closer to $2 \times 0.999960\ V_{DC}$ or $1.999920\ V_{DC}$ ($80\ \mu V$ too low). (To arrive at the actual values, we normally would have to iterate our guesses concerning the magnitude of V_{OUT}, but in this case, our first guess was close enough to the actual values that can be calculated for this circuit.)

So feedback is a way of using an op amp to force a small input voltage to exist at the tap point of a divider or attenuator network (which is now called *the feedback network*). We take advantage of *the larger voltage that now must exist* at the *top* of this attenuator and *let this be our desired amplified output voltage.*

This transfer function is not as accurate as that of the attenuator by itself, because the op amp must always have some small differential input voltage to support a non-zero output voltage. This is called the *gain error* of the op amp. For a high gain op amp, this input error signal will be quite small and gives rise to the concept of a "virtual short" between the inputs of an op amp. We will have more to say about this later on in this chapter in Section 2.3, DC Closed-Loop Gain Dependence on DC Open-Loop Gain.

As a hint to the possibilities, if a small capacitor were placed across only R1 (of Figure 2-2), the gain would fall at high frequencies. Placing a small capacitor across only R2 would, instead, cause the gain to rise at high frequencies. Audio applications use complex RC feedback networks to provide *frequency equalization*. This is compensation for the frequency distortions that are introduced in the audio signal source, whether phonograph records or magnetic tape.

Many useful op amp application circuits result by replacing R1 and R2 by general impedances Z1 and Z2. This allows frequency dependent gain as given by

$$A_V = \frac{v_{OUT}}{v_{IN}} = 1 + \frac{Z1}{Z2}$$

and, as we will see later in this chapter, the "1" in this gain expression can be removed, if desired, if we use an inverting op amp application circuit.

In the inverting op amp application circuit, an integrator results if Z1 is simply a capacitor, the *integration capacitor*. The general use of a few or even many Rs, Cs, and even Ls for Z1 and/or Z2 can provide a wide variety of application circuits. We will consider many of these in Chapter 5.

2.2 DERIVING THE KEY EQUATION FOR FEEDBACK CONTROL SYSTEMS

A large body of knowledge has been developed in support of the design of servomechanisms and feedback control systems. We will show that the results of this analysis also apply to the closed-loop op amp application circuits.

There are three basic blocks in a system diagram of a feedback control system, Figure 2-3. (We will initially restrict this discussion to dc to simplify the analysis.) At the input, the circle with an "X" in it is a symbol that is used to represent a voltage differencing circuit. The rectangle with the "A," sometimes "μ" or "G," in it represents a noninverting voltage amplifier — the open-loop voltage gain of the system. The rectangle with the "β," sometimes "f" or "H," in it represents *the feedback network. The value of β is defined to be the fraction of the output voltage that is fed back to the input. Therefore, β can range from 0 (no feedback) to 1 (100% feedback).*

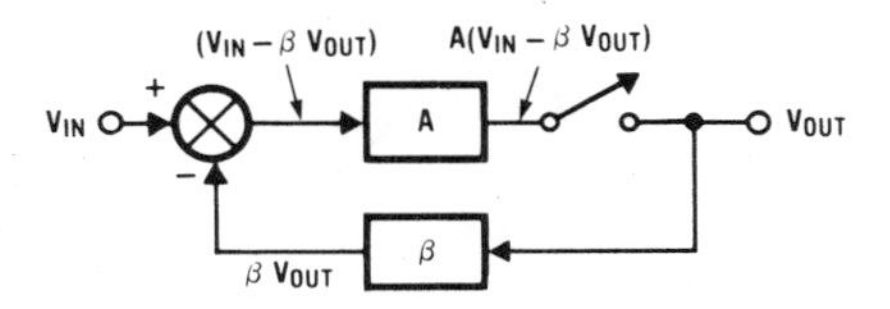

Fig. 2-3. The Classic Feedback Control Loop

If we open the switch at the output, Figure 2-3, and consider what would happen if we assumed an output voltage were somehow to exist, we would find that voltages would be established, as is indicated at each node on this diagram. The important thing to notice is that when we chase through the symbols of the diagram and arrive at the output voltage point (the left side of the opened switch), we don't really find the output voltage to exist, as we may have hoped. Instead, we get an expression that involves every element of the system diagram.

If we were to now close the switch, we would force the voltages on each side of the switch to be equal. Forcing this condition mathematically provides the starting equation for analyzing a feedback control system, or

$$V_{OUT} = A\,(V_{IN} - \beta\, V_{OUT})$$

Collecting terms, this becomes

$$V_{OUT}\,(1 + A\,\beta) = A\,V_{IN}$$

or, the dc closed-loop gain, V_{OUT}/V_{IN}, is given by the very important feedback equation:

$$A_{CL} = \frac{V_{OUT}}{V_{IN}} = \frac{A}{1 + A\beta}$$

This is the key equation of feedback control systems and we will be continuously making use of this equation. We see that the closed-loop gain depends on both the open-loop gain, A, and also the feedback factor, β. The product, $A\beta$, that appears in the denominator, has been called *loop-gain* because this can be thought of as a signal propagating around the loop that consists of the A and the β networks. If $A\beta$ is large, it can be said that we have a *tight loop* or lots of feedback; conversely, if $A\beta$ is small, we would not have much feedback. This $A\beta$ product is also sometimes called *loop transmission* and is represented by the symbol "T."

We can graphically represent how the available open-loop gain of an op amp is split up into the closed-loop gain and the loop gain, Figure 2-4. The closed-loop response cannot have any higher gain at any frequency than what is available in the open-loop response. As expected, when we have a small A_{CL}, we have a large loop gain. In the limit, for $A_{CL} = +1$ (0 dB), all of our available dc open-loop gain becomes loop gain. This is usually the largest problem for closed-loop stability. We will look at stability problems in Chapter 4. Alternatively, when the dc closed-loop gain is equal to the dc open-loop gain, we no longer have any loop gain. There is no β network. The amplifier is operating open loop. There is no stability problem for this case because there is no feedback loop.

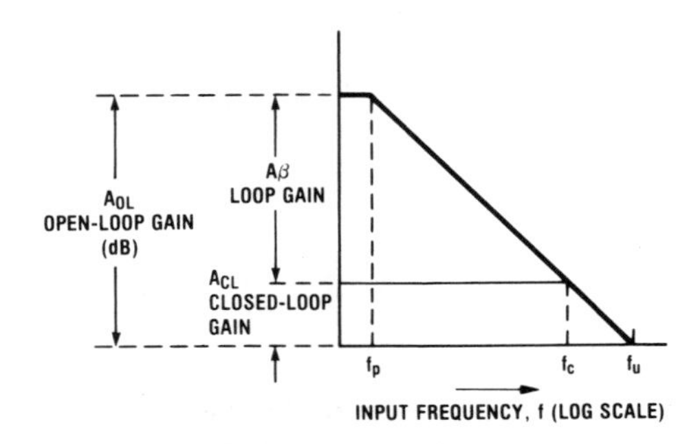

Fig. 2-4. The Gains of a Feedback Amplifier

We can now ask what would happen to the dc closed-loop gain if the dc open-loop gain were to approach infinity, $A \rightarrow \infty$. Before we let A approach infinity, we will divide both numerator and denominator by A to provide the feedback equation in an equivalent form, or

$$A_{CL} = \frac{1}{\frac{1}{A} + \beta}$$

Now, if A were to approach infinity, $1/A \rightarrow 0$ and the closed-loop gain expression approaches a value that is given by

$$A_{CL} \cong \frac{1}{\beta} \text{ (as } A \rightarrow \infty)$$

This is the important result that demonstrates the benefits of feedback — especially when the open-loop gain is large. For this noninverting amplifier case, *the closed-loop gain is simply the inverse of the feedback factor, β*. Because β is usually established by passive components (and many times with only two resistors) we now have an amplifier with a well-controlled value of voltage gain. It is important to notice that this value of *closed-loop voltage gain depends only on the values of components that are external to our amplifier, A: those components that establish β*. Therefore, we can plug in many amplifiers or even change the temperature of an amplifier, which usually changes the magnitude of its open-loop voltage gain, and still have the closed-loop gain essentially independent of all of this uncertainty and under the complete control of the β network.

Problems result when you go to your engineering stockroom and ask for *a circle with an X in it*, as was shown in Figure 2-3, so you can build a feedback circuit. This is where the op amp comes to your rescue; it gives you this circle with the X and also the rectangle with the A (both in the one triangle, as shown in Figure 2-5). Now, you add your own β network and you have a closed-loop control system: an op amp application circuit.

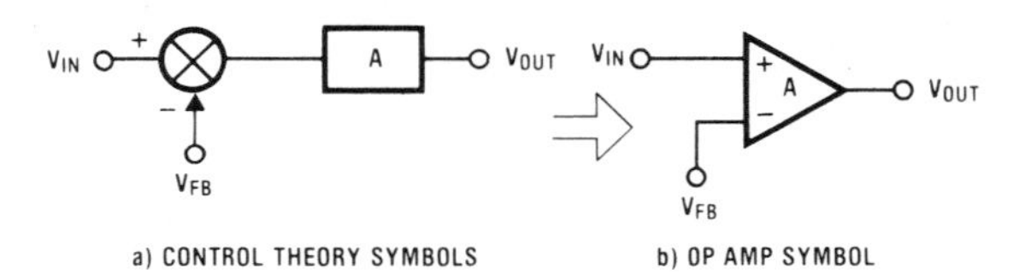

Fig. 2-5. Op Amps Easily Allow for Feedback

As we stated earlier, this basic equation for the closed-loop gain of a feedback amplifier (the feedback equation) that we have developed in this section, is very important. We will now see how we can make use of this feedback equation in many different ways (using mathematically equivalent forms) to obtain most of the numerical measures for the performance of op amp application circuits.

2.3 DC CLOSED-LOOP GAIN DEPENDENCE ON DC OPEN-LOOP GAIN

We can make use of the standard feedback equation to obtain a numerical measure of the effects of changes in the magnitude of the dc open-loop gain on the resulting changes in the magnitude of the dc closed-loop gain. The additional problems of ac gain will be discussed in the next chapter. Many times this is derived by making use of calculus. Starting with the feedback equation (2-1),

$$A_{CL} = \frac{A}{1 + A\beta}$$

we can take the derivative of A_{CL} with respect to A and then substitute A/A_{CL} for one of the $(1 + A\beta)$ terms in the denominator to obtain

$$\frac{d\,(A_{CL})}{dA} = \frac{(1 + A\beta) - A\beta}{(1 + A\beta)^2} = \frac{1}{(1 + A\beta)}\frac{(A_{CL})}{A}$$

or

$$\left[\frac{d\,(A_{CL})}{A_{CL}} \times 100\right] = \left[\frac{dA}{A} \times 100\right]\left[\frac{1}{1 + A\beta}\right]$$

$$\Delta A_{CL}\,(\%) = \Delta A\,(\%)\left[\frac{1}{1 + A\beta}\right]$$

which shows that the percentage change in the dc open-loop gain is reduced by the factor within the brackets. *This relation is only valid for small changes in A because it is based on calculus and the derivative at a specific point is not representative for large changes.*

Notice that it is $(1 + A\beta)$, called the *desensitivity factor, that is, desensitizing the gain of a closed-loop application circuit to changes in the open-loop gain.* To obtain a large desensitivity factor, obtain an op amp with a large value for A. Unfortunately, the application circuit reduces the desensitivity factor when β is small by reducing the $A\beta$ product. For example, an application circuit with a closed-loop gain of + 1000 ($\beta = 10^{-3}$ because, as we found in the last section, $A_{CL} \cong 1/\beta$) reduces the desensitivity benefits when compared with a unity-gain voltage follower application (where $\beta = 1$) by 1000:1. This can be seen if we have an op amp with an open-loop gain of 100,000 by looking at these desensitivity factors, DF. For $\beta = 1$

$$DF_{+1} = (1 + A\beta) = 1 + 10^5\,(1) = 1 \times 10^5$$

but for $\beta = 10^{-3}$

$$DF_{+1000} = (1 + A\beta) = 1 + (10^5)\,(10^{-3}) \cong 1 \times 10^2$$

A more useful expression that relates the dc closed-loop gain change to the dc open-loop gain change that is valid for large changes in A can be derived. Consider the differences in closed-loop gain that would result from two different values of dc open-loop gain, A1 and A2. This can be found from differencing the two feedback equations

$$A_{CL1} = \frac{A1}{1 + A1\beta}$$

and

$$A_{CL2} = \frac{A2}{1 + A2\beta}$$

now,

$$\Delta A_{CL} = (A_{CL2} - A_{CL1}) = \frac{A_2}{(1 + A2\beta)} - \frac{A_1}{(1 + A1\beta)}$$

combining, we get

$$\Delta A_{CL} = \frac{A2\,(1 + A1\beta) - A1\,(1 + A2\beta)}{(1 + A1\beta)\,(1 + A2\beta)}$$

or,

$$\Delta A_{CL} = \frac{A2 + A1A2\beta - A1 - A1A2\beta}{(1 + A1\beta)\,(1 + A2\beta)}$$

also,

$$\Delta A = (A2 - A1)$$

so,

$$\Delta A_{CL} = \frac{\Delta A}{(1 + A1\beta)(1 + A2\beta)}$$

and for the form of percentage changes, we divide both sides of this equation by A_{CL1} (and also use the previous equation that relates A_{CL1} to A1 and β) to obtain,

$$\frac{\Delta A_{CL}}{A_{CL1}} = \frac{\Delta A}{(1 + A1\beta)(1 + A2\beta)} \frac{(1 + A1\beta)}{A1}$$

or

$$\frac{\Delta A_{CL}}{A_{CL1}} = \frac{\Delta A}{A1}\left[\frac{1}{(1 + A2\beta)}\right]$$

To get percentage changes, we multiply both sides by 100 to give

$$\left[\frac{\Delta A_{CL}}{A_{CL1}} \times 100\right] = \left[\frac{\Delta A}{A1} \times 100\right]\left[\frac{1}{(1 + A2\beta)}\right]$$

or

$$\Delta A_{CL}\,(\%) = \Delta A\,(\%)\left[\frac{1}{1 + A2\beta}\right] \tag{2-2}$$

We now have an equation, similar in form (we use A2 instead of the A1), *that can be used for large changes in A*.This is useful to provide a numerical answer to the following situation:

Given: We want a dc closed-loop gain of 100 ($\beta = 0.01$). The A at +25°C is 700,000 and at +125°C is 130,000. (Note how much A can decrease at high temperatures.)

To Find: What change will this cause in our dc closed-loop gain? (Neglect, for now, any problems with the β network.)

Solution: $\beta = 0.01$

A1 = 700,000 and A2 = 130,000

$$\Delta A_{CL}\,(\%) = \Delta A\,(\%)\frac{1}{[1 + (1.3 \times 10^5)(10^{-2}]}$$

$$\Delta A_{CL}\,(\%) = \Delta A\,(\%)\,[7.7 \times 10^{-4}]$$

and the percentage change in the open-loop gain is

$$\Delta A\,(\%) = \frac{\Delta A}{A1} \times\cdot 100 = \frac{-570{,}000}{700{,}000} \times 100 = -81.4\%$$

so

$$\Delta A_{CL}\,(\%) = (-81.4\%)(7.7 \times 10^{-4}) = -0.063\%$$

Now that we have a numerical measure we can decide if this amount of change in the dc closed-loop gain is acceptable in our application; and, if not, we can determine what should be done about it. For example, we have these choices: select an op amp with a higher dc open-loop gain; increase β; or find an op amp with less temperature change in dc gain. Also consider adding a locally fed-back op amp in an overall loop, Figure 2-6, so that the open-loop amplifier is composed of two op amps. This raises the dc open-loop gain and also reduces the temperature change in the open-loop gain. But the stability of this overall loop must be carefully considered.

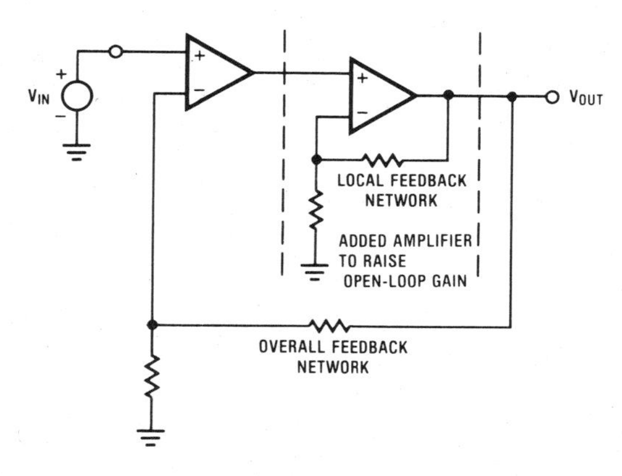

Fig. 2-6. Cascading Op Amps to Raise the Open-Loop Gain

It is always interesting to determine the effects of changes in the open-loop gain on an application circuit so you can become aware, in advance, of the magnitude of impending production or environmental problems. Alternatively, for a less demanding application, you may be able to obtain more dc closed-loop gain and thereby reduce the number of op amps that are required in an application.

We will now return to the *key equation* of feedback control systems to determine what value of A is required for a predetermined accuracy in dc closed-loop gain. (The β network is considered perfect and is not included in this analysis.) We again start with the feedback equation (2-1):

$$A_{CL} = \frac{A}{1 + A\beta}$$

and now divide each term of the numerator and denominator by $A\beta$, to provide an equivalent, but more useful form

$$A_{CL} = \frac{\frac{1}{\beta}}{\frac{1}{A\beta} + 1}$$

As $\frac{|1|}{|A\beta|} \ll 1$, we can approximate this equation as

$$A_{CL} \cong \frac{1}{\beta}\left[1 - \frac{1}{A\beta}\right] \qquad (2\text{-}3)$$

by making use of the small value approximation

$$\frac{1}{1 + \Delta} \cong 1 - \Delta \text{ for } \Delta \ll 1, \text{ where } \Delta = \frac{1}{A\beta}$$

Equation (2-3) is in a convenient form as $1/\beta$ is the ideal dc closed-loop gain and the error term, $1/A\beta$, is compared to 1. For example, if $1/A\beta = 0.01$, this will cause a 1% error in the accuracy of the dc closed-loop gain. This is the *dc gain error* of this application circuit.

This equation (2-3), can be used to calculate the required open-loop gain, when a known dc closed-loop gain is to be provided, for a specified *dc gain error*. For example, if we want to use an op amp at a closed-loop gain of +1000, $\beta = 10^{-3}$, we can calculate the open-loop dc gain (A_0) that is required to keep the dc gain error, ϵ_{DC}, to less than 1% (0.01) as

$$\epsilon_{DC} = \left|\frac{1}{A_0\beta}\right| \leq 0.01$$

$$A_0 \geq \frac{1}{0.01\beta} = \frac{1}{(10^{-2})(10^{-3})} = 10^5$$

A few test examples, using this equation, rapidly indicate the problem of providing high dc gain-accuracy at large values of closed-loop gain.

The noninverting gain application is seen to fit standard feedback control theory. We will now consider the changes that are necessary when we have an inverting gain application. The differences between these application circuits can create much confusion. The next section will clarify this problem.

2.4 THE INVERTING-GAIN APPLICATION IS DIFFERENT

The standard feedback equation is well suited to the noninverting gain application. When the op amp is used in an inverting-gain mode, the relationship between the feedback factor and the closed-loop gain is no longer the same.

The inverting amplifier is useful for: polarity reversal, fast-settling and high-speed amplifiers, the elimination of CMRR errors, multiple-input signal summation, and the general isolation benefits that result because of the "virtual ground" that exists at the inverting input.

Consider an inverting amplifier with both resistors equal, and equal to 10 kΩ, (Figure 2-7). We can understand the operation of this circuit by noticing that an input voltage of 10V will cause a current of 1 mA to enter the input resistor, R_{IN}. Under the assumption that the input current to the op amp is essentially zero, this 1 mA must therefore flow through the feedback resistor, R_F, and be absorbed by the output of the op amp. Actually, a small error voltage at the inverting input is required to cause the output voltage of the op amp to swing far enough negative so the feedback resistor can absorb this 1 mA. With this feedback resistor also equal to 10 kΩ, an output voltage of −10V is provided. This is the detailed operation of this unity-gain inverter.

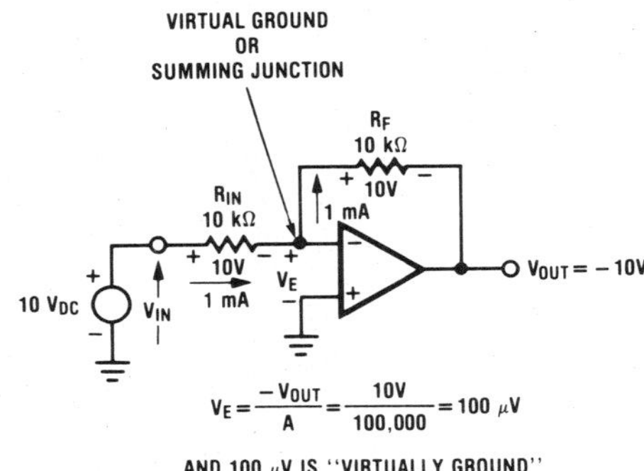

Fig. 2-7. The Inverting Gain Application and Virtual Ground

The assumption of large open-loop gain for the op amp means that the voltage at the inverting input will never be far away from ground. In this example it is only + 100 μV. This has allowed the inverting input to be called a *virtual ground.* This concept is very useful in many op amp circuits. For example, even if two (or more) input resistors are connected together at this common virtual ground, they will not cause any significant voltage coupling between the input signal sources because *there is essentially (or virtually) no voltage present at this common connection point. The virtual ground isolates the input signal sources from interaction with each other.*

This isolation also exists between the feedback impedance and the input impedance and is basic to the op amp integrator. For the integrator, Figure 2-8, the feedback impedance is a capacitor. The voltage that exists on this capacitor does not affect the magnitude of the charging current that is supplied through the input resistor. This is why this op amp RC integrator provides a linear ramp as opposed to the exponential charging voltage that results with simply a passive RC integrator circuit.

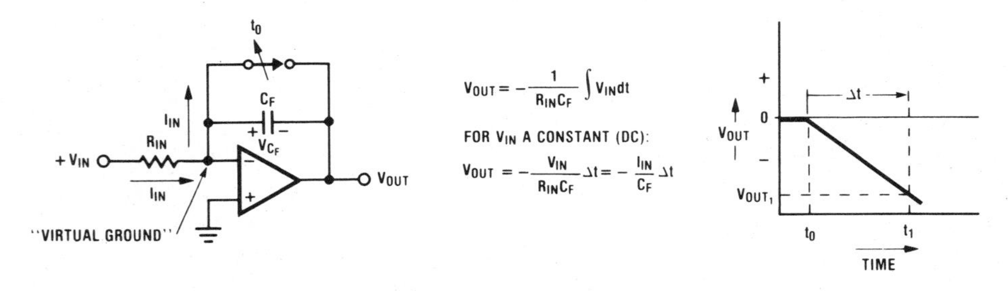

Fig. 2-8. The Integrator Makes Use of Virtual Ground

If we look at the unity-gain inverting circuit from the feedback standpoint and solve for β by superposition, we still see that one-half of the output voltage is fed back to the inverting input. To determine β, V_{IN} is grounded. We have no knowledge of where V_{IN} will eventually be applied. The β of this inverting application circuit has to be the same as that of the non-inverting application circuit that uses these same resistors. The result is

$$\beta = \frac{1}{2}$$

We can see why this connection is more stable. The $A\beta$ product is one-half that of a unity-gain noninverting amplifier, the voltage follower, where $\beta = 1$. It is also the reason for the decrease in the small-signal bandwidth of the inverting amplifier, $f_c = f_u/2$.

The dc gain of this circuit is given by

$$A_V \cong -[1/\beta - 1]$$

or, since $\beta = \dfrac{R_{IN}}{R_F + R_{IN}}$

$$A_V \cong -[1 + R_F/R_{IN} - 1] = -R_F/R_{IN}$$

and, with both resistors equal, the dc voltage gain is -1.

If we use these same resistors, but ground R1 and enter the signal at the (+) input of the op amp, the dc gain becomes

$$A_V \cong +(1/\beta) = 1 + R_F/R_{IN} = +2$$

The difference of 1 that exists in the magnitude of the dc closed-loop gain between this inverting and a noninverting gain application that uses the same resistors is sometimes confusing, because both circuits have the same feedback factor. There is no change in the resistor values that are used, only the way in which the input signal is applied, Figure 2-9.

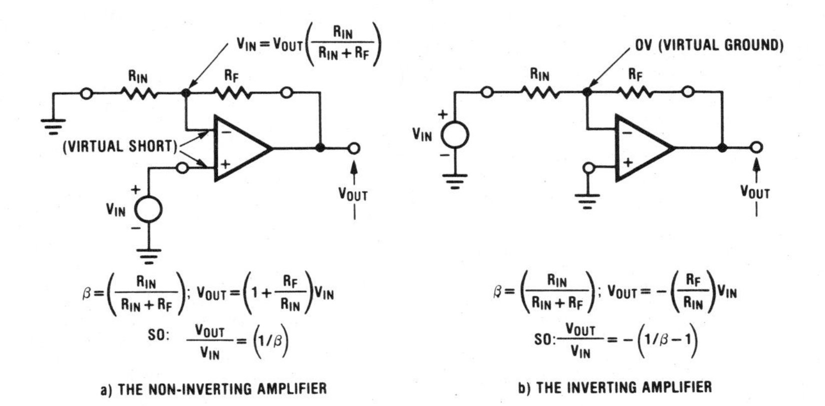

Fig. 2-9. Closed-Loop Voltage Gain of the Noninverting and Inverting Amplifiers

In general, the bandwidth, f_c, of an application circuit actually depends on β as

$$f_c = f_u\beta$$

and not A_{CL}. This also causes confusion, because for the noninverting amplifier,

$$A_{CL} = 1/\beta$$

so f_c can also be expressed as

$$f_c = \frac{f_u}{A_{CL}} \quad \text{(for the noninverting amplifier)}$$

For the inverting amplifier, we have found

$$|A_{CL}| = 1/\beta - 1$$

or

$$1/\beta = [|A_{CL}| + 1]$$

Therefore, the bandwidth, f_c, of the inverting amplifier relates to the dc closed-loop gain as

$$f_c = \frac{f_u}{[|A_{CL}| + 1]} \quad \text{(for the inverting amplifier)}$$

A unity-gain voltage follower ($\beta = 1$, $A_V = +1$) therefore has a bandwidth equal to the f_u of the op amp, but a unity-gain inverting amplifier ($\beta = 1/2$, $A_V = -1$) has a bandwidth equal to only $1/2\ f_u$.

If we try to set up a control system type of block diagram (as we used earlier for the *non-inverting* amplifier) for this *inverting* amplifier, we end up with the diagram of Figure 2-10, and some rather strange results. Notice that we have used the same blocks, but have changed the sign of the amplifier block to $-A$. The overall feedback loop, $A\beta$, must still be negative. Where the signal being fed back is entered into the summing block we also have to change the sign to (+). This provides $(+) \times (-) = (-)$, and not $(-) \times (-) = (+)$ for $A\beta$. Further, we must leave the (+) sign at the V_{IN} input, so the negative sign of A is not cancelled in providing V_{OUT}. We want to end up with a closed-loop gain that is inverting.

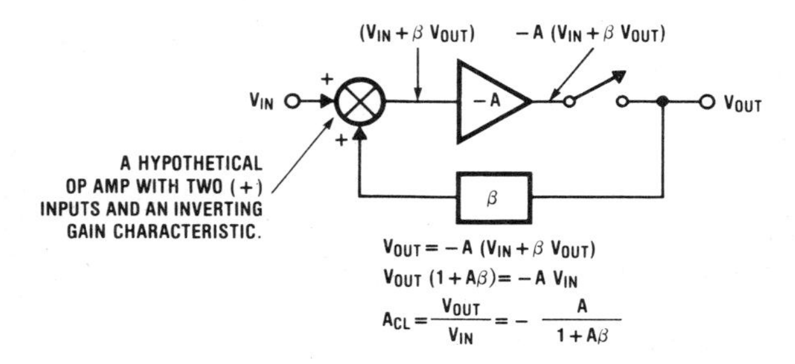

Fig. 2-10. Obtaining an Inverting Closed-Loop Gain with the Standard Feedback Form

A similar analysis of this block diagram shows we have achieved our goal. An inverting amplifier is provided. It is interesting that this is the only difference. Otherwise we have the same equation. The problem comes when we try to buy or build an op amp that has the input-output characteristic indicated by the symbol shown in Figure 2-11a. This new op amp, if it could be built, would use a feedback network, as shown in Figure 2-11b, *to provide an inverting closed-loop gain.* We could, instead, have left the gain block positive and then changed both signs of the differencing element to (−) to achieve the same inverting closed-loop gain. The symbol for this equally difficult to realize op amp can be obtained by changing both of the input signs on the op amp symbol of Figure 2-11a.

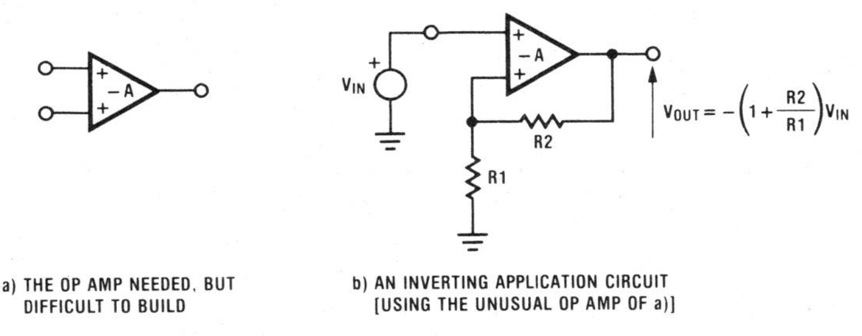

Fig. 2-11. The Unusual Op Amp Needed for Providing an Inverting Amplifier Using the Standard Feedback Form

Fortunately for the op amp circuit designers, the op amp users have found it easier to simply ground the noninverting input of a standard op amp and to thereby create all of the confusion over the differences that result between this inverting amplifier and a noninverting amplifier.

The input signal isolation that is provided by the virtual ground of the *commonly used inverting amplifier,* with $V_{IN}(+)$ simply grounded, allows a summing amplifier connection, where none of the input signal sources will interact with each other (Figure 2-12). The output voltage of the op amp will swing to whatever voltage is necessary to cause the current flow through the feedback resistor to absorb or balance the algebraic residue of all of the input currents. In so doing, this circuit provides the proper action for a multiple-input summing amplifier that is useful to algebraically combine many input signals.

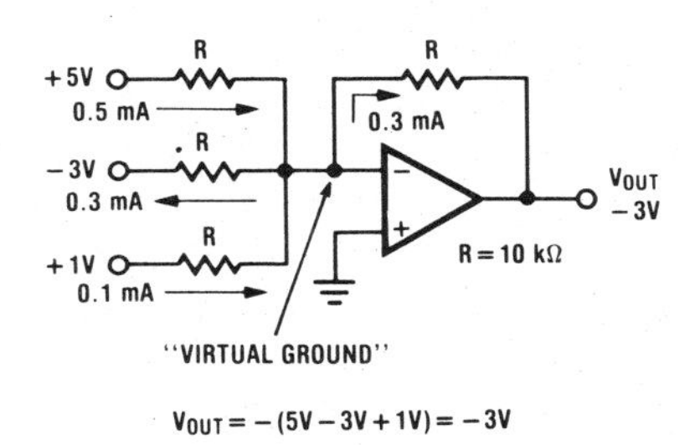

Fig. 2-12. Virtual Ground Isolates Multiple Inputs

Some of the other mathematical operations that op amps can perform are: gain (multiplication by a constant); difference amplifier (subtraction); attenuator (division by a constant); integration; and differentiation. Also, many nonlinear functions can be performed such as in a log amplifier (the output voltage varies as the logarithm of the input voltage), multipliers, dividers, and raising inputs to various powers such as squaring, and extracting square roots.

2.5 THE FOUR BASIC FEEDBACK CONFIGURATIONS

We can generalize the feedback amplifier by considering the *open-loop amplifier* as *a two-port network. There is an input port and an output port. The feedback, or β network* is also *a two-port network*, as shown in Figure 2-13. This will allow a systematic way to consider the four basic feedback configurations. These basic circuits result from the four possible ways of connecting a feedback network to the open-loop amplifier.

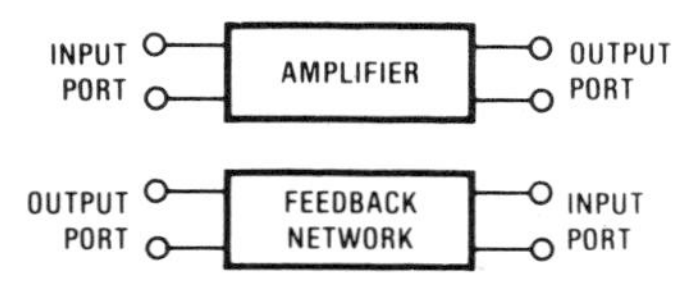

Fig. 2-13. The Two-Port Representation of the Open-Loop Amplifier and the Feedback Network

Another way to anticipate these four basic circuits is to note that both the input and the output parameters of a closed-loop amplifier can be either voltages or currents. This provides two dimensionless transfer functions, V_{OUT}/V_{IN} (voltage gain) and I_{OUT}/I_{IN} (current gain). There are two that have dimensions: V_{OUT}/I_{IN} (with the dimension of ohms; or, in general, "transimpedance") and I_{OUT}/V_{IN} (which becomes Mhos; or, in general, "transadmittance").

We will briefly look at these circuits to familiarize you with some new names for op amp applications and to show that feedback-stabilized transfer functions are not always "voltage gain." Large magnitudes for any of those transfer functions can also be obtained so we can have "high gain" circuits even if the input and output parameters are different. The effects of these various feedback configurations on the input and output impedances of the closed-loop amplifier will also be stated for each. We will then look at this more closely in the next section.

The input and output ports of the amplifier and feedback network can be connected in series or parallel. Some of the old literature on feedback theory names the resulting circuits based on how the feedback network is connected to the amplifier. If the input to the feedback network is wired in parallel with the output of the amplifier, it is called *shunt* feedback; otherwise it is *series* feedback. A second name results from how the output of the feedback network is wired to the input of the amplifier: here again, it can be either shunt or series. Four feedback circuit names result: shunt-series, series-shunt, series-series, or shunt-shunt, as shown in Figure 2-14. We will consider each of these shortly.

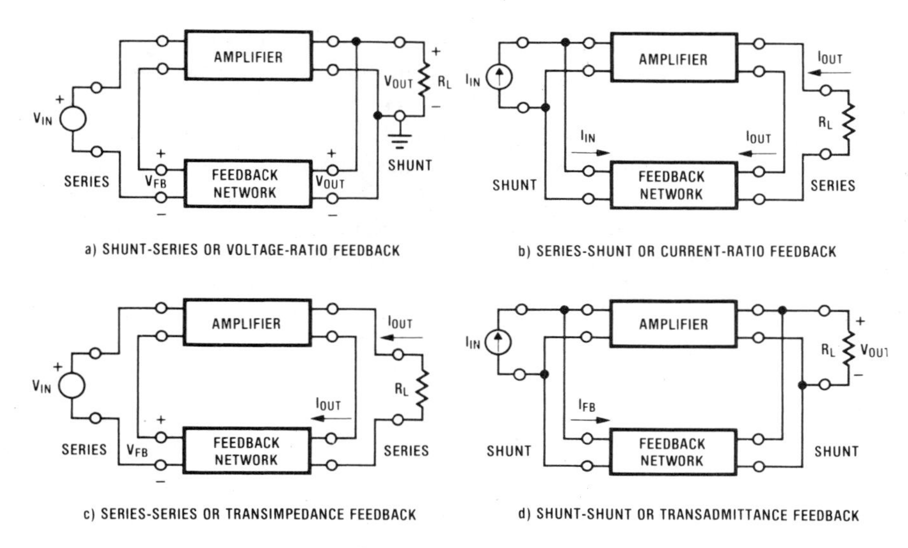

Fig. 2-14. The Four Feedback Possibilities

Confusion also results from the fact that the closed-loop "gain" or transfer function of a feedback system is the inverse of the feedback factor (because, as we have previously seen, $A_{CL} \cong 1/\beta$). For example, if the feedback network were to sample the output current of the amplifier and from this produce a voltage to be placed in series with the input voltage to the amplifier, the feedback would be called *transimpedance feedback* (V/I). But the stabilized closed-loop transfer function will be the reciprocal of this, (I/V). A *transadmittance amplifier* will result from this *transimpedance feedback.* Hopefully, this double-talk will become clearer as we consider the four feedback examples.

Voltage-Ratio Feedback

When the feedback network is shunting the output of the amplifier and is wired in series with the input to the amplifier, Figure 2-15, we have *shunt-series or voltage ratio feedback.* This typical schematic, Figure 2-15b, is recognized to be the standard noninverting amplifier.

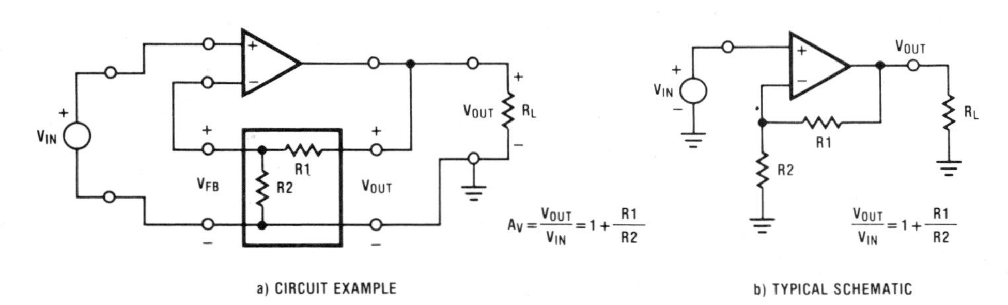

Fig. 2-15. Shunt-Series or Voltage-Ratio Feedback

A convenient thing about this feedback circuit is that it has a non-dimensional voltage-ratio feedback factor. The feedback network is providing V_{FB}/V_{OUT}. The closed-loop transfer function is the reciprocal of this, or V_{OUT}/V_{IN}. Because we are using a high gain op amp, the required input voltage of the amplifier can usually be neglected so $V_{IN} \cong V_{FB}$. This stabilized closed-loop transfer function is also a nondimensional voltage ratio: the voltage gain of the circuit.

An idealized voltage-ratio feedback amplifier, Figure 2-16, provides a high input impedance, a low output impedance, and a stabilized, dimensionless transfer function that is a voltage ratio or a voltage gain, A_V. This is one of the most common application circuits for an op amp.

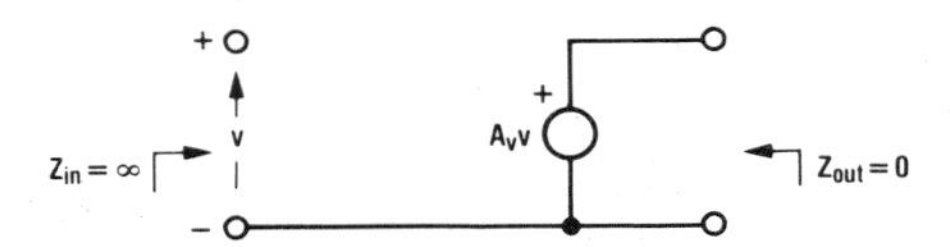

Fig. 2-16. The Idealized Equivalent Circuit for Voltage-Ratio Feedback

Current-Ratio Feedback

We now consider the interconnection shown in Figure 2-17. This *series-shunt feedback configuration* is called *current-ratio feedback* and *provides a stabilized current gain.*

The typical schematic, Figure 2-17b, shows a practical problem that results from the series connection at the output. The load is no longer ground referenced. It must float. This would present no problems if the load was, for example, a D'Arsonval meter — if you remember what that was.

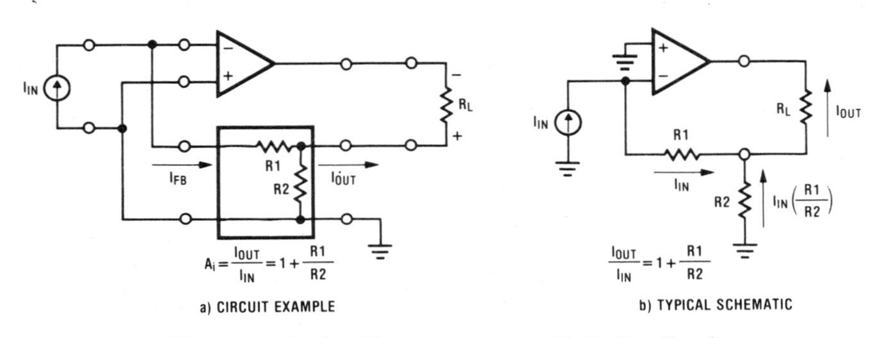

Fig. 2-17. Series-Shunt or Current-Ratio Feedback

This idealized current-ratio feedback, Figure 2-18, provides a low input impedance, a high output impedance, and a stabilized, dimensionless transfer function that is a current ratio or a current gain, A_i. Because of the requirement for a floating load (R_L), this circuit is not very popular.

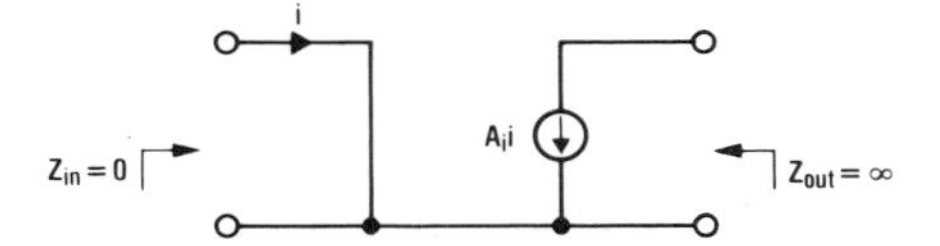

Fig. 2-18. The Idealized Equivalent Circuit for Current-Ratio Feedback

Transimpedance Feedback

When the feedback network samples the output current and provides an input voltage, the feedback factor becomes V_{FB}/I_{OUT} and has the dimensions of resistance, or (to be more general) impedance. This *transimpedance feedback* is shown in Figure 2-19. A typical schematic is shown in Figure 2-19b. Notice that the series interconnection at the output again requires a floating load.

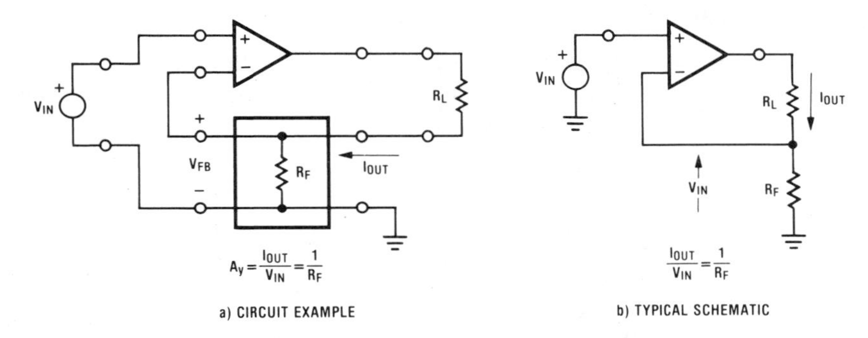

Fig. 2-19. Series-Series or Transimpedance Feedback

The closed-loop transfer function or *gain* now has dimensions: 1/R, conductance or (the more general) admittance. So *transimpedance feedback provides a transadmittance amplifier* because the closed-loop transfer function is the reciprocal of the feedback factor.

This idealized amplifier, Figure 2-20, has a high input impedance; a high output impedance; and a transfer function; or "gain," A_y, that has the dimensions of admittance. It provides a stabilized output current in response to an input voltage. The floating load also reduces the popularity of this application circuit but an adaptation, shown in Figure 2-21, is popular for providing current sources. We will discuss these in Chapter 5.

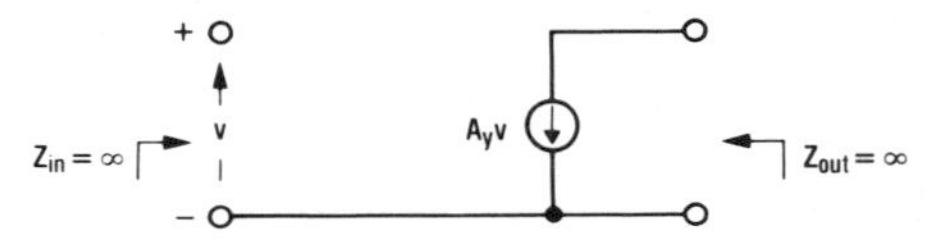

Fig. 2-20. The Idealized Equivalent Circuit for Transimpedance Feedback

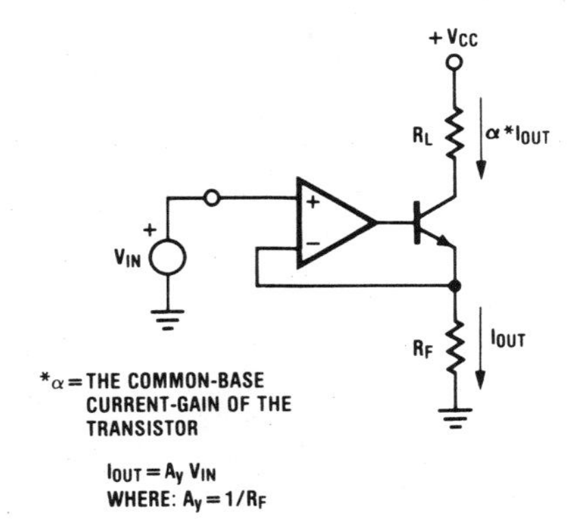

Fig. 2-21. A Practical Adaptation of a Transadmittance Amplifier

Transadmittance Feedback

The last way to interconnect the feedback network and the amplifier is shown in Figure 2-22. This *shunt-shunt* configuration also has dimensions associated with the feedback factor (I_{FB}/V_{OUT}): *transadmittance.*

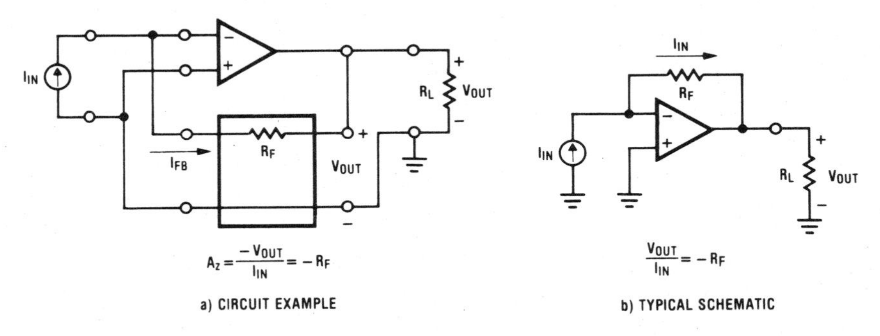

Fig. 2-22. Shunt-Shunt or Transadmittance Feedback

This idealized amplifier, Figure 2-23, has a low input impedance, a low output impedance, and provides a stabilized transfer function (V_{OUT}/I_{IN}) or *gain,* A_z, *that has the dimensions of impedance.* This circuit is often called a *current-to-voltage converter* or a *transimpedance amplifier.* This transimpedance amplifier is also the basis of many additional, very useful, op amp application circuits. For example, an integrator results when R_F is replaced by a capacitor, as shown in Figures 2-24a and 2-24b. The standard inverting voltage amplifier results when an input resistor is added to convert from an input voltage to an input current, as shown in Figure 2-24c.

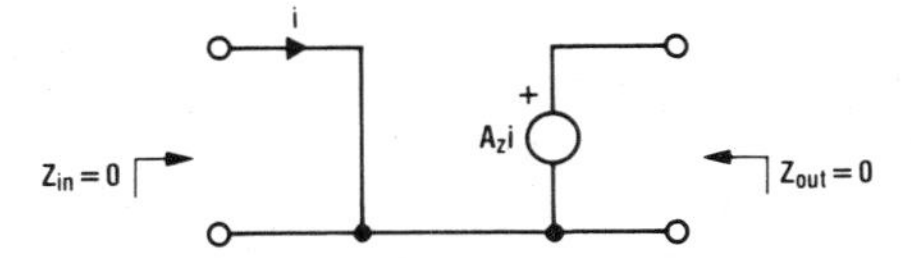

Fig. 2-23. The Idealized Equivalent-Circuit for Transadmittance Feedback

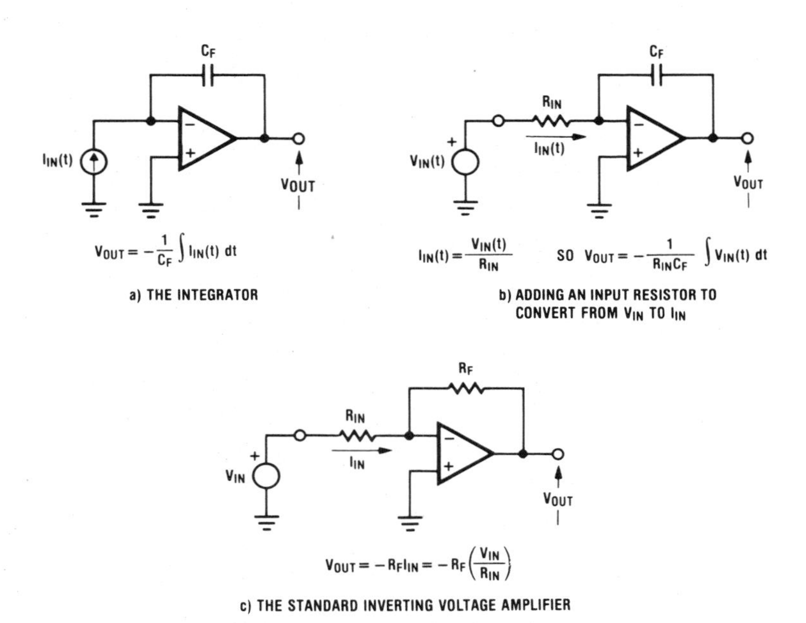

Fig. 2-24. Some Common Transimpedance Amplifiers

This transimpedance amplifier is perhaps the most common of the op amp application circuits because it includes all of the inverting applications of op amps that make use of *virtual ground (the summing junction).* In many of these circuits, an input impedance is added to convert an input voltage into an input current. The resulting voltage gain of an inverting voltage amplifier is actually not stabilized by the use of feedback, as it is for the noninverting voltage amplifier, and this is the reason for the difference in the equations for voltage gains of these two types of voltage amplifiers.

We will now look more closely at the reason that feedback affects the input and output resistance levels of the closed-loop circuits. We will also develop the powerful concept of *putting undesirable resistances "within-the-loop"* of a feedback circuit *in order to eliminate them.*

2.6 THE EFFECT OF FEEDBACK ON INPUT AND OUTPUT RESISTANCE

We can provide some insight into the output resistance to expect in a closed-loop amplifier by stating that *whatever the feedback network "samples" at the output resists being changed.* If

V_{OUT} is sampled, then V_{OUT} essentially can't be changed. For example, adding a load (R_L) will not cause V_{OUT} to drop, so we have a low closed-loop output impedance.

This active lowering of the output resistance results from *amplifier action*. Any small changes in V_{OUT} that result from changes in I_{OUT} are passed on to the input of the amplifier by way of the feedback network. This causes the output of the amplifier to respond by sourcing or sinking any current that may be necessary to keep V_{OUT} constant. This amplifier action is made use of in providing a low output impedance for power supply voltage regulator circuits.

If, instead, we had used a series interconnection of the input to the feedback network and the amplifier output, then I_{OUT} would resist change and a high output impedance would therefore result.

At the input of the closed-loop amplifier, a high impedance results if a series interconnection of the amplifier and feedback network is used. This is because the voltage fed back keeps the actual input voltage that is applied to the amplifier negligibly small. Conversely, a shunt or parallel interconnection of the amplifier and the feedback network provides essentially a zero input impedance for the closed-loop circuit.

Lowering the output impedance and raising the input impedance are important benefits obtained with op amp application circuits. Unfortunately, these impedance changes involve the action of a feedback loop and quantitative results must be obtained by an actual analysis of the circuit. Intuition is rather difficult with feedback, especially where a numerical measure of an effect is desired. We will, however, also show how you can quickly estimate whether a high or low impedance results at the input or the output of an op amp application circuit.

The impedance, or resistance, to simplify the analysis, will be determined by calculating the terminal voltage that results at either the input or the output of the closed-loop amplifier when we force in 1 amp of current at that terminal. This calculated voltage then will be numerically equal to the value of the input or output resistance we are looking for. This computational trick is *permitted* by the circuit models *but would burn out a real op amp*. If 1 amp bothers you, 1 mA can be used if you keep track of the 10^{-3} scaling factor that is introduced.

Output Resistance with Shunt Feedback

When the feedback network is placed in shunt with both the output and the input of the amplifier, the shunt-shunt circuit of Figure 2-25 results. The resistor shown as R_o is used to represent the open-loop output resistance of the op amp.

The current, $V_{OUT}/(R_{IN} + R_F)$, that enters the feedback network, via R_F, produces a voltage, V_ϵ, at the input to the op amp that is given by,

$$V_\epsilon = \frac{V_{OUT}}{(R_{IN} + R_F)} R_{IN}$$

This is amplified by $-A$ to provide a voltage at the internal side of R_o that is given by,

$$-AV_\epsilon = -AR_{IN} \frac{V_{OUT}}{(R_{IN} + R_F)}$$

The output voltage, V_{OUT}, can now be found by adding to this the voltage that is dropped across R_o when the current that is indicated on the figure enters the output of the amplifier, or

$$V_{OUT} = R_{OUT} = -AR_{IN}\left[\frac{V_{OUT}}{(R_{IN} + R_F)}\right] + \left[1 - \frac{V_{OUT}}{(R_{IN} + R_F)}\right]R_o$$

collecting terms, this becomes

$$R_{OUT}\left[1 + \frac{AR_{IN}}{R_{IN} + R_F} + \frac{R_o}{R_{IN} + R_F}\right] = R_o$$

or

$$R_{OUT} = \frac{R_o}{1 + \dfrac{AR_{IN}}{R_{IN} + R_F} + \dfrac{R_o}{R_{IN} + R_F}}$$

but,

$$\beta = \frac{R_{IN}}{R_{IN} + R_F}$$

so, substituting this, we find

$$R_{OUT} = \frac{R_o}{1 + A\beta + \dfrac{R_o}{R_{IN}}\beta}$$

and, as $A \gg \dfrac{R_o}{R_{IN}}$

$$R_{OUT} \cong \frac{R_o}{1 + A\beta}$$

This equation shows that the open-loop output resistance of the op amp, plus any external resistance that may be placed in series with R_o, is reduced by the desensitivity factor. This is often called placing resistance within the loop. For example, if $R_o = 100\Omega$; $A = 50{,}000$; and $\beta = 0.01$; R_{OUT} becomes

$$R_{OUT} \cong \frac{100\Omega}{1 + (5 \times 10^4)(10^{-2})} \cong 0.2\Omega$$

As a way to remember this result, we can now apply intuition to quickly analyze this shunt-feedback circuit (Figure 2-25). We still imagine forcing current into the output terminal and simply determine whether a large or a small output voltage (and therefore a large or a small output impedance) would result. In this case, notice that the output voltage would initially (before any feedback action takes place) tend to rise in response to this current that is entered into the output terminal. This increase in voltage is fed to the inverting input of the op amp, so V_{OUT} would tend to drop to oppose this increase in V_{OUT}. Therefore, we would correctly conclude that a low output impedance results.

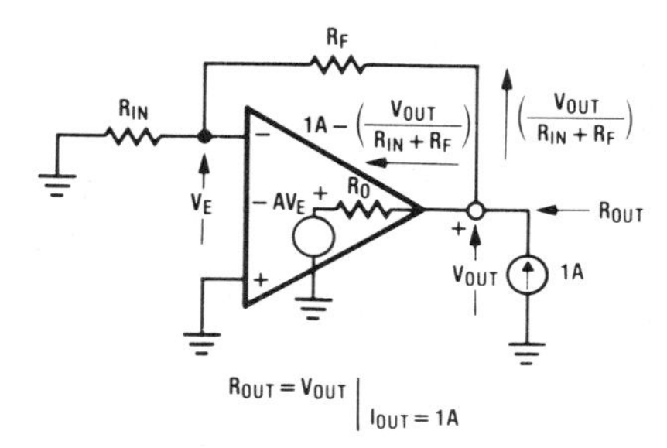

Fig. 2-25. Shunt Feedback Lowers the Output Resistance

This is the circuit action used to provide a low output impedance for a series voltage regulator. The β network is altered to provide a desired V_{OUT} from a given fixed V_{REF}. Changing the value of β changes the closed-loop dc gain. Because of these changes in the value of β, the closed-loop output resistance is not constant. Specifications for voltage regulators therefore are given in terms of percent regulation.

The change in output resistance occurs because there is less feedback when V_{OUT} is large because more closed-loop gain is needed to amplify the reference voltage. Consequently, β is smaller. Percentage regulation specifications allow more absolute V_{OUT} change when V_{OUT} is large. This accommodates the smaller β that exists when V_{OUT} is large.

Output Resistance with Series Feedback

The effect of series feedback at the output can be determined by analyzing the circuit shown in Figure 2-26. We enter 1 amp at the designated output terminals and calculate the output voltage that results between these terminals.

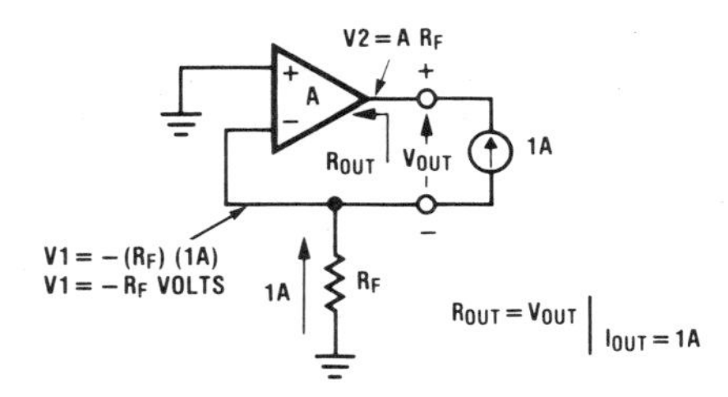

Fig. 2-26. Series Feedback Raises the Output Resistance

The 1 amp creates a drop across R_F, shown as V1 on the figure, of $-(R_F)(1A)$ or $-(R_F)$ volts (note the numerical simplification and also the confusion that is caused by our 1A test current). This voltage is amplified by $-A$ to produce the voltage V2 at the output of the op amp. The output resistance, R_{OUT} therefore is,

$$R_{OUT} = V_{OUT} = V2 - V1$$

$$R_{OUT} = AR_F + R_F = (A + 1) R_F$$

The output resistance is essentially the open-loop gain times the feedback resistance. As an example, if $R_F = 100\Omega$ and $A = 50{,}000$,

$$R_{OUT} \cong (50{,}000)(100) = 5\ M\Omega$$

To remember this result, we can apply intuition by noticing (in Figure 2-26) that if we forced the 1 amp into the output, the resulting negative voltage drop across R_F is applied to the inverting input of the op amp. This causes a much larger V_{OUT} to result. We get a large V_{OUT} in response to our 1 amp test current and therefore, we would correctly conclude that this circuit has a high output impedance.

Obtaining a Stabilized Output Resistance

As expected, if the voltage fed back depends on both the output voltage and the output current, a controlled output resistance can be obtained. A circuit that accomplishes this is shown in Figure 2-27. The R_{OUT} of this circuit can be evaluated by using the circuit shown in Figure 2-28. Here, 1 amp is forced into the output of the amplifier, V_{IN} is replaced by a short-circuit, and the resulting output voltage, V_{OUT} (where $R_{OUT} = V_{OUT}$), is calculated.

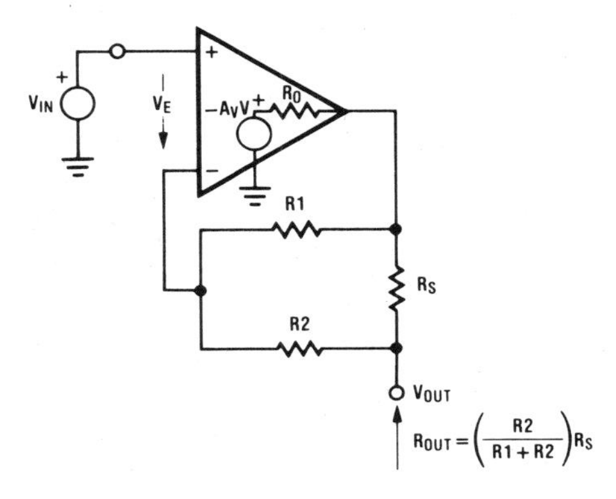

Fig. 2-27. Stabilizing the Output Resistance

From this figure, we can equate V_{OUT} with the expression that is calculated for V_{OUT}. First, assume that a V_{OUT} exists. Then chase this around the loop to derive an expression for V_{OUT} in terms of the parameters of the circuit. This provides the starting equation. The 1A cur-

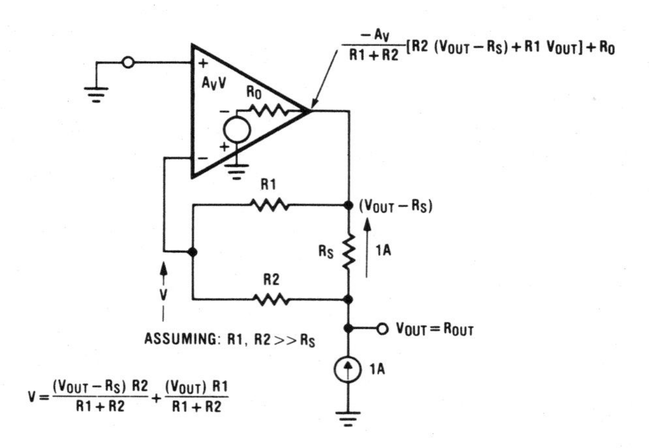

Fig. 2-28. Calculating R_{OUT} for the Stabilized Output-Resistance Circuit

rent flow through both R_o and R_s, we are assuming R1, R2 >> R_s, creates voltage drops of (R_o) (1A) and R_s (1A) or simply R_o and R_s as shown in the equation.

$$V_{OUT} = (R_o + R_s) - \left(\frac{A_V}{R1 + R2}\right) [R2\,(V_{OUT} - R_s) + R1\,V_{OUT}]$$

where the second term is $-A_V\,V_E$.

Collecting terms

$$V_{OUT} = \left[1 + \left(\frac{A_V\,R2}{R1 + R2}\right) + \left(\frac{A_V\,R1}{R1 + R2}\right)\right] = (R_o + R_s) + \frac{R2\,R_s\,A_V}{R1 + R2}$$

$$V_{OUT} = R_{OUT} = \frac{(R_o + R_s) + \left(\dfrac{R2\,R_S\,A_V}{R1 + R2}\right)}{\left[1 + \left(\dfrac{A_V\,R1}{R1 + R2}\right) + \left(\dfrac{A_V\,R2}{R1 + R2}\right)\right]}$$

after simplifying,

$$R_{OUT} = \frac{\left[\dfrac{R_o\,R1 + R1\,R_s + R_o\,R2}{1 + A_V}\right] + R2\,R_s}{(R1 + R2)}$$

For large values of A_V,

$$R_{OUT} \cong \left(\frac{R2}{R1 + R2}\right) R_s$$

This provides a value for R_{OUT} that is smaller, or at most equal to, R_s. For R2 much larger than R1, we approach the limiting case of a voltage follower and now R_s is completely outside the feedback loop, so $R_{OUT} = R_S$. This is often the way a prescribed output resistance is realized in practice.

Consider an op amp with a stabilized output resistance driving a load resistor that is equal in value to this output resistance, as shown in Figure 2-29. In this example we have let R1 = R2 = R so that R_{OUT} is $R_s/2$ (2 kΩ/2 = 1 kΩ). The output voltage is reduced by one-half. This *flat loss* of one-half (−3 dB) is sometimes overlooked and can cause some confusion. Designing for a particular output resistance is usually restricted to those applications where a signal is transmitted over a cable with a prescribed characteristic impedance. One example is the 600Ω standard that is used for audio and telephone lines where both the source and load resistance have the same values as the characteristic impedance. The relatively poor high frequency response of the standard op amp and the limited output current does not allow them to be used to drive the 50Ω coaxial cable transmission lines that are used for high frequency signals. As we will see in Chapter 5, at low frequencies these cables are not resistive loads, they are capacitive.

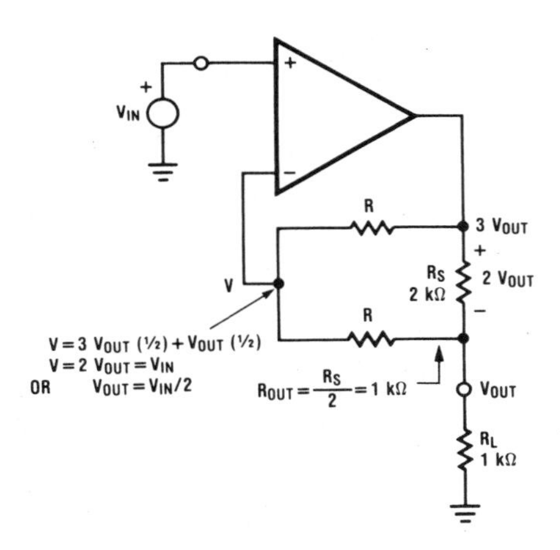

Fig. 2-29. V_{OUT} is Reduced by One-Half when $R_L = R_{OUT}$

Input Resistance with Series Feedback

The circuit shown in Figure 2-30 will be used to show the effects of series feedback on the input resistance (R_I, the open-loop differential input resistance) of the op amp. To evaluate R_{IN}, we force 1A in and then calculate the V_{IN} that results.

We see that the input voltage for the op amp, V_ϵ, is given by

$$V_\epsilon = V_{IN} - V2$$

$$V_\epsilon = V_{IN} - \beta\, V_{OUT}$$

and the output voltage of the op amp must be the open-loop gain times V_ϵ, or

$$V_{OUT} = A\,(V_{IN} - \beta V_{OUT})$$

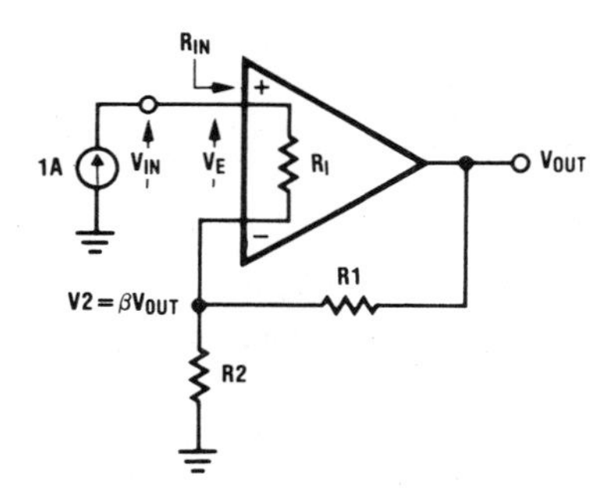

Fig. 2-30. Raising the Input Resistance with Series Feedback

collecting terms,

$$V_{OUT}\,(1 + A\beta) = A\,V_{IN}$$

or

$$V_{OUT} = \left(\frac{A}{1 + A\beta}\right) V_{IN}$$

The input voltage is given by

$$V_{IN} = V_{\epsilon} + V2$$

or

$$V_{IN} = V_{\epsilon} + \beta V_{OUT}$$

Substituting for V_{OUT} from the previous equation and noting that $V_{\epsilon} = (R_I)(1A) = R_I$, we have

$$V_{IN} = R_I + \beta\left[\frac{A}{1 + A\beta}\right] V_{IN}$$

collecting terms

$$V_{IN}\left[1 - \frac{A\beta}{1 + A\beta}\right] = R_I$$

or

$$V_{IN} = R_{IN} = \frac{R_I}{1 - \left[\frac{A\beta}{1 + A\beta}\right]}$$

Simplifying

$$R_{IN} = \frac{R_I}{\frac{1 + A\beta - A\beta}{1 + A\beta}} = (1 + A\beta)\, R_I$$

So, the differential input resistance of the op amp is seen to be raised by the desensitivity factor. The result is an effective increase in input resistance. This increase is often referred to as "bootstrapping the input resistance." (The term "boostrap" is derived from the rather fanciful concept of "lifting yourself up by pulling on your own bootstraps.") The increased input resistance also exists at the inverting input. Consequently, the op amp does not load the β network. There is, however, a common-mode input capacitance that does load the β network. This capacitance will be considered in the next chapter where we will look at the effects of common-mode impedance on the op amp.

A simple way to remember this result is to notice that the feedback network is returning a signal to the inverting input of the op amp that is essentially equal to V_{IN}. Therefore, only a very small part of the input voltage actually exists directly across the input terminals of the op amp. This very small voltage across the open-loop input resistance of the op amp causes a very small input current to flow from the input signal source. Consequently, the input impedance of the op amp "appears" unusually large.

Input Resistance with Shunt Feedback

A circuit with shunt feedback at the input is shown in Figure 2-31. We can calculate the V_{IN} that results from our 1A test current by noticing that

$$V_{OUT} = V_{IN} - (1A)\,(R_F) = V_{IN} - R_F$$

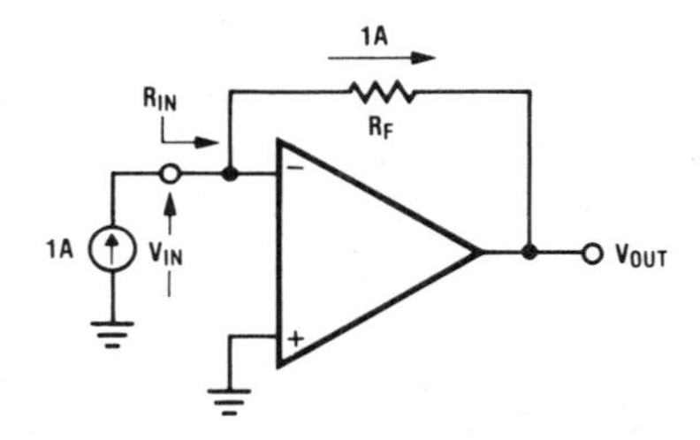

Fig. 2-31. Reducing the Input Resistance with Shunt Feedback

but; also, as a result of the op amp gain,

$$V_{OUT} = -A\, V_{IN}$$

therefore, these two equations must be equal, or

$$-A\, V_{IN} = V_{IN} - R_F$$

collecting terms

$$V_{IN}(1 + A) = R_F$$

or, knowing $V_{IN} = R_{IN}$,

$$R_{IN} = \frac{R_F}{1 + A} \cong \frac{R_F}{A}$$

(this approximation is valid because, for an op amp, $A \gg 1$) and the input resistance is therefore the value of the shunt feedback resistor, R_F, divided by the open-loop gain. For example, with $R_F = 10\ \mathrm{k\Omega}$ and $A = 50{,}000$

$$R_{IN} \cong \frac{10^4}{5 \times 10^4} \cong 0.2\Omega$$

To intuitively appreciate this result, notice that the 1 amp current that is forced into the input tends to cause V_{IN} to increase. This slight increase in voltage is directly at the inverting input of the op amp and so the op amp responds by providing an amplified (and negative) V_{OUT}. This amplifier action causes this forced input current to be diverted through the feedback resistor, where it will be absorbed by the output of the op amp, which reduces the input voltage, and therefore a small input resistance results.

These calculations have demonstrated the *amplifier action* that takes place and modifies the terminal impedances in an amplifier with feedback. We will now consider an unexpected internal thermal feedback mechanism that modifies the dc open-loop voltage gain of an IC op amp.

2.7 THERMAL FEEDBACK EFFECTS

All electronic circuits that are in thermal proximity, whether simply in the same box, the same module, or the same IC package, respond to the heat that is dissipated by the circuit components. This changing thermal environment introduces an additional feedback mechanism to the circuitry.

Thermal feedback is limited to dc and low frequencies because of the relatively large thermal inertia of the components in the thermal feedback path. Therefore, some strange things can be noticed: dc output resistance can be negative (the output voltage can actually increase under load) *and the dc gain of some IC op amps can be of either sign and "any magnitude it wants to be."* A plot of the dc gain of an IC op amp that shows some "interesting" thermal effects is shown in Figure 2-32. These thermal effects are made worse when the op amp is driving a relatively small valued (2 kΩ) load resistor.

The usual condition that results with thermal feedback is that the dc performance parameters of the op amp are drastically changed. Thermally-induced changing error magnitudes and signs result that are related to the magnitude of the output voltage. A changing gain error results because this relates to the input voltage that is instantaneously needed to support an output voltage. This output-dissipation dependent *gain error* is important because it *can cause dc gain linearity-errors* in a precise application. For these applications, an external emitter follower transistor or a unity-gain op amp buffer should be added at the output of the first op

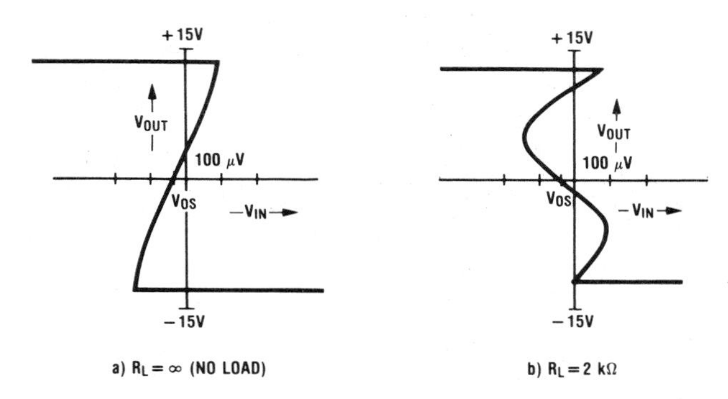

Fig. 2-32. Thermal Coupling Can Drastically Affect the DC Voltage Gain

amp to unload this op amp and thereby reduce the heat generated. This buffer should be placed within the overall feedback loop. The buffer will improve the gain linearity.

Thermally-symmetric IC chip layouts are used to reduce this thermal feedback problem by preventing differential thermal errors between critical elements of the circuit (changes in die temperature then become a uniform or *common-mode effect*). The more accurate component placement that exists on an IC chip can provide a better and more repeatable solution to this thermal feedback problem than can be achieved with other high-density electronic circuit packaging techniques. Best results are obtained in IC chip layouts by making use of as much thermal isolation (physical separation) as is possible between the output stage and the input stage.

Now that we have discussed the basic idea of feedback, we will consider the factors that make an amplifier with feedback depart from an ideal amplifier. This requires that we look at some of the error sources that exist with real op amp circuits.

CHAPTER III

Op Amp Error Sources

In the previous chapter we considered the dc gain error that results from the finite gain of an op amp. There are many other sources of error in application circuits. An understanding of these op amp imperfections is essential for designers of high accuracy systems. This knowledge will also benefit all designers by making them aware of the limitations of their circuits. In many cases, however, errors caused by the op amp are negligible when compared with other parts of the circuit, such as the initial and temperature induced errors that result from the passive components used for the feedback network.

Many, perhaps even most, of the common applications of op amps will tolerate errors of a few percent. As more accuracy is required, the details of the application circuit, the choice of external components, and the specs of the op amp must be more carefully checked. In this section, the sources of error will be discussed, errors will be calculated, and examples showing the general magnitude of each will be given.

3.1 PROBLEMS WITH THE FEEDBACK NETWORK

The benefits of feedback and high performance op amps are lost if little attention is paid to the selection of the critical feedback components. If we achieve our goal, and have the closed-loop performance depend *only* on these external components, the next question is, did we use components of high enough quality to meet our performance objectives? (We will consider the problems of passive component selection in Chapter 6.)

When the feedback network consists of only two resistors, we have the ideal case. Performance will then depend on the ratio of these resistors and this ratio dependence usually provides at least an order of magnitude improvement over that of the individual components in regard to temperature stability, for example. In very high accuracy applications, the large signal-voltage drop across the feedback resistor can cause nonlinear gain errors because of the unequal power dissipation in the resistors of the feedback network, even if the temperature coefficients of all of the resistors are exactly matched. High power rating resistors that are kept in good thermal contact or thin film resistors on the same substrate, can reduce this error source. Also, when working with high accuracy analog circuits, the critical feedback resistors and even the PC board should be stabilized by use of a burn-in cycle prior to the calibration of a new electronic board.

The problem of the initial tolerances of the resistors of the feedback network will cause a corresponding initial tolerance in the gain provided. This can be improved upon in only two ways: specify a low tolerance on the initial component values (such as 1% or better, if needed) or provide a way to adjust the initial tolerance of the resistors that establish the feedback factor. For highest accuracy, this last, more costly approach must be used. The tradeoff is between the costs of trimming and the added costs for a tighter initial tolerance specification on the resistors. Some high volume assembly techniques use film resistors deposited on a substrate and then use computer controlled individual resistor trimming to solve this initial tolerance problem.

The low cost and usually adequate performance of 1% tolerance metal-film resistors makes these popular for use in feedback networks. Not all the passive components associated with a given application circuit have a direct effect on β, the feedback factor. Some are used as biasing or frequency compensating elements. These less critical resistors usually do not require 1% tolerance or exceptional stability with time or temperature changes.

Another consideration when selecting the scaling factors is the magnitudes of the resistances that are used in the feedback network. Large values of resistance can create problems as a result of the flow of the input current of the op amp even when the input resistance is matched on both of the input leads of the op amp. A second factor that is aggravated with large-valued resistors is the problem of stray pickup. These considerations limit the maximum values of the resistors. As these resistors are made small in value, a lower limit in value also exists because the output of the op amp must be able to drive the paralled combination of its feedback resistor, perhaps a separate load resistor (not used when op amps only drive other op amps) and sometimes the input resistor of a following inverting-gain op amp stage. Most op amps are specified to drive a minimum load resistance of 2 kΩ, but some op amps will only handle a 10-kΩ load.

Special problems exist when an R and a C (as in an integrator) must be used. Temperature tracking of different types of passive components is much more difficult. The popularity of polystyrene capacitors results because of the availability of film resistors with a temperature coefficient of resistance, TCR, of + 100 ppm/°C to temperature compensate these capacitors. Capacitors, in general, are not capable of the same performance that resistors can provide in regard to low initial tolerance, wide selection of values offered, temperature independence, and the general lack of undesired parasitic effects. Although, high-cost capacitors can achieve 0.1% tolerance and ± 30 ppm/°C.

A large problem with capacitors is dielectric absorption. This results from physical characteristics of the dielectric material that cause capacitors to have an undesired *memory effect*. This can most vividly be demonstrated by charging a capacitor to a high voltage, momentarily shorting the leads (in an attempt to discharge it) and then handing this *discharged* capacitor to an innocent bystander. This shocking experience represents the dielectric storage problem that plagues many of the op amp application circuits that use capacitors. More will be said about this in Chapter 6.

3.2 DYNAMIC ERRORS

Dynamic errors are those additional errors that exist because the input signal is no longer dc. Some of these error problems are the natural, but often overlooked, amplitude inaccuracies of the transfer function, V_{OUT}/V_{IN}, of a single-pole, low-pass filter, Figure 3-1.

The Bode plot (named after Dr. Hendrik W. Bode of the Bell Telephone Laboratories) in this figure shows how the actual smoothly shaped magnitude response is simply approximated by

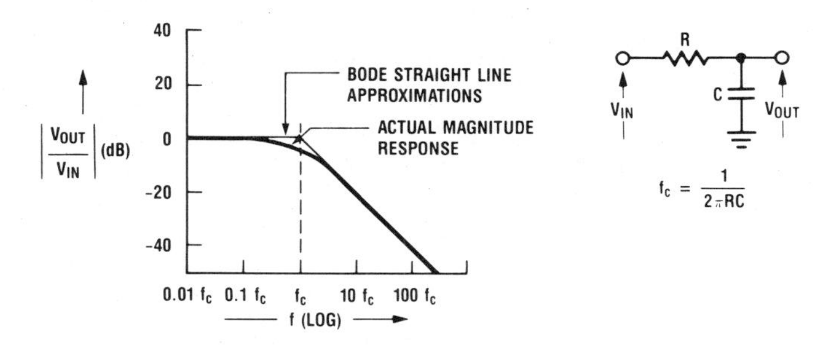

Fig. 3-1. A Low-Pass Filter Transfer Function

two intersecting straight lines. The actual magnitude response is down by 3 dB (0.707) at the corner frequency f_C.This value of −3 dB is accepted as a useful limit in audio applications because of the logarithmic response of the human ear (1 dB change in sound level is barely perceptible to the ear), but *this is an intolerable error for a high performance measurement system. Frequencies of interest* therefore *must be kept below this corner frequency.*

If we normalize the magnitude of this low-pass transfer function, $|v_{out}/v_{in}|$, to unity (at low frequencies), we can then express this magnitude response, M, as

$$M = \left|\frac{v_{out}}{v_{in}}\right| = \left|\frac{1}{1 + j\left(\frac{\omega}{\omega_C}\right)}\right| = \left|\frac{1}{1 + j\left(\frac{\omega}{\omega_C}\right)} \bullet \frac{1 - j\left(\frac{\omega}{\omega_C}\right)}{1 - j\left(\frac{\omega}{\omega_C}\right)}\right|$$

$$M = \left|\frac{1 - j\left(\frac{\omega}{\omega_C}\right)}{1 + \left(\frac{\omega}{\omega_C}\right)^2}\right|$$

or

$$M = \sqrt{\frac{1}{1 + \left(\frac{\omega}{\omega_C}\right)^2} + \frac{\left(\frac{\omega}{\omega_C}\right)^2}{1 + \left(\frac{\omega}{\omega_C}\right)^2}} = \sqrt{\frac{1}{1 + \left(\frac{\omega}{\omega_C}\right)^2}}$$

For a 1% magnitude error, $|V_{OUT}/V_{IN}| = 0.99$, and the largest value for ω, ω_{MAX}, that can be passed through this filter can be found as

$$0.99 = \frac{1}{\sqrt{1 + \left(\frac{\omega_{MAX}}{\omega_c}\right)^2}}$$

$$0.98 = \frac{1}{1 + \left(\frac{\omega_{MAX}}{\omega_c}\right)^2}$$

$$\left(\frac{\omega_{MAX}}{\omega_C}\right)^2 = \frac{1}{0.98} - 1$$

$$\omega_{MAX} = \sqrt{\frac{1}{0.98} - 1}\ \omega_C = 0.14\ \omega_C$$

or

$$f_C = \frac{1}{0.14} f_{MAX} = 7\ f_{MAX}$$

This means, that for a 1% amplitude error in a 1 kHz signal, the corner frequency must be at least moved out to 7 kHz.

A 0.01% magnitude error would limit the largest ω to

$$\omega_{MAX} = \sqrt{\frac{1}{0.9999} - 1}\ \omega_C = 0.01\ \omega_C$$

or

$$f_C = \frac{1}{0.01} f_{MAX} = 100\ f_{MAX}$$

This dramatically shows the magnitude errors that are introduced by filters. This accuracy-versus-frequency limit exists for the simplest, passive, RC, low-pass filter. It is not a unique op amp problem. It also indicates that to achieve small values of dynamic error requires careful compensation for this amplitude error of a filter. Furthermore, it is very difficult to measure the amplitude of a sinewave to within 0.01%.

We can restate the above equation in a general way by again considering the normalized magnitude, M, of the response with an error, ϵ, as

$$M = (1 - \epsilon) = \frac{1}{\sqrt{1 + \left(\frac{f_{MAX}}{f_C}\right)^2}}$$

$$1 - 2\epsilon + \epsilon^2 = \frac{1}{1 + \left(\frac{f_{MAX}}{f_C}\right)^2}$$

and, as $\epsilon \ll 1$, $\epsilon^2 \cong 0$, and can therefore be neglected to give

$$(1 - 2\epsilon) \cong \frac{1}{1 + \left(\frac{f_{MAX}}{f}\right)^2}$$

$$\left(\frac{f_{MAX}}{f_C}\right)^2 \cong \frac{1}{1 - 2\epsilon} - 1$$

and as

$$\frac{1}{1 - 2\epsilon} \cong 1 + 2\epsilon, \text{ for } 2\epsilon \ll 1$$

$$f_{MAX} \cong \sqrt{2\epsilon}\, f_c$$

which is a very useful approximation for calculating the signal loss when the operating frequency is much less than the corner frequency. Don't forget that the approximation used in obtaining this simple form requires $2\epsilon \ll 1$.

Dynamic Gain Errors

Our previous discussions of gain error were restricted to dc operation. As the frequency is increased, the magnitude of $A\beta$ drops and this also causes dynamic gain errors.

Conceptual problems often exist in handling this particular dynamic error of a fed-back amplifier. Many application mistakes are made because of improperly (or not even) considering the phase relationships that exist with A or β. In many applications, the phase associated with $A\beta$ greatly reduces the resulting error.

To investigate this we will return, once again, to the feedback equation

$$A_{CL} = \frac{A}{1 + A\beta} = \frac{1}{\beta}\left(\frac{A\beta}{1 + A\beta}\right)$$

and we will again assume, for simplicity, that β is independent of frequency. The ideal closed-loop voltage gain is $1/\beta$, so any departure of the second term of the above equation from the value of unity will represent the gain error that we are now interested in. In Chapter 2 we considered only dc gain errors. Now we will see what happens when we consider the ac case and account for the phase shift that is associated with A.

To provide a comparison, we will start with a dc example and assume $A\beta = 100$ (at dc). The second term in the above equation becomes

$$\frac{100}{1 + 100} = \frac{100}{101} = 0.99$$

and the dc gain error is therefore 1%.

Now, if we consider operation at a frequency higher than the dominate pole of the open-loop response, the phase lag of A (and therefore also of $A\beta$) will be $-90°$ and the phasor sum of $1 + A\beta$ means that we are adding two phasors ("vectors"), one with a length of 1 and the other with a length of 100 and there is an angle of 90° between them. This phasor or sum can be expressed as

$$1 + A\beta = \sqrt{(1)^2 + (|A\beta|)^2} = \sqrt{1 + (100)^2} = 100.005$$

and has a phase angle of approximately $-90°$ ($-89.4°$).

The error term now becomes

$$\frac{A\beta}{1 + A\beta} = \frac{100\angle -90°}{100.005\angle -89.4°} = 0.99995\angle -0.6°$$

[For our assumed condition of no phase shift associated with the β network (only resistors used), the phase of this error term is also the phase of the closed-loop gain: the amount of phase lag of the output voltage with respect to the input voltage.]

The ac gain error is the departure of this error term from unity, or

$$\text{ac gain error} = 1 - 0.99995 = 0.00005$$

and, expressed as a percent, this ac gain error is only 0.005%!

This is *considerably smaller* than the 1% error that we found to exist at dc where, strangely enough, we have the largest open-loop gain. This improvement is often confusing. A smaller ac gain error results because of the very significant difference that exists between a conventional numerical summation (that applies for a dc signal) and the quite different phasor summation (that must be used for an ac signal). When the signal of interest is no longer simply dc, this strange advantage of phasor addition must be kept in mind.

Another often overlooked problem also lurks in the determination and specification of this ac gain error. Note that the output voltage will be essentially in-phase with the input voltage for the previously assumed conditions, because the 90° phase lag of $(1 + A\beta)$ is nearly the same as the phase lag of $A\beta$. This will not be the case at higher frequencies.

As the operating frequency is increased, the phase lag of A initially remains essentially at -90°, although this phase lag will increase as f_u is approached. But the magnitude of A falls. Therefore, at higher frequencies the phasor sum $(1 + A\beta)$ will have an angle less than -90°. This will no longer balance the -90° phase lag of the $A\beta$ term in the numerator of our error expression and the output voltage waveform will now have a larger phase lag with respect to the input voltage waveform. Even though feedback is used, *the phase of high frequency output voltages will lag the input voltage.*

This phase relationship between the ideal output voltage (with no phase lag) and the actual output voltage (that includes phase lag) is shown in Figure 3-2. At each point in time, the instantaneous dynamic error is the difference between these two phasors or *vectors*. This is called the *vector error*, ϵ_V, and is the important error for application circuits that depend on providing output voltages that reproduce the instantaneous values of the input voltage or that require preservation of the phase shift of the input voltage, such as sample-and-hold circuits, instrumentation amplifiers, phase sensitive applications, multipliers, and multiplying digital-to-analog converters.

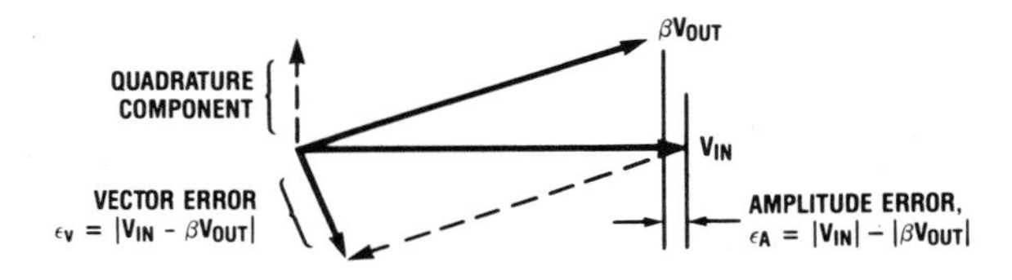

Fig. 3-2. The Relationship Between Amplitude Error and Vector Error

The strange thing about these output voltage waveforms is that a relatively large vector error can exist and yet the magnitude or amplitude error, ϵ_A, can be relatively small. This provides an unexpected benefit for those op amp applications that measure errors as simply how well the output *signal amplitude* meets the *ideal amplitude* expectations. Therefore, the performance of active filters and audio amplifiers is more precise because these applications depend more on ϵ_A than they do on ϵ_V.

We can appreciate this difference between ϵ_A and ϵ_V by relating each of these errors to the input frequency, f, and the upper corner frequency of the closed-loop response, f_c, (we will consider the derivation of these approximate equations for these errors shortly), as

$$\epsilon_V \cong \frac{f}{f_c}$$

and

$$\epsilon_A \cong 1/2 \left(\frac{f}{f_c}\right)^2$$

As an example, Figure 3-3, with f_c = 100 kHz and an input frequency of 10 kHz, these respective errors would be

$$\epsilon_V \cong \frac{f}{f_c} = \frac{10 \text{ kHz}}{100 \text{ kHz}} = 0.1$$

$$\epsilon_V \cong 10\%$$

and

$$\epsilon_A \cong 1/2 \left(\frac{f}{f_c}\right)^2 = 1/2\,(0.1)^2 = 0.005$$

$$\epsilon_A \cong 0.5\%$$

The vector error is alarming, but the amplitude error is not too bad, considering that *the input frequency is only a decade away from the upper corner frequency* (100 kHz).

To calculate the magnitude of these dynamic errors we first define error as

$$\text{error} \equiv \frac{\text{actual } v_{OUT} - \text{ideal } v_{OUT}}{\text{ideal } v_{OUT}}$$

The equation for the dynamic error therefore becomes

$$\text{error} = \frac{\left(\dfrac{A}{1 + A\beta}\right) v_{IN} - (1/\beta)\, v_{IN}}{(1/\beta)\, v_{IN}}$$

or

$$\text{error} = \left[\frac{A\beta}{1 + A\beta} - 1\right] \tag{3-1}$$

and this error is expressed as two error components: the vector error

$$\epsilon_V = \left|\frac{A\beta}{1 + A\beta} - 1\right| \tag{3-2}$$

and the amplitude error

$$\epsilon_A = \left| \frac{A\beta}{1 + A\beta} \right| - 1 \tag{3-3}$$

If we consider the vector error first, we can rewrite equation (3-2) as

$$\epsilon_V = \left| \frac{1}{1 + \frac{1}{A\beta}} - 1 \right|$$

and for $A\beta \gg 1$

$$\epsilon_V \cong \left| 1 - \frac{1}{A\beta} - 1 \right|$$

where we have used the small magnitude approximation

$$\frac{1}{1 + \Delta} \cong 1 - \Delta$$

with $\Delta = \dfrac{1}{A\beta}$

Now

$$\epsilon_V \cong \frac{1}{|A\beta|}$$

The magnitude of the vector error, ϵ_V, can be easily determined from the familiar plots of open-loop gain and closed-loop gain (as was shown in Figure 3-3) because $|A\beta|$ is the loop gain.

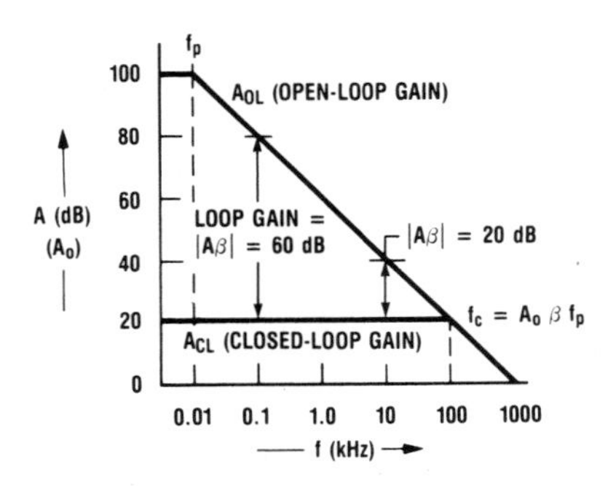

Fig. 3-3. Loss of Loop Gain at High Frequencies

Notice that vector errors start prior to 10 Hz for this example and at a frequency of 100 Hz the magnitude of the loop gain is 60 dB (1000), so the vector error at this frequency is

$$\epsilon_v \text{ @ } 100 \text{ Hz} = \frac{-1}{|A\beta|} = \frac{-1}{10^3} = -0.001 = -0.1\%$$

If we restrict β to be frequency independent (only resistors used) then we can develop an equation for the frequency dependence of this vector error by first calculating $|A|$ starting with a single-pole approximation for the op amp, or

$$A_{(\omega)} = \frac{A_o}{1 + j\left(\frac{\omega}{\omega_p}\right)}$$

where: A_o = open-loop dc gain
ω_p = dominate pole of the open-loop response

or

$$A_{(f)} = \frac{A_o}{1 + j\left(\frac{f}{f_p}\right)} \tag{3-4}$$

and similar to an earlier development, we can find

$$|A_{(f)}| = \frac{A_o}{\sqrt{1 + \left(\frac{f}{f_p}\right)^2}}$$

so, now our vector error, ϵ_V, becomes

$$\epsilon_v = \frac{1}{|A_{(f)}\beta|} = \frac{1}{\left|\frac{A_o\beta}{\sqrt{1 + \left(\frac{f}{f_p}\right)^2}}\right|} = \sqrt{1 + \left(\frac{f}{f_p}\right)^2}\ \epsilon_{dc}$$

where, ϵ_{dc}, is the dc gain error ($1/A_o\beta$).

From this equation, we can determine that at a frequency equal to f_p (10 Hz in this example) the vector error is up by $\sqrt{2}$ or 40% over the dc gain error ($\epsilon_{dc} = 10^{-4}$ or 0.01% in this example). For $f > f_p$,

$$\epsilon_v \cong \left(\frac{f}{f_p}\right)\epsilon_{dc} = \frac{f}{A_o\beta f_p}$$

$$\epsilon_v \cong f/f_c$$

and the vector error can be approximated by this simple equation.

To derive an expression for the amplitude error we can start with equation (3-3) and write this as

$$\epsilon_A = \left| \frac{1}{1 + \dfrac{1}{A\beta}} \right| - 1$$

and again, for $A\beta >> 1$, this becomes

$$\epsilon_A \cong \left| 1 - \frac{1}{A\beta} \right| - 1$$

using the expression for A from equation (3-4)

$$\epsilon_A \cong \left| 1 - \frac{\dfrac{1}{A_o\beta}}{1 + j\left(\dfrac{f}{f_p}\right)} \right| - 1$$

$$\epsilon_A \cong \left| 1 - \frac{1 + j\left(\dfrac{f}{f_p}\right)}{A_o\beta} \right| - 1$$

$$\epsilon_A \cong \left| \left(1 - \frac{1}{A_o\beta}\right) - j\left(\frac{f}{A_o\beta f_p}\right) \right| - 1$$

For $A_o\beta >> 1$, and again writing $A_o\beta f_p$ as f_c

$$\epsilon_A \cong \left| 1 - j\left(\frac{f}{f_c}\right) \right| - 1$$

$$\epsilon_A \cong \sqrt{1 + \left(\frac{f}{f_c}\right)^2} - 1$$

For $\left(\dfrac{f}{f_c}\right)^2 << 1$, we can use the small magnitude approximation

$$\sqrt{1 + \Delta} \cong 1 + \frac{\Delta}{2}$$

for $\Delta << 1$, where $\Delta = \left(\dfrac{f}{f_c}\right)^2$, so this becomes

$$\epsilon_A \cong 1 + \frac{1}{2}\left(\frac{f}{f_c}\right)^2 - 1$$

$$\epsilon_A \cong \frac{1}{2}\left(\frac{f}{f_c}\right)^2$$

which is the useful approximation for the amplitude error that we made use of earlier. Note that because f/f_c is less than 1, squaring this small value provides an even smaller number, which we then further reduce by one-half to provide an estimate of the amplitude error.

Rate Errors

When an op amp is used to amplify an input ramp voltage, a steady-state rate error exists. The resulting amplified output voltage ramp will therefore always be slightly less in voltage than an ideal output voltage ramp, as shown in Figure 3-4. (This creates an instantaneous error, not an error in the output rate-of-change.) At the start, the output voltage undergoes an initial exponential change (with $\tau_c = \frac{1}{2\pi f_c}$) and then converges to a steady-state ramp. This steady-state ramp lags the ideal output ramp by an error; ϵ_r, that is given by

$$\epsilon_r = \left(\frac{\Delta V_{OUT}}{\Delta t}\right)\left(\frac{1}{2\pi f_c}\right) \text{ volts}$$

where f_c is the bandwidth of the closed-loop circuit and, for a noninverting amplifier, is given by

$$f_c = \frac{f_u}{(1/\beta)} = \beta f_u$$

The reason for this rate, or ramp, error ϵ_r, is that the open-loop transfer characteristic of an op amp is basically the same as that produced by an integrator circuit. The integration capacitor for the op amp is provided by the internal frequency compensation capacitor.

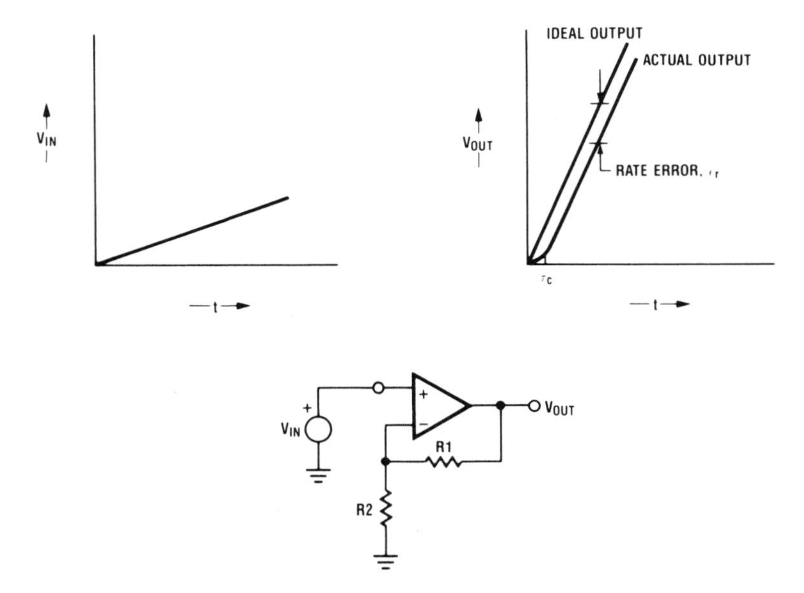

Fig. 3-4. The Rate Error of a Noninverting Gain Stage

It is useful to keep in mind that a dc differential input voltage *must always exist* to allow the input differential stage of an op amp to provide an output current. In Chapter 1 it was seen that to achieve the large-signal slew-rate spec of an op amp requires a relatively large differential input voltage, or *an input error voltage* that is large enough to completely switch the input differential amplifier. It is not surprising that to provide any output ramping voltage will require a continuous dc input error voltage. This error voltage must be large enough to provide for the required current flow out of the input stage that is needed to change the voltage across the comp cap at the desired rate. This is the reason for the rate error of an op amp. It is also now more obvious why the generation of a slow output ramp voltage causes a smaller rate error than if a very high speed output ramp voltage was required.

As an example of the magnitude of the rate error, we can consider a 1 MHz op amp, operating with a gain of 100 ($\beta = 0.01$), and producing an amplified output ramp voltage of 10V/ms. For these conditions the rate error, ϵ_r, would be

$$\epsilon_r = -\left(\frac{\Delta V_{OUT}}{\Delta t}\right)\left(\frac{1}{2\pi\beta f_u}\right)$$

$$\epsilon_r = -\left(\frac{10 \text{ volts}}{1 \times 10^{-3} \text{ sec}}\right)\left[\frac{1 \text{ sec}}{2\pi\,(0.01)\,(1 \times 10^{6})}\right]$$

$$\epsilon_r = -160 \text{ mV}$$

which is 1.6% of a 10V sweep range.

This type of error affects sample-and-hold circuits and other applications that involve time-changing output voltage signals, such as the active-trimming control electronics for adjusting the values of thin-film resistor networks. This error should be kept in mind whenever a ramping output voltage exists.

This rate error also shows up in high speed integrators, Figure 3-5, where opening the shorting switch across the integration capacitor at $t = t_0$ initiates the output ramp. This switch location avoids the output voltage glitch that would result from the abrupt application of an input voltage. The output voltage response is again an exponential approach to the steady-state ramp, but now the time constant, shown as τ_u, depends directly on the unity-gain cross frequency of the op amp, f_u. The rate error, ϵ_r, in this case is given by

$$\epsilon_r = \left(\frac{V}{RC}\right)\left(\tau_u\right) = \left(\frac{V}{RC}\right)\left(\frac{1}{2\pi f_u}\right) \text{ volts}$$

As an example; if $f_u = 1$ MHz, $R = 10$ kΩ, $C = 0.1\ \mu$F, and the input voltage was 10 volts, this error would be

$$\epsilon_r = \left[\frac{10 \text{ volts}}{(10^4)\,(10^{-7}) \text{ sec}}\right]\left[\frac{1 \text{ sec}}{2\pi \times 10^{6}}\right]$$

$$\epsilon_r = 1.59 \text{ mV}$$

This rate error would be even less for a slower output ramp.

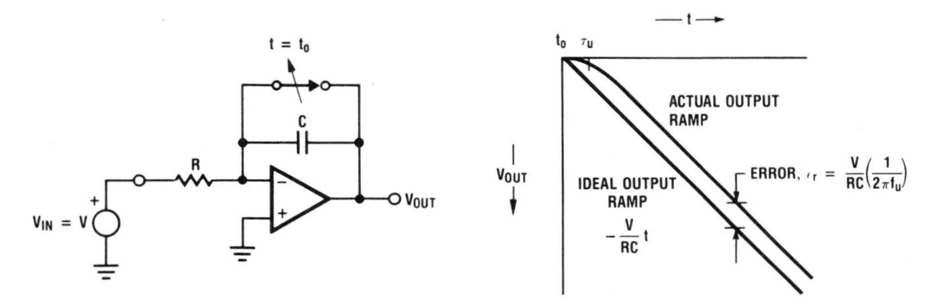

Fig. 3-5. The Rate Error of an Integrator

For a faster output ramp, created with R = 3 kΩ and C = 0.01 μF, the rate error can be found by scaling the previous result, or

$$\epsilon_r = 1.59 \text{ mV} \left[\frac{(10^4)(10^{-7})}{(3 \times 10^3)(10^{-8})}\right] = 53 \text{ mV}$$

An op amp with a 5 MHz unity-gain frequency would reduce this error by a factor of five and the use of this higher frequency op amp would probably also insure that the $\Delta V_{OUT}/\Delta t$ is not further limited by the slew rate of the slower op amp.

The required slew rate, SR, to provide for an output ramp is given by

$$SR > \frac{\Delta V_{OUT}}{\Delta t} = \frac{I}{C} = \frac{V}{RC}$$

In the above example, this becomes

$$SR > \frac{10 \text{ volts}}{(3 \times 10^3)(10^{-8}) \text{ sec}} = 0.33 \text{ V}/\mu\text{sec}$$

3.3 RESPONSE TO THE COMMON-MODE INPUT SIGNAL

We have previously stated that an ideal op amp will not respond to a common-mode input voltage. Actual op amps unfortunately do respond, and often have a gain of unity to these common-mode inputs. This sounds alarming, but is not too bad when it is remembered that the voltage gain provided for a differential input signal can be 100,000 or more. A further complication that is introduced by actual op amps is that the values for both of these gains usually vary nonlinearly with voltage levels (the input common-mode voltage and the output voltage).

A measure of the quality of an op amp is the ratio of the voltage gain for a differential-mode input signal to the voltage gain for a common-mode input signal. This is one definition of the Common-Mode Rejection Ratio, CMRR, and is stated as

$$CMRR = \frac{A_{DM}}{A_{CM}} = \frac{\left(\dfrac{\Delta V_{OUT}}{\Delta V_{IN_D}}\right)}{\left(\dfrac{\Delta V_{OUT}}{\Delta V_{IN_{CM}}}\right)}$$

or to provide the same ΔV_{OUT}

$$\text{CMRR} = \frac{\Delta V_{IN_{CM}}}{\Delta V_{IN_D}}$$

With this form we can include the effects of CMRR in our standard feedback circuit, Figure 3-6. Notice that we have converted the CMRR effect to a differential equivalent signal by dividing the common-mode input change by the CMRR or

$$\frac{\Delta V_{IN_{CM}}}{\text{CMRR}} = \Delta V_{IN_{CM}} \left(\frac{\Delta V_{IN_D}}{\Delta V_{IN_{CM}}} \right)$$

or

$$\frac{\Delta V_{IN_{CM}}}{\text{CMRR}} = \Delta V_{IN_D}$$

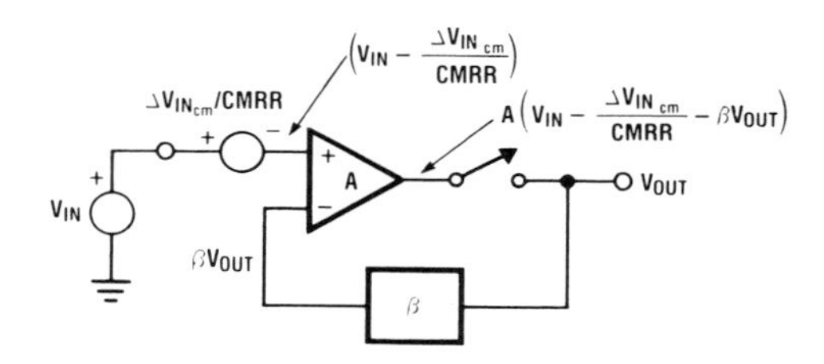

Fig. 3-6. Including CMRR in the Feedback Analysis

We can now chase signals around this block diagram to find an expression that includes both the CMRR error and the previously determined gain error.

For a noninverting amplifier:

$$\Delta V_{IN_{CM}} = V_{IN}$$

therefore

$$A \left(V_{IN} - \frac{V_{IN}}{\text{CMRR}} - \beta V_{OUT} \right) = V_{OUT}$$

$$A_{CL} = \frac{V_{OUT}}{V_{IN}} = \left[\frac{A}{1 + A\beta} \right] \left(1 - \frac{1}{\text{CMRR}} \right)$$

The first term in brackets, is the standard feedback equation and in Chapter 2, Equation (2-3), we found that this can be approximated as

$$\left[\frac{A}{1 + A\beta} \right] \cong \frac{1}{\beta} \left(1 - \frac{1}{A\beta} \right)$$

The closed-loop gain therefore becomes (after multiplying the terms)

$$A_{CL} \cong \frac{1}{\beta}\left[1 - \frac{1}{A\beta} - \frac{1}{CMRR} + \frac{1}{(A\beta)(CMRR)}\right]$$

and as both (1/Aβ) and (1/CMRR) are both small compared to 1, the last product term in this equation becomes even smaller and can therefore be neglected to provide a convenient approximate form for this equation where an additional error term appears in our closed-loop response, or

$$A_{CL} \cong \frac{1}{\beta}\left[1 - \left(\frac{1}{A\beta} + \frac{1}{CMRR}\right)\right]$$

and this dc CMRR term is seen to be as important as the dc gain error term.

Noninverting gain applications typically have a common-mode input voltage that is equal to the input voltage. The *inverting amplifier applications have* a very small common-mode input voltage and *CMRR effects can therefore be neglected* (except for the additional dc offset voltage component that is introduced if the dc bias voltage at the noninverting input is not 0V).

If we look at the specs of a high-performance hybrid op amp, we can find A = 120 dB and CMRR = 120 dB. Notice that this is stating that the common-mode voltage gain = 1 because CMRR is the ratio of differential voltage gain, A, to common-mode voltage gain, A_{CM}.

We can now include the effects of the CMRR error term in a typical example:

Given: $A_{CL} = 100$ ($\beta = 10^{-2}$)
A = 50,000
and CMRR = 95 dB

To Find: The size of the dc CMRR error term and the dc gain error term.
Solution: These dc error terms are

$$\left(\frac{1}{CMRR} + \frac{1}{A\beta}\right) = \left[\frac{1}{10^{95/20}} + \frac{1}{(5 \times 10^4)(10^{-2})}\right]$$

$$1.8 \times 10^{-5} + 2 \times 10^{-3} \cong 0.2\%$$

Here we see that the CMRR error term is two orders of magnitude smaller, and is therefore not contributing to the overall 0.2% error. This also shows that to improve performance you shouldn't buy an op amp with a better CMRR spec; instead, increase A or increase β.

We can also find out how much dc closed-loop gain we can take for a given error. For example:

Given: A = 120 dB (10^6)
CMRR = 120 dB

To Find : $A_{CL\,MAX}$ ($A_{CL} \cong 1/\beta$) for (CMRR error + dc gain error) $\leq 0.1\%$

Solution:

$$10^{-3} \geq \left(\frac{1}{\text{CMRR}}\right) + \left(\frac{1}{A\beta}\right)$$

$$10^{-3} - \left(\frac{1}{\text{CMRR}}\right) \geq \frac{1}{10^6\beta}$$

or

$$A_{CL} \cong \left(\frac{1}{\beta}\right) \leq 10^6\,[10^{-3} - (1 \times 10^{-6})] \leq 999$$

For smaller errors, this much closed-loop voltage gain couldn't be taken. The low-cost op amps, with the following specifications: CMRR = 65 dB_{MIN} and A = $25{,}000_{MIN}$, would only allow a maximum closed-loop gain of 11 for this same total error of 0.1%.

3.4 DIFFERENTIAL AND COMMON-MODE INPUT IMPEDANCE

Standard feedback control theory has indicated some of the dc errors (gain error and CMRR error) of op amps. We will now consider the errors that are produced by the input impedance of a real op amp.

There are *two components of the input impedance* of an op amp: that seen by a differential input signal *(the differential input impedance,* C_D and R_D) and that seen by a common-mode input signal *(the common-mode input impedance,* C_{CM} and R_{CM}). These are modeled at the input of an op amp as shown in Figure 3-7. In the noninverting amplifier applications, the differential input impedance of the op amp is increased by the use of feedback, but the common-mode input impedance remains unaffected.

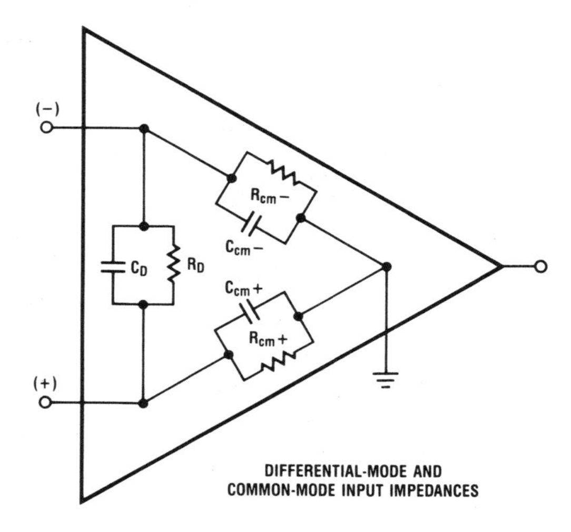

Fig. 3-7. Modeling the Input Impedance of an Op Amp

For an FET input op amp, the differential and the common-mode input resistance, both approximately $10^{12}\Omega$, can usually be neglected. Only the effect of the common-mode input capacitance (3 to 5 pF, where approximately 1 pF is from the package) needs to be considered. (The bipolar input op amps usually have a common-mode input capacitance of approximately 2 to 3 pF and both op amp types have a differential input capacitance of approximately 2 pF.)

The differential input resistance of a bipolar op amp (listed as *input resistance* on the data sheet) can range from approximately 0.3 MΩ to 100 MΩ. The common-mode input resistance is usually several orders of magnitude larger.

In the previous chapter we saw how feedback increased the input impedance that otherwise would have resulted from the differential input impedance of the op amp. We recall that the high open-loop gain of the op amp keeps the differential input voltage of the op amp at a very small value. Thus, the full input voltage is not impressed across the differential input impedance of the op amp. Only the very small input differential voltage necessary to provide the required output voltage exists, which greatly reduces the input current that must be supplied by the signal source, thereby *increasing the effective input impedance.*

Let's now consider the effects of the common-mode input impedance of the op amp.

DC Errors Resulting from Common-Mode Input Resistance

A simple resistive attenuator is formed by the signal source resistance, R_s, when it is loaded by $R_{CM}{}^{+}$, as shown in Figure 3-8. This has a voltage transfer, or dc gain, that is given by

$$\frac{V_{OUT}}{V_{IN}} = \frac{R_{CM}{}^{+}}{R_s + R_{CM}{}^{+}}$$

so, in this case; the dc error, ϵ_0, is given by

$$\epsilon_0 = \frac{\left(\dfrac{R_{CM}{}^{+}}{R_s + R_{CM}{}^{+}}\right) - 1}{1}$$

$$\epsilon_0 = \frac{R_{CM}{}^{+} - R_s - R_{CM}{}^{+}}{R_s + R_{CM}{}^{+}}$$

or

$$\epsilon_0 = -\frac{R_s}{R_s + R_{CM}{}^{+}}$$

and, as $R_{CM}{}^{+} >> R_s$,

$$\epsilon_0 \cong -\frac{R_s}{R_{CM}{}^{+}}$$

As a numerical example of this error magnitude, we can consider a bipolar op amp with $R_{CM}{}^{+} = 300$ MΩ that is operating with an $R_s = 1$ MΩ. The dc error for this application is

$$\epsilon_0 \cong -\frac{R_s}{R_{CM}{}^{+}} = -\frac{10^6}{3 \times 10^8} = -3 \times 10^{-3}$$

or

$$\epsilon_o \cong -0.3\%$$

For comparison, an FET input op amp with $R_{CM}^+ = 10^{12}\Omega$ would reduce this error to

$$\epsilon_o \cong -\frac{10^6}{10^{12}} = -10^{-6} = -0.0001\%$$

In both of the above examples, the dc errors that result from the V_{OS} and the input current of the op amp will usually dominate and therefore dc errors owing to R_{CM} can be ignored.

The Effects of the Common-Mode Input Capacitance

The *input capacitance* spec for an op amp is generally the *input common-mode capacitance.* This was previously modeled as a capacitor that is shunting R_{CM}^+, the common-mode input resistance. When operating with an input source resistance, R_s, the equivalent circuit that was shown in Figure 3-8 results. The signal source resistance causes a loss of signal owing to the loading by R_{CM}^+ and C_{CM}^+.

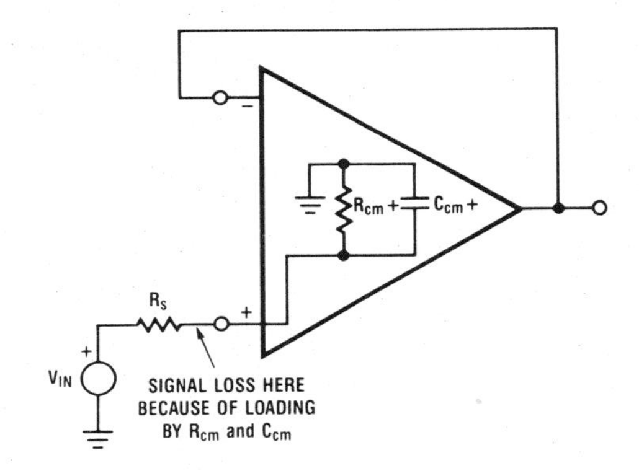

Fig. 3-8. Input Capacitance Causes Signal Loss

This circuit, redrawn in Figure 3-9a, has a dc voltage loss because of R_{CM}^+ (as previously discussed) and a dynamic voltage loss because of the resulting low-pass filter, Figure 3-9b. If we identify the corner frequency as $f_{c\ CM}$, we have

$$f_{c\ CM} = \frac{1}{2\pi\, C_{CM}^+\, (R_{CM}^+ \| R_s)}$$

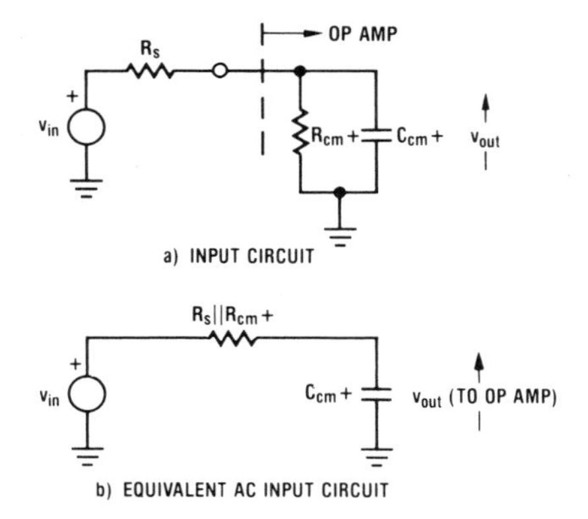

Fig. 3-9. Common-Mode Input Impedance Causes AC Signal-Loss

A dynamic amplitude error can be caused by this filter. For example, if we choose an FET input op amp with $R_{CM}{}^{+} = 10^{12}\Omega$ and $C_{CM}{}^{+} = 3$ pF, and use it as a voltage follower that is driven from a signal source resistance of 10 MΩ, we find

$$f_{c\ CM} = \frac{1}{2\pi\,(3 \times 10^{-12})\left[\frac{(10^{12})\,(10^{7})}{10^{12} + 10^{7}}\right]}$$

$$f_{c\ CM} = 5.3 \text{ kHz}$$

For a dynamic amplitude error of 0.1% ($\epsilon = 0.001$) we are limited to input frequencies that are less than

$$f_{max} \cong \sqrt{2\epsilon}\;\; f_c$$

$$f_{max} \cong \sqrt{0.002}\;\; 5.3 \text{ kHz}$$

$$f_{max} \cong 240 \text{ Hz}$$

Where we have used the approximate relation that was derived earlier in this chapter. If we picked up an additional 10 pF at this input node because of stray wiring capacitance, we would find (by scaling the previous result)

$$f_{c\ CM} = 5.3 \text{ kHz} \left(\frac{3 \text{ pF}}{13 \text{ pF}}\right) = 1.2 \text{ kHz}$$

and

$$f_{max} \cong 240 \text{ kHz} \left(\frac{12 \text{ kHz}}{53 \text{ kHz}}\right) = 55 \text{ Hz}$$

This indicates why an op amp should be kept physically close to the signal source.

Low impedance (53Ω), coaxial cable (RG 58A/U) is often used for shielded wire. The capacitance of this cable is 29 pF per foot and can cause severe dynamic or ac amplitude errors because of signal loss at surprisingly low frequencies. Even the lowest capacitance shielded cable must be kept very short to avoid adding excessive capacitance at this sensitive input node.

3.5 THE DC NOISE SOURCES: OFFSET VOLTAGE AND INPUT CURRENT

In our previous discussions of the ideal op amp (Chapter 1), we stated that the output voltage would go to exactly zero volts when the input differential voltage was zero volts. Real op amps usually don't provide such idealized performance because of both internal circuit imbalances that unavoidably result during the manufacture of the IC chip (this causes an input offset voltage, V_{OS}) and the effects of the dc input currents (I_B^+ and I_B^-, the base current of bipolar input transistors or the gate leakage current of Bi-FETs). Amplifier application circuits therefore don't necessarily produce an output of zero volts for an input of zero volts. With a real op amp, to get the output to go to zero, some small voltage (called the offset voltage) must be supplied at the input. This is a result of the dc noise sources of the op amp. (Noise, in general, is any undesired signal.)

In dc-coupled applications, these dc noise sources of the op amp cannot be distinguished from the dc output voltage that would result from a valid dc input signal. If gain is provided by the application circuit, then these dc noise sources will also be amplified in their effects on the output signal.

DC noise sources of the op amp contaminate all dc-coupled application circuits. In ac-coupled applications, we can neglect this problem because these dc errors simply shift the bias (or quiescent operating point) of the output voltage and will usually only slightly restrict the magnitude of the peak output voltage swing that can be provided. In general, every dc application of an op amp should be investigated to determine the effects of these dc errors on overall circuit performance.

Two possibilities exist to reduce V_{OS}, a major contributor to dc error: specify an op amp that has a guaranteed small value of V_{OS}, or adjust the V_{OS} of the op amp to zero as part of the PC board production procedure. Offset voltage is one of the key specs of an op amp and special products are available that guarantee low values of V_{OS}. If the dc input currents (I_B^+ and I_B^-) of the op amp are causing problems, a Bi-FET op amp (that has much lower input currents) should be considered. If you were already using a Bi-FET op amp, then an LM11 should be considered.

Modeling the DC Noise Sources

A circuit model for these dc noise sources is shown in Figure 3-10. This is identical (in form) to the model which we will use for the op amp ac noise sources in the next section. *These dc noise sources* will be modeled and handled in the same way as the ac noise sources. This similarity aids the analysis of both types of noise in op amps.

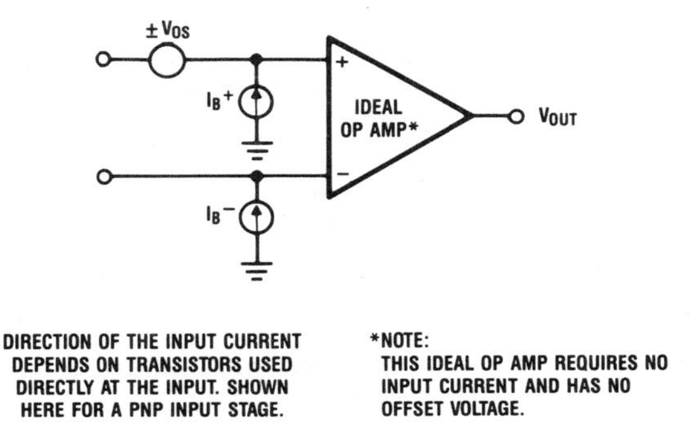

Fig. 3-10. Modeling the DC Error-Sources of an Op Amp

A major difference between handling the ac and the dc noise sources is that for the dc noise sources *we will keep track of the algebraic signs of the current sources.* The ones shown in the figure are for a PNP input stage, where the dc bias current flows out of the op amp. These current directions should be reversed for an NPN-input or Bi-FET op amp where the dc bias current (or leakage current for the Bi-FET op amps), instead, enters the op amp. The uncertainty of the algebraic sign of V_{OS} requires that both signs be carried in the analysis.

Once these dc noise sources are externally modeled at the inputs, the remaining op amp is ideal. It has no V_{OS} and also requires no dc input current. If desired, to simplify a calculation, or to aid your thinking, the V_{OS} generator can therefore be moved over to be in series with the $V_{IN}(-)$ lead.

The dc effects at the output of an op amp application circuit can be found by making use of superposition; each of these three dc error sources can be considered separately and the effects that each produce at V_{OUT} can then be algebraically added.

Matching the DC Resistance at Each Input

There is an old rule with op amps that states; *match the input resistors.* We can now use our dc noise model to determine why this helps and to discover a few more things about reducing the dc errors in an op amp application circuit.

As an example, consider the voltage follower shown in Figure 3-11. To analyze this circuit, we will start by first considering only the effects of the input current noise sources and therefore, both the V_{OS} generator and V_{IN} will be temporarily replaced with short circuits, as shown in Figure 3-12.

Notice that we have included a resistor, R, in series with the $V_{IN}(-)$ lead. The benefits of this added resistor can now be seen by inspection; $I_B{}^+$ will create a dc drop across R_S and, without R, $I_B{}^-$ would simply flow into the output of the op amp with no dc effects on V_{OUT} or V_{IN}. What we hope to accomplish by adding the extra resistor R is to cause the dc voltage drop across R_S, that is caused by $I_B{}^+$, *to be matched by a similar dc voltage drop across the added resistor, R.* When we do this, rather than creating a dc output voltage error because of the flow of $I_B{}^+$ back through R_S (this current direction will increase V_{OUT}) we hope to also use the fact that $I_B{}^-$ flows through R to create a matching dc voltage drop. Both input voltages of the op

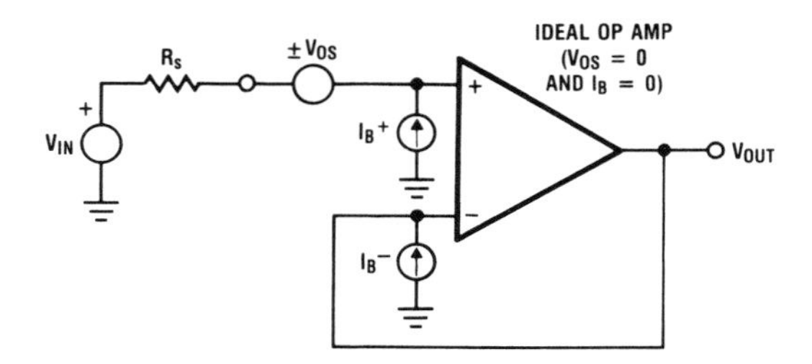

Fig. 3-11. Model for DC Errors of a Voltage Follower

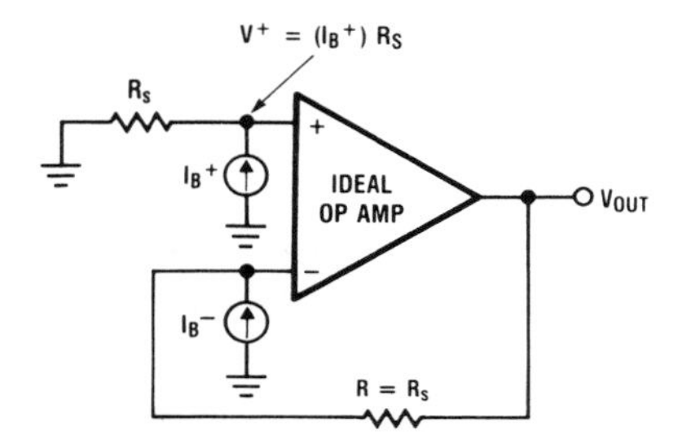

Fig. 3-12. Considering the Effects of Input Current

amp will therefore equally rise (a common-mode shift that the op amp can reject) and there is no longer any dc error in V_{OUT}. [In calculations with this ideal op amp, we can easily add up the dc voltages around the circuit, even right across the input terminals, because $V_{IN}(+) = V_{IN}(-)$.] As expected, a good choice is $R = R_S$. This has the additional benefit of eliminating the effects of the gain error that would otherwise result owing to R_{CM}.

If we write an equation for V_{OUT} we find

$$V_{OUT} = (I_B^+)(R_S) - (I_B^-)(R)$$

or for $R = R_S$

$$V_{OUT} = R_S(I_B^+ - I_B^-)$$

and

$$(I_B^+ - I_B^-) = \pm I_{OS}$$

so

$$V_{OUT} = \pm I_{OS} R_S$$

The term $\pm I_{OS}$ is the input offset current for the op amp. It can be of either sign and represents the difference between the magnitudes of the input currents. Most op amps provide a factor of 5 to 20 reduction in I_{OS} as compared to I_B. This matching of the input resistors therefore, provides a big benefit in circuit performance for the small cost of adding only a single resistor.

To eliminate the ac thermal noise that is associated with this input matching resistor (we will consider the ac noise problem later in this chapter) and to prevent oscillations, which we will consider in the next chapter, a capacitor should be placed across this added resistor. A large valued (0.01 μF), low cost disc-ceramic capacitor is a good choice. It also provides good noise reduction. Smaller valued capacitors (5 to 15 pF) can also be used, but the intermediate range (100 to 1000 pF) can disturb the settling characteristics (long "tails" may result) due to interaction with $C_{IN_{CM}}$.

We can now consider the effects of V_{OS} acting alone by making use of the simplified circuit of Figure 3-13. (For more complex circuits, it is often a benefit to consider three separate schematics: one for $I_B{}^+$, one for $I_B{}^-$, and one for V_{OS}.)

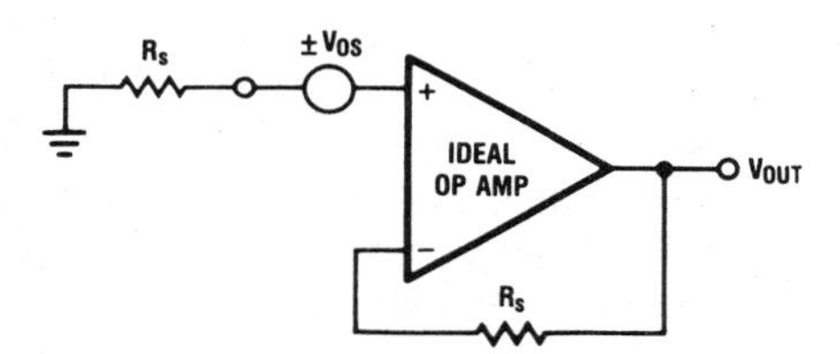

Fig. 3-13. Considering the Effects of Offset Voltage

Remember that the ideal op amp we are working with has no input current and also no V_{OS}. The effects of the external V_{OS} noise source therefore simply provides

$$V_{OUT} = \pm V_{OS}$$

We now add this to the V_{OUT} that we previously found to result from the input offset current to find

$$V_{OUT_{TOTAL}} = \pm (I_{OS}R_S) \pm V_{OS}$$

This result is for a unity-gain application, dc gain in an application circuit will correspondingly increase these dc errors in the output voltage. If this first term is too large, a Bi-FET op amp can be used, but the eventual limit on dc performance will depend on the V_{OS} of the op amp; therefore, trim, specify, or let it be, but be aware.

All three of the terms in the above equation are temperature dependent. To determine the changes to expect in $V_{OUT_{TOTAL}}$ we can take the partial derivative of the above equation with respect to temperature to find

$$\frac{\partial V_{OUT}}{\partial T}_{TOTAL} = \pm I_{OS} \frac{\partial R_S}{\partial T} \pm R_S \frac{\partial I_{OS}}{\partial T} \pm \frac{\partial V_{OS}}{\partial T}$$

If the two resistors R_S and R (where R = R_S) have different temperature coefficients, they should be handled separately using the I_B values instead of I_{OS}.

Problems usually exist because the temperature drifts of I_{OS} and V_{OS} are not simple straight lines that fit nicely into the above equation. For the high volume op amps, these drift figures are usually not guaranteed because the extra temperature testing adds excessive costs. (A V_{OS} max limit that applies over the complete temperature range that is specified for the part is guaranteed for some op amps.)

For more precise dc applications, low cost op amps that specify a maximum V_{OS} drift (10 μV/°C) are available (LF411 and LF412). Special low offset voltage op amps can be used, and the LM11 or an instrumentation op amp (or an instrumentation amplifier) should be used for the most critical applications.

DC Noise Gain

We will consider a second example to introduce the concept of *dc noise gain*. Actually, if the previously shown noise source model is used, no consideration has to be given to the concept of dc noise gain because the correct gain, the dc noise gain, (and therefore the correct answer) will result.

The best example to demonstrate this gain problem is the unity-gain inverting amplifier, Figure 3-14, where R1 = R2. If the signal source has a resistance, shown as R_S, it should be included in selecting the value for R1 as shown on the figure. For now, we will assume that either R1 is compensating for R_S or R_S = 0. Again, in looking at this schematic, we wonder if a matching resistor should be used; this time in the $V_{IN}(+)$ lead.

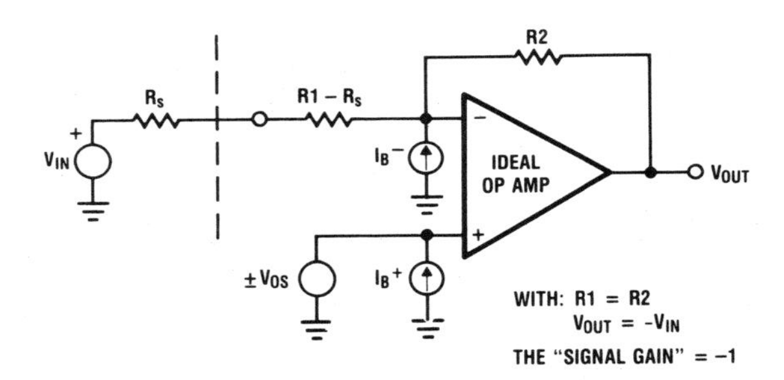

Fig. 3-14. The Unity-Gain Inverting Amplifier

We can redraw this circuit to separately consider the effects of the I_Bs as shown in Figure 3-15, where we have added a general resistor, R, in the $V_{IN}(+)$ lead, because; intuitively, if both inputs *see* the same resistance, their I_Bs should create the same voltages at each input and this common-mode voltage can then be rejected by the op amp.

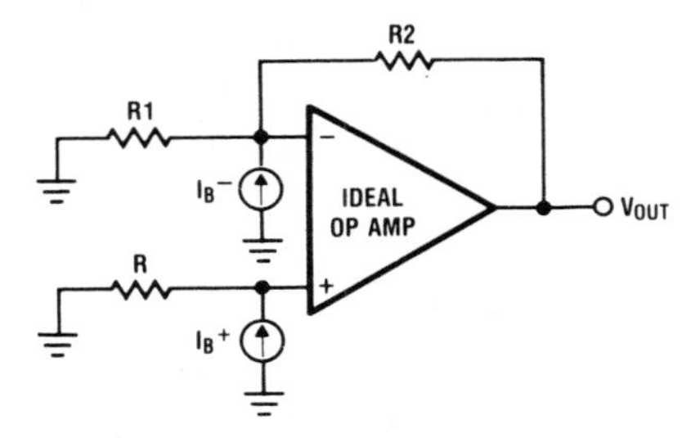

Fig. 3-15. The Input-Current DC Errors of the Unity-Gain Inverting Amplifier

We can now further simplify the schematic by only considering the effects of I_B^- by redrawing the circuit as shown in Figure 3-16. This current source has been converted to the equivalent voltage source, V_{IN}, shown. The output voltage now becomes,

$$V_{OUT} = -V_{IN} \left(\frac{R2}{R1}\right)$$

$$V_{OUT} = -[(I_B^-)(R1)] \left(\frac{R2}{R1}\right)$$

or

$$V_{OUT} = -(I_B^-) R2$$

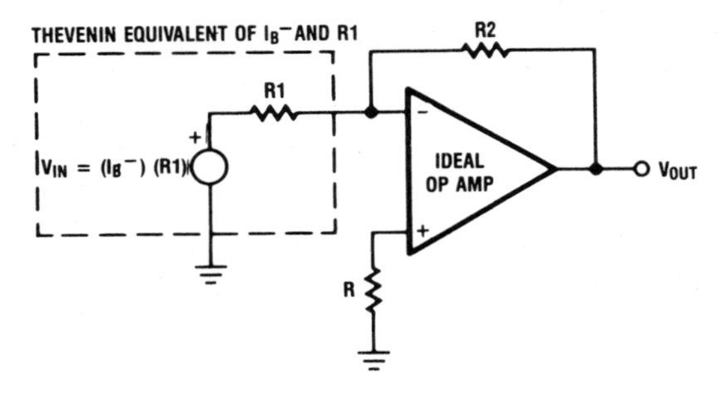

Fig. 3-16. The Effects of I_B^- Acting Alone

This result could have been written by inspection from the previous figure by noticing that there is no voltage across R1 (I_B^+ is considered equal to zero when we evaluate I_B^- by superposition), there is no V_{OS} of the op amp, so $V_{IN}(-)$ equals zero, and therefore no current can flow in R1. This then means that all of I_B^- must flow up into R2 to produce V_{OUT}.

To consider the effects of I_B^+, we redraw the circuit as shown in Figure 3-17; where we have, again, substituted an equivalent voltage source, V_{IN}, for the current source. The expression for V_{OUT} now becomes

$$V_{OUT} = V_{IN} \left(\frac{1}{\beta}\right)$$

or

$$V_{OUT} = [(I_B^-)(R)] \left[1 + \frac{R2}{R1}\right]$$

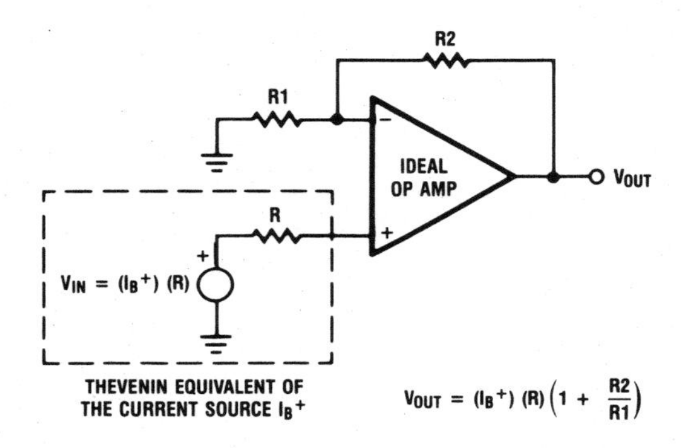

Fig. 3-17. The Output DC-Error That Results from I_B^+

A final circuit is used to determine the effects that result from V_{OS} acting alone, Figure 3-18. Notice that V_{OUT} becomes

$$V_{OUT} = \pm V_{OS} \left(1 + \frac{R2}{R1}\right)$$

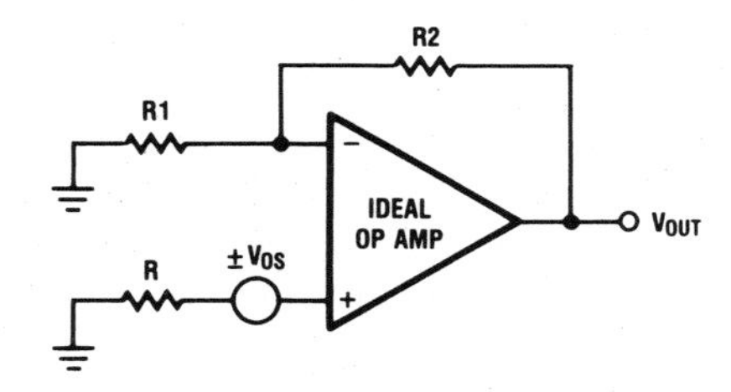

Fig. 3-18. The Effects of V_{OS} Acting Alone

So, again, we can collect the results of these three separate calculations to provide an overall $V_{OUT_{TOTAL}}$ that becomes

$$V_{OUT\ TOTAL} = \pm V_{OS}\left(1 + \frac{R2}{R1}\right) + (I_B^+)(R)\left(1 + \frac{R2}{R1}\right) - (I_B^-)(R2)$$

To achieve the benefits that result because $I_B^+ \cong I_B^-$, we can force the last two terms to equal zero and use this result to select a value for R, or

$$(I_B^+)(R)\left(1 + \frac{R2}{R1}\right) = (I_B^-)(R2)$$

from which

$$R = \frac{R2}{\left(1 + \frac{R2}{R1}\right)} = \frac{R1\ R2}{R1 + R2}$$

or

$$R = R1||R2$$

Using this value for R in the previous equation we find,

$$V_{OUT\ TOTAL} = \pm V_{OS}\left(1 + \frac{R2}{R1}\right) + (I_B^+)\left(\frac{R1\ R2}{R1 + R2}\right)\left(\frac{R1 + R2}{R1}\right) - (I_B^-)(R2)$$

$$= \pm V_{OS}\left(1 + \frac{R2}{R1}\right) + R2\,(I_B^+ - I_B^-)$$

$$= \pm V_{OS}\left(1 + \frac{R2}{R1}\right) \pm I_{OS}\,(R2)$$

This has been a rather lengthy path to such a simple result. But the techniques that have been used are general and can be used to evaluate any circuit, if you can keep from making simple algebraic mistakes. The benefit of considering three, separate, simplified circuits is that errors are less likely to creep into the analysis.

The important thing to notice from the last equation is that the V_{OS} of the unity-gain inverting amplifier is amplified by 2 (not −1) when it appears in V_{OUT}. In this example there is a large difference between the *signal gain* (−1) and the *dc noise gain* (+ 2).

With a Bi-FET op amp, input resistor matching can usually be omitted, the extra resistor saved, and (as we will see in the next section), omitting this resistor will produce less ac noise in V_{OUT}. The previous analysis can be used to indicate the dc output voltage errors that result and whether or not input resistor matching should be used. An input matching resistor could be important for a Bi-FET op amp, especially when working with a large source resistance, and at elevated temperatures, where I_B increases in value.

Most ac applications of op amps have less concern with exact dc performance. The matching of the resistors is generally of no importance. Actually, ac noise considerations may dominate, and matching the input resistances is not the best policy for lowest ac noise performance. There will be more on this in Section 3.6.

Nulling V_{OS} and the Effects on Drift

Changes in the operating temperature of the op amp will change the magnitude of V_{OS}. This is called V_{OS} drift. The magnitude of this drift can range from 10's of $\mu V/°C$ to less than 1 $\mu V/°C$, depending on the particular op amp selected (a value of 10 $\mu V/°C$ is considered a typical spec for the high volume low cost op amps).

For a differential amplifier that consists of only two transistors (a non-Darlington input stage as we considered in Chapter 1 for our basic op amp), we can derive an expression that shows how the V_{OS} drift is affected by the existence of an initial V_{OS}. (Unfortunately, most op amps use a more complex input circuit and V_{OS} can therefore result from many other on-chip circuit imbalances, so this simple relation no longer holds. But the LM725 and the LM321 do make use of a basic input differential amplifier and this simple theory therefore applies for these op amps.)

To determine an equation for V_{OS} drift, we start with the ideal diode equation, that relates the base-emitter voltage, V_{BE}, to the collector current, I_C, of a bipolar transistor as

$$I_C = I_S \exp\left(\frac{q\,V_{BE}}{kT}\right)$$

where:

I_S = The reverse saturation current
q = The charge of an electron
k = Boltzman's constant
and T = The temperature of the transistor (in °C).

For our differential amplifier (using transistors Q1 and Q2) we then have for Q1

$$I_{C1} = I_{S1} \exp\left(\frac{q\,V_{BE1}}{kT}\right)$$

and for Q2

$$I_{C2} = I_{S2} \exp\left(\frac{q\,V_{BE2}}{kT}\right)$$

The ratio of these two collector currents is given by

$$\frac{I_{C1}}{I_{C2}} = \frac{I_{S1} \exp\left(\frac{q\,V_{BE1}}{kT}\right)}{I_{S2} \exp\left(\frac{q\,V_{BE2}}{kT}\right)}$$

and, if we assume matched IC transistors, we will have

$$I_{S1} = I_{S2}$$

Therefore

$$\frac{I_{C1}}{I_{C2}} = \exp\left[\frac{q\,(V_{BE1} - V_{BE2})}{kT}\right]$$

and, as

$$(V_{BE1} - V_{BE2}) = V_{OS}$$

this becomes

$$\frac{I_{C1}}{I_{C2}} = \exp\left(\frac{q\,V_{OS}}{kT}\right)$$

If we take the natural log of both sides,

$$\ln\left(\frac{I_{C1}}{I_{C2}}\right) = \frac{q\,V_{OS}}{kT}$$

or the way in which a V_{OS} causes an imbalance in collector currents (and vice versa) is given by

$$V_{OS} = \frac{kT}{q}\ln\left(\frac{I_{C1}}{I_{C2}}\right)$$

and the temperature drift of V_{OS} becomes

$$\frac{\partial V_{OS}}{\partial T} = \frac{k}{q}\ln\left(\frac{I_{C1}}{I_{C2}}\right)$$

If we substitute for the last term by making use of the previous equation, we find

$$\frac{\partial V_{OS}}{\partial T} = \frac{V_{OS}}{T}$$

and therefore, if we had an initial V_{OS} of 1 mV we would find

$$\frac{\partial V_{OS}}{\partial T} = \frac{1\text{ mV}}{298°\text{C}} = 3.35\ \mu\text{V}/°\text{C}$$

or we could expect 3.4 μV/°C of V_{OS} drift per mV of initial V_{OS}. Also, if we adjusted V_{OS} to zero (by adjusting I_{C1}/I_{C2}), the V_{OS} drift also goes to zero.

Only a few IC op amps use a basic differential input stage and therefore only these op amps follow this theory for the correlation between V_{OS} and V_{OS} drift. In a more complex op amp input stage, many initial circuit imbalances cause an initial V_{OS}. To adjust this V_{OS} to zero, a different circuit imbalance is usually introduced by the V_{OS} adjusting circuit. It would be too complex to try to identify and separately cure each of the possible causes of an initial V_{OS}. Therefore, there is the possibility of many separate circuit imbalances, and this increases the V_{OS} drift. In most IC op amps, when the initial V_{OS} is close to zero, the temperature drift of V_{OS} is also very low.

Data sheets recommend circuits to be used to adjust the V_{OS} to zero. It is important to use these circuits to prevent degrading the V_{OS} drift. The zeroing benefits obtained at room temperature can otherwise quickly be lost if a wide temperature range of operation is required.

Degradation in V_{OS} drift usually results from the large mismatch in the temperature drift of the resistance element of the nulling pot as compared with the temperature drift of the on-chip diffused resistors that are often used in the V_{OS} adjust circuit. This problem exists, for example, with the 741 op amp. The LF356 nulling circuit is a large improvement: V_{OS} adjustment does not significantly degrade V_{OS} drift or CMRR. For example, the effects of V_{OS} nulling on V_{OS} drift are typically 0.4 μV/°C per mV of V_{OS} adjustment for the LF356 op amp.

Additional major V_{OS} drift problems can result if the V_{OS} zeroing adjustment procedure should accidentally also trim out errors that result from input current flow through the external resistors, as shown in Figure 3-19. The additional component of input offset voltage that is created by the dc input current flow, (Figure 3-19b), will cause an improper V_{OS} nulling for the op amp. The introduced circuit imbalance from the offset adjustment can result in a large V_{OS} drift. Keep the V_{OS} that results from input current negligibly small by using a small value for R1. In general, if voltage drop owing to the input current of the op amp cannot be compensated by equalizing the resistance at each input, then two separate adjustments should be provided: one to null V_{OS} and one to compensate for I_B. The natural matching and the temperature tracking of the I_Bs of an op amp generally provide better performance than can be obtained from attempts to match the I_Bs with external circuitry. For example, at high temperature I_B gets smaller for most bipolar op amps but much larger with the Bi-FET op amps. This is not the case if I_B is essentially independent of temperature, as exists with the LM324; or, in general, if the current that is produced by the external compensation circuitry has the same temperature change as I_B.

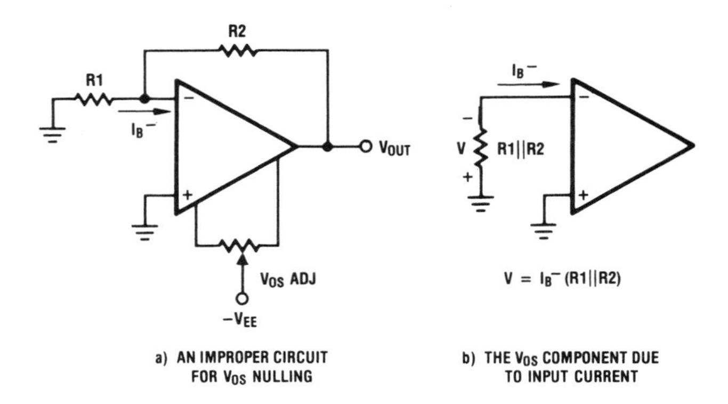

Fig. 3-19. V_{OS} Adjustment Should Not Include I_B Effects

If V_{OS} adjust pins are not available on the IC package, as is the case for most duals and quads, V_{OS} adjust can still be achieved by *carefully* introducing small dc correcting voltages into the resistance networks of the application circuit. For inverting amplifiers, this type of V_{OS} adjustment can be achieved by introducing a small dc voltage in the normally grounded noninverting input lead. In the case of a noninverting application, the introduction of a small dc voltage at the normally grounded resistor at the inverting input will slightly alter the closed-loop gain if the value of the total resistance from the inverting input to ground changes as a result of this V_{OS} adjustment.

The best solution to the V_{OS} problem of the op amp in critical applications is for the V_{OS} adjustment to be made by the IC manufacturer during the wafer probing step. This automatic high speed technique is the most cost-effective solution and is very effective with the modern high performance, and even some of the low cost, IC op amps.

Thermoelectric Voltages as Sources of V_{OS}

In very precise dc applications, there are many other problems to be considered. One of these is the thermoelectric potentials that are generated by the electrical connections that are made between different metals, when these connections are exposed to different localized temperatures. For example, the copper PC board electrical contacts to the kovar* input pins of an IC in a TO-5 can create an offset voltage of 40 $\mu V/°\Delta T$, where ΔT is the temperature difference between the two dissimilar metal contacts. The common lead-tin solder, when used with copper, creates a thermoelectric voltage of 1 to 3 $\mu V/°\Delta T$. (Special cadmium-tin solders are available that reduce this to 0.3 $\mu V/°\Delta T$.)

These small thermoelectric voltages are also associated with the physical construction of discrete electronic components. Fortunately, such small voltages are usually disregarded in common applications, but they complicate the construction of *μV analog circuitry*. Generally, the solution to this problem is to insure that the external components connected to the inputs of the op amp are constructed with the same materials (identical components) and are symmetrically located so that corresponding components will be at the same temperature.

After we have taken care of all of the dc problems of the op amp, a small signal will still exist at the output. This remaining signal results from ac noise and establishes a limit on the smallest magnitude of an ac input signal that can be resolved. We will now consider this ac noise and see how to predict the amount of this noise that we can expect to find at the output of an application circuit.

3.6 THE AC NOISE SOURCES

With the present low cost of op amps, it is certainly economical to cascade four stages, each locally fed back for a gain of 100, to make one *super amplifier*, with an overall gain of 100,000,000. It would seem that this cascaded amplifier should then be easily able to amplify ac input signals that are considerably less than 1 μV and, from these extremely low-level input signals, provide any desired output peak-to-peak voltage level, such as 10V.

The thing that is missing in the above discussion is that *every amplifier, in addition to providing gain, also provides a few signals of its own*. These ever present, or inherent, ac signals of real amplifiers place a limit on the smallest magnitude of a desired input signal (and therefore also the maximum gain) that can be used. We have to be able to determine that our intended signal *appears* at the output of the amplifier; that it is not completely hidden within this minimum background of signals that naturally exists and establishes a "noise floor."

These undesired background signals are called *ac noise*. They result from physical mechanisms within the resistors and transistors that are used to fabricate the amplifier. We will now look at these sources of ac noise and determine how to predict an equivalent input, and output, noise magnitude that results from all of these separate noise sources so we can get an idea of just how small the magnitude of an ac input signal can be.

*Kovar is a trademark of Westinghouse Electric Corporation.

The purpose of this section on noise is to create a general awareness of many concepts that will assist your thinking about noise. Many applications of op amps deal with relatively large signals and noise is therefore not a factor. Occasionally, op amps are used to receive or process low-level signals — noise considerations then become very important.

Equivalent Input AC Noise Sources

AC noise calculations are made easier by using the concept of equivalent input ac noise sources. This is an artificial way to simultaneously represent all of the individual sources of ac noise within the op amp and to lump them all into equivalent ac noise generators that are placed at the input to a, now, *noiseless op amp*. These artificial noise generators will produce the same ac noise as the actual op amp; and, being at the input, will provide increased ac noise in the output voltage as more gain is taken in the application circuit.

This is similar to our previous discussion of the dc noise sources of the op amp, except; now, these are the ac noise sources. They will be handled in a similar manner, but we also have to consider that these ac noise sources may be frequency dependent, as they are, and the bandwidth of our application circuit now becomes important. There is now no way to cancel one ac noise source with another, because both are assumed completely random and uncorrelated, and they also have to be added up in a peculiar way (using rms addition).

In addition to this *inherent ac noise* of the op amp, *interference noise* can also exist. Sources of this type of noise include all of the outside disturbances that can contaminate the output signal of an amplifier: ac power line coupling, power supply noises, signals from unintentional ground loops, radiations from radio stations, and others. Even if these interference noise sources are "cleaned up," *the inherent ac noise sources remain*. These result from the discrete nature of electric current because current flow is made up of moving charges, and each charge carrier transports a definite value of charge (the charge of an electron, 1.6×10^{-19} coulombs).

At the atomic level, current flow is very erratic. The motion of the current carriers resembles popcorn popping. This was chosen as a good analogy for current flow and has nothing to do with the "popcorn noise" that we will discuss at the end of this chapter. Just as you can't say for sure when the next kernel of corn will pop, you can't be sure of the minute details of current flow. For example, when a PN junction is biased for forward conduction, the details of exactly when each electron will cross the junction are unknown and unknowable. For this reason, statistics must be used to describe these random processes and this makes the descriptions of noise voltages and noise currents very strange; things become probabilistic instead of definite.

The uncertainty of carrier movement is compounded by the thermal energy that these carriers possess because they are at room temperature. We are apt to find room temperature comfortable and to our liking. Particles in the physical world find room temperature is not simply +25°C; it is, instead, +298°K! The thermal energy associated with such a high temperature causes very erratic movements of the charge carriers. In a resistor, for example, these random thermal motions cause a *Johnson noise* to result.

J.B. Johnson concluded that since thermal noise has no cyclic character, it is best considered in terms of random noise power or of effective mean-square voltage, $\overline{e^2_n}$, in his classic paper, "Thermal Agitation of Electricity in Conductors", Phys. Rev., July, 1928.

As expected, this thermal noise increases with increases in temperature and the mean-square value; $\overline{e^2_n\ (t)}$ (or simply $\overline{e^2_n}$) can be calculated from

$$\overline{e^2_n} = 4kTR\Delta f \text{ volts}^2/\text{Hz}$$

where

- $k = 1.38 \times 10^{-13}$ Joules/°K (Boltzmann's constant)
- T = Thermal equilibrium temperature, in °K
- R = Resistance value, in ohms
- Δf = Bandwidth over which the measurement is made, in Hz

It may seem strange that a common resistor is a signal generator. *Even a resistor suspended on a silk thread in free space is a source of ac noise voltage.* The energy that provides for this noise is the thermal energy that the resistor picks up from ambient heat sources. Deep-space radio receivers, for example, use refrigeration to keep the electronic components of the receiver as cold as possible to minimize this thermally generated noise, so that smaller signals can be detected.

Johnson noise is frequency independent over a broad range of frequencies and, therefore, is often called *white noise* because of the similarity to *white* light, which is composed of all the frequencies within our visual spectrum.

Johnson noise has a uniform spectral density that is the square root of the previous mean-squared value that can be simplified as

$$e_n\ (R) = \frac{4\ \text{nV}}{\sqrt{\text{Hz}}} \times \sqrt{\frac{R}{1\ \text{k}\Omega}}$$

where we have replaced the 4 kT term by the appropriate constants, used T = 298°K, and normalized to a 1 kΩ resistance. This equation is now easier to use. For example, a 10 kΩ resistor would produce more noise as given by

$$e_n\ (10\ \text{k}\Omega) = \frac{4\ \text{nV}}{\sqrt{\text{Hz}}} \times \sqrt{\frac{10\ \text{k}\Omega}{1\ \text{k}\Omega}}$$

$$e_n\ (10\ \text{k}\Omega) = 12.7\ \text{nV}/\sqrt{\text{Hz}}$$

When a resistor is placed at the input to an op amp, the ac noise voltage that it produces at the output of the op amp increases directly as the square root of the bandwidth that is allowed to exist in the op amp application circuit. *To reduce ac noise, the bandwidth of an application circuit should therefore be kept small, only wide enough to pass the required signal information.* (Bandwidths of only 1 or a few Hz are often used to obtain low noise in the receivers for the space exploration satellites and this restricts the allowed rate of information transfer. Therefore, a relatively long time is taken to transmit a single picture from outer space and this has generated interest in removing all redundant information prior to transmission.)

We will perform all of our ac noise calculations using the individual spectral densities of the ac noise sources as $nV/\sqrt{Hz}$ (pronounced "nano-volts per root-Hertz") and then account for the equivalent bandwidth, BW′, of the circuit by multiplying by the square root of this equivalent bandwidth, $\sqrt{BW'}$, after the various ac noise sources have been properly combined. (This dependence of ac noise on the square root of the bandwidth implies that if we reduced the bandwidth by a factor of 100:1, the ac noise voltage at the output would be reduced by 10:1.)

We have considered only one source of ac noise in op amp circuits: external resistors. This will be used to represent the ac noise that is generated by the signal source resistance and also by the resistors that are used in the feedback network. These resistors are all external to the op amp and the values that are used depend on the application.

Even if an op amp is perfect from the standpoint of noise and is entirely free of noise, the resistors that are used around the op amp will cause noise at the output of the application circuit. As an example, the value of the resistance of the input signal source will be added to the parallel equivalent of the feedback resistors to give an equivalent total resistance that is the source of Johnson noise. If this equivalent resistance has a value of 100 k, the Johnson noise voltage per root hertz at the input to the op amp can be determined from

$$e_n(100\ k\Omega) = \frac{4\ nV}{\sqrt{Hz}} \times \sqrt{\frac{100\ k\Omega}{1\ k\Omega}}$$

$$e_n(100\ k\Omega) = 40\ nV/\sqrt{Hz}$$

Further, if the bandwidth of the application circuit is assumed to be 100 kHz (as we will see later, a correction to this bandwidth needs to be made to account for the added noise that is allowed to pass through to the output when the high-frequency cut-off is not sharp and noise at frequencies beyond the cut-off frequency gets through), the total noise voltage becomes

$$e_n\ (total) = 40\ nV/\sqrt{Hz} \times \sqrt{100\ kHz}$$

$$e_n\ (total) = 12.7\ \mu V$$

If the application circuit provides voltage gain, the noise at the ouput of the op amp will also be increased. To complete this example, we will assume a closed-loop voltage gain of 100, so the overall total rms noise voltage to expect at the output (neglecting the bandwidth correction mentioned above) becomes

$$e_n\ (output) = 12.7\ \mu V \times 100$$

$$e_n\ (output) = 1.27\ mV\ rms$$

The inherent ac noise sources of the op amp are modeled as shown in Figure 3-20, where all of the ac noise sources within the op amp are brought outside and are represented by the ac noise generators that are shown. The remaining op amp is now considered as being *free of ac noise*, or *noiseless*.

The normal sign convention of the symbols that are used for voltage sources and current sources doesn't apply with ac noise. (+ or – doesn't apply, because, as we will see later, rms addition will be used for ac noise sources — not algebraic addition. Unfortunately, this also means that uncorrelated ac noise sources can never cancel. They are each individually too

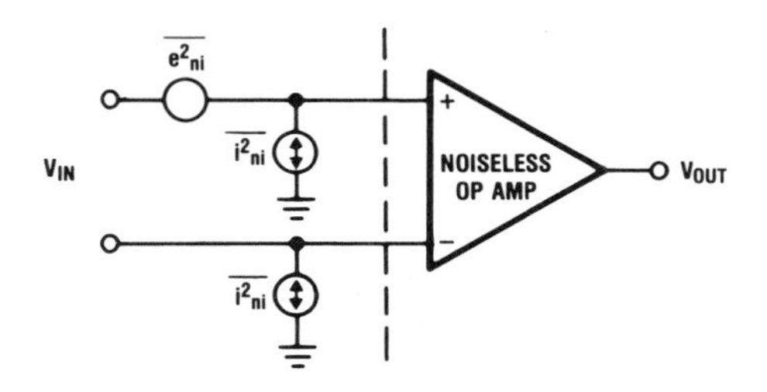

Fig. 3-20. Equivalent Input AC Noise-Sources of an Op Amp

erratic to have the proper detailed symmetry required for cancellation). The symbols used for the ac noise sources are therefore modified so they do not specify an algebraic sign.

The ac noise current generators, shown at each input, represent the ac noise that results because of the dc current flow, I_B, at each of the inputs to the op amp. This "*shot*" *ac noise is associated with all dc current flow.* (This name was chosen because the noise sounded like "a hail of shot striking a target.") It is also *a white noise* and has a mean-square value, $\overline{i^2_n}$, when observed over a frequency band, Δf, that is given by

$$\overline{i^2_n} = 2qI_B\Delta f \text{ Amps}^2/\text{Hz}$$

[This equation shows why Bi-FET op amps have low ac noise current. The small I_B (50 pA) causes this ac shot noise current to be extremely small (16×10^{-30} Amps^2/Hz or 4×10^{-15} Amps/$\sqrt{\text{Hz}}$ = 4 fA/$\sqrt{\text{Hz}}$) and very difficult to actually measure.]

Some op amp designs provide an internally generated dc current that is equal to the expected value of I_B. This current is then fed directly into the input lead of the op amp to make the external *apparent dc input bias current equal to zero,* or nearly so. For op amps that use this input current compensation, the I_B that finally exists does not determine the ac noise current magnitude. In fact, although the dc input currents can be arranged to cancel, the ac noise is increased by $\sqrt{2}$ (40%) over the uncompensated I_B in most of these designs because of the increased ac noise of the added I_B.

The magnitude of the input ac noise voltage source, e_{ni}, depends on the internal design of the op amp. In general, if a bipolar op amp has large dc input currents, it will have a small ac noise voltage spec. This is the reason for the unusually low ac noise voltage that can be obtained from one of the early IC op amps, the 709. As we will see later on, in applications with a low value of dc source resistance, this op amp can provide very low ac noise performance.

Predicting AC Noise In The Output Signal

We will now use this ac noise model of the op amp to predict the ac noise in the output signal of a typical application, Figure 3-21, where we have added our signal source (with a dc source resistance of R_s ohms) and the resistors R1 and R2 of the β network.

For optimum dc stability, we have learned to match the input resistors. In the case of a design for low ac noise, we would like to avoid using large valued resistors at the input because of the extra ac noise they add. For this low ac noise example we will, somewhat

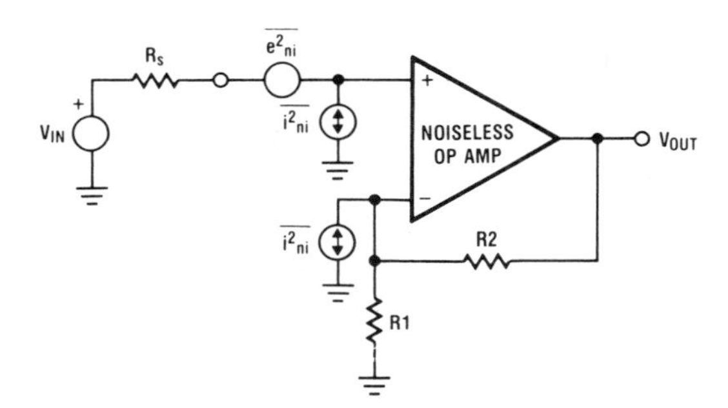

Fig. 3-21. Incorporating the Op Amp Noise Model in an Application Circuit

arbitrarily, choose

$$R1||R2 = R_P = \frac{R_S}{2}$$

Selecting large enough values for R1 and R2 to prevent amplifier loading presents problems for low values of R_S. It is relatively easy to have $R_P \ll R_S$ at large values of R_S. We can see the effects of altering this choice for R_P later on, but using $R_P = R_S/2$ could reduce the total ac noise by as much as 20% to 30% when compared to the condition of matching the resistors ($R_S = R_P$). Further reductions in ac noise level may be possible if even smaller resistance values are used in the β network. If this input resistor mismatch causes a dc problem, a Bi-FET op amp should be considered.

Accounting for All of the AC Noise Sources. The next problem is how to properly combine or add up all these ac noise sources to arrive at an *equivalent input ac noise voltage.* We will start by considering the noise sources at the inverting input, shift these over to the noninverting input side of the op amp, and then find the sum of all of these individual noise sources.

Two ac noise voltages result at the inverting input: the Johnson noise due to R_p and the voltage that results from the flow of $\overline{i^2_{ni}}$ through R_p. We can represent these as shown in Figure 3-22, where the resistors of the β network are now considered noiseless.

We can now simply move these ac noise voltages over into the noninverting input lead. The ac noise sources that were at the noninverting input can be handled in a similar manner. There is Johnson noise associated with R_S and also $\overline{i^2_{ni}}$ flows through R_S.

Combining all this, we arrive at a string of ac noise voltage sources as shown in Figure 3-23. The question now is how do we properly add up all of these individual ac noise voltages to arrive at the equivalent ac input noise?

Because all of these ac noise voltages are considered to be uncorrelated, they should be added in an rms fashion. This provides the mean-square value of the equivalent ac input noise, $\overline{e^2_{n\,eq}}$, as

$$\overline{e^2_{n\,eq}} = \overline{e^2_{ni}} + \overline{e^2_n (R_S)} + \overline{e^2_n (R_P)} + [\overline{i^2_{ni}}]\,[(R_S)^2] + [\overline{i^2_{ni}}]\,[(R_p)^2]$$

where e^2_n (R_S) implies the mean-square value of the Johnson noise of resistor R_S.

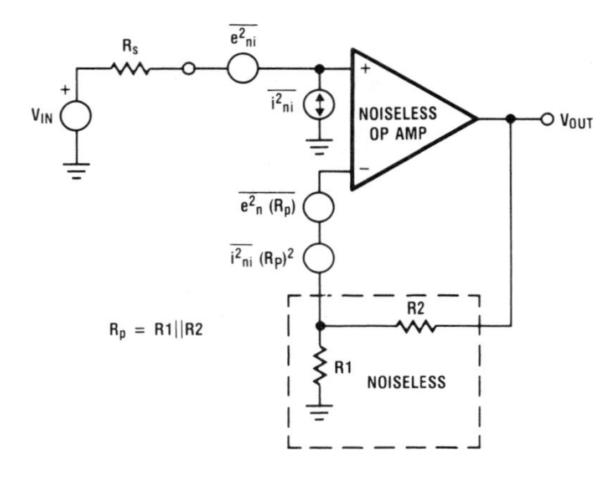

Fig. 3-22. Converting to Noise Voltages at the Inverting Input

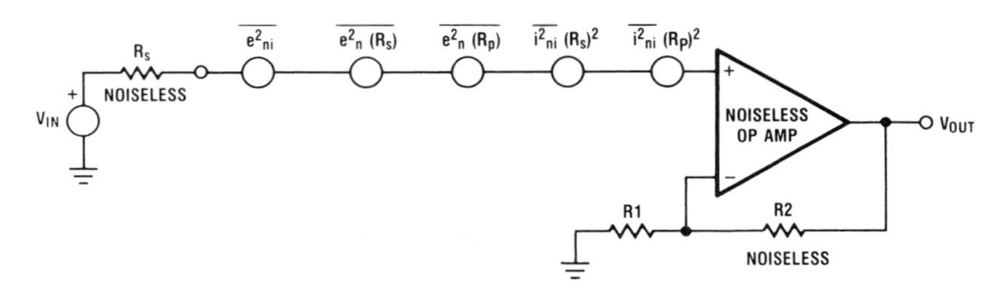

Fig. 3-23. All the Noise Voltages Grouped at the Noninverting Input

We can now factor this, as

$$\overline{e^2_{n\,eq}} = \overline{e^2_{ni}} + \overline{e^2_n (R_S + R_P)} + [\overline{i^2_{ni}}]\,[(R_S)^2 + (R_p)^2]$$

Because we have enforced the condition that

$$R_P = \frac{R_S}{2}$$

then

$$\overline{e^2_{n\,eq}} = \overline{e^2_{ni}} + \overline{e^2_n}\left(R_S + \frac{R_S}{2}\right) + \overline{[i^2_{ni}]}\left[(R_S)^2 + \left(\frac{R_S}{2}\right)^2\right]$$

or

$$\overline{e^2_{n\,eq}} = \overline{e^2_{ni}} + \overline{e^2_n}\left(\frac{3\,R_S}{2}\right) + \overline{[i^2_{ni}]}\left(\frac{5\,R_S2}{4}\right)$$

The interesting thing to notice is that we have three quite different terms here: the first term is independent of R_S, the second term increases directly with R_S, and the last term increases as $(R_S)^2$. We will make use of this observation shortly. For now, simply notice that if *$R_S = 0$, we would still get the ac noise voltage that is inherent to the op amp. This is the best, or lowest ac noise, we can achieve with an op amp.* As R_S increases in value, it can also happen that the second term (and eventually even the third term) will dominate, and then $\overline{e^2_{ni}}$ will have essentially no effect.

Before we move on, we will pause to look a little more closely at rms addition. You may have used this concept when observing the rise time response of a circuit to a pulse input excitation (with a known rise time) on an oscilloscope (that, also, has a known rise time). The rise time to expect for the resulting display is achieved by the addition of all three rise times in an rms fashion. This form of addition has the peculiar property that the rms sum quickly follows the larger term. For example: the rms sum, S, of two numbers A and B is given by

$$S^2 = A^2 + B^2$$

or

$$S = \sqrt{A^2 + B^2}$$

and if $B = A/2$

$$S = \sqrt{A^2 + \frac{A^2}{4}} = \sqrt{\frac{5A^2}{4}} = 1.12A$$

which states that only a 12% error would be made in S if we neglected B entirely, even though it is as large as one-half of A!

In the case of our ac noise sources, similarly strange results occur. For example, two equal valued ac noise sources increase the total ac noise by only $\sqrt{2}$ or 40% (not 100%). And, as we have just seen, if one ac noise voltage is only twice the magnitude of a second one, we can usually neglect the second one. In other words, we should expect to find that the total ac noise of a given application will depend essentially on only one of the three ac noise terms that appear in our final ac noise equation.

It is interesting to plot this final equivalent ac input noise equation as a function of R_S, because we often must work with a given value of R_S (unless ac coupling and transformers are allowed in the application). This has been done in Figure 3-24. We again notice that we can do no better than the e^2_{ni} spec of the op amp, even if $R_S = 0$.

We have plotted the noise voltage specs of three different op amps on this same graph: a low noise voltage op amp (6 nV/$\sqrt{\text{Hz}}$), a medium noise voltage op amp (12 nV/$\sqrt{\text{Hz}}$), and a relatively high noise voltage op amp (40 nV/$\sqrt{\text{Hz}}$). If you recall, a 10 kΩ resistor produces approximately 13 nV/$\sqrt{\text{Hz}}$ and this is twice the ac noise voltage of a low noise op amp! This is why large valued resistors at the input of an op amp can easily degrade the overall ac noise performance. For example: if you must work with an R_S = 20 kΩ, don't buy a special low noise voltage op amp. Also notice that if an op amp has an input current of 50 nA, the noise current is only a problem for source resistances of 1 MΩ or greater. (The analysis, so far, has considered only the flat band, or white noise. The situation is different when low frequencies are involved, as we will see shortly.)

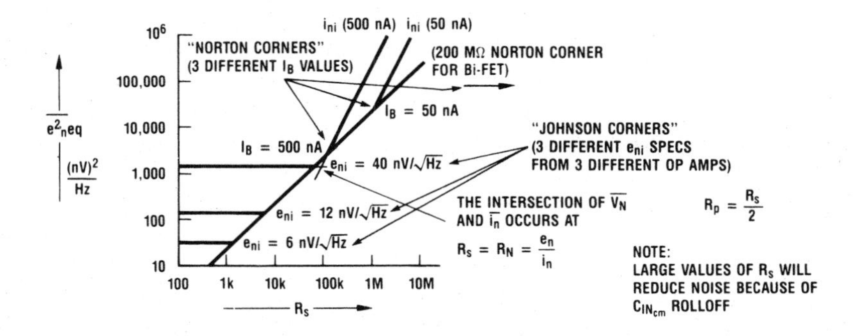

Fig. 3-24. Seeing the White-Noise Tradeoffs

On this graph, we can identify two *break points* (similar to those on a Bode plot). The total ac noise dependence on R_S starts with a horizontal line for small values of R_S (showing that ac noise is, at first, independent of R_S). This horizontal line represents the ac noise voltage spec of the op amp. As we move to higher values of R_S, we next intersect a line with a slope of + 1 (one logarithmic unit per logarithmic unit); this represents the *Johnson corner* because we are now starting to become limited by the Johnson noise of the input resistors. As R_S is further increased, we continue along this slope of + 1 until we come to a second intersection, where the ac noise current spec of the op amp is finally starting to dominate; the *Norton corner.* Now the ac noise increases as $(R_S)^2$, so the slope changes to + 2.

We have plotted the Norton corner of two different op amps on this figure, one with an I_B of 500 nA and the other with an I_B of 50 nA. (The Bi-FET op amp, with an I_B of 50 pA, has a Norton corner of 200 MΩ and is off this graph — this clearly shows the benefits of the Bi-FETs at large values of R_S.)

This plot also shows that we will probably not be limited by both the ac noise voltage and the ac noise current spec of an op amp at the same time (that is, for the same R_S). Further, as we have previously seen, even exactly at the break points, the ac noise voltage is only up 40% due to the simultaneous contributions from the two equal valued ac noise sources.

Problems can exist when ac noise reduction of an op amp application circuit is based on minimizing the *noise figure* of the circuit. For this case, *the total ac noise of the circuit is referenced to the ac noise that is produced only by the signal source resistance, R_S, without the amplifier.* Now, *if R_S were to equal zero, the noise figure would become infinite and this would be interpreted as a poor noise figure and therefore R_S = 0 would not be allowed.* The investigations based on noise figure usually find that an optimum source resistance, R_S opt, should, instead, be used that is given by

$$R_{S\ opt} = \frac{e_{ni}}{i_{ni}} = R_n$$

This ratio of the two equivalent noise sources of an op amp has also been called the *noise resistance*, R_n, of the op amp. For the noninverting gain application, *the lowest ac noise is achieved with an R_S = 0* (or at least low enough to be below the *Johnson noise corner*). [Note that if R_S were equal to zero (or even a small value), the resistors of the β network will still present noise problems.] Most of the time we don't have the luxury of an $R_S = 0$, so we do the best we can with the R_S that we must work with.

When operating with large values of R_S, this analysis can predict larger ac noise voltage than is actually measured. As we have previously seen, with a large value of R_S, the common-mode input capacitance, $C_{IN\ cm}$, can severely limit the bandwidth of the Johnson noise of the external resistors and thereby provide an unexpectedly low output ac noise voltage.

A final consideration is to check the small-signal frequency response of the application circuit to insure that there is no peaking near the upper band edge. This peaking, if present, can produce an unexpected increase in the output ac noise of the op amp circuit.

We have, so far, evaluated the mean-square value of the equivalent ac input noise per unit bandwidth. The square root of this provides the rms ac noise voltage per square root bandwidth, often called *root bandwidth* or *root Hz*. The question now is how do we determine the bandwidth to use in a given application circuit? Once we have the proper bandwidth, we can multiply the equivalent input ac noise voltage per root Hz by the square root of this bandwidth to arrive at the equivalent rms input noise voltage.

To determine the proper bandwidth, we must first consider the concept of *ac noise bandwidth* and also introduce an additional ac noise mechanism that increases only the low frequency noise (the "flicker" or 1/f noise problem that we will consider shortly), in case the bandwidth in our application should extend to low frequencies or to dc.

Up to this point, all the ac noise we have considered has been white noise. As we will soon see, this is not the case in the 1/f noise region.

AC Noise Bandwidth. If we limit the bandwidth of an op amp by using a single-pole low-pass filter, we would find that more ac noise would appear at the output of the op amp than the *signal bandwidth* of the filter would indicate. This problem is caused by the relatively gradual high frequency attenuation characteristic of this simple filter. Additional high frequency noise is admitted by the slowly falling *skirt* or roll-off of the filter response and this increases the noise.

AC noise bandwidth is a conceptual way to eliminate the dependence of the resulting ac noise measurement on the type of filter that is used. AC noise bandwidth is defined to be the equivalent *brickwall* filter response, Figure 3-25, that will provide the same rms noise as a real filter where contributions from the decreasing amplitudes of high frequency noise are allowed because of the unavoidable skirt response. The signal bandwidth of a practical filter, therefore, has a multiplier that is associated with it to represent the increase in the ac noise bandwidth beyond the signal bandwidth of the filter. This is called the *ac noise bandwidth ratio*. The closer the attenuation characteristic of the filter approaches a brickwall response, the more the ac noise bandwidth ratio becomes approximately equal to 1. This concept of ac noise bandwidth only applies for constant ac noise voltage (or current) versus frequency; therefore, it is limited to the considerations of only white noise sources.

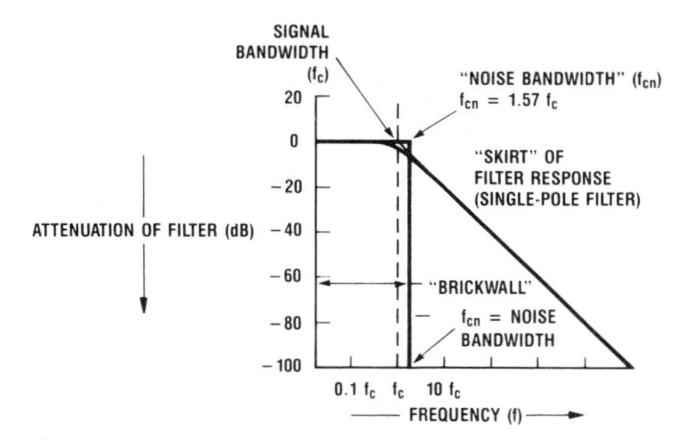

Fig. 3-25. Noise-Bandwidth Increase Over Signal-Bandwidth

If we cascade two single-pole filters (and prevent interaction by including an isolating unity-gain buffer between these filters) that have the same signal bandwidth, f_C, then a new overall effective signal bandwidth, f_C', results that is smaller because of bandwidth shrinkage. We now are down by -3 dB at f_C in both filters, or -6 dB total, so the -3 dB point now occurs at a frequency f_C' that is smaller than f_C. This shrinkage can be calculated for a cascade of n low-pass filters (all with a cutoff frequency of f_C) as

$$f_C' = f_C \sqrt{(2^{1/n} - 1)}$$

The bandwidth shrinkage factor, f_C'/f_C, and the resulting ac noise bandwidth ratios, f_{cn}/f_C', for a few values of n (the number of identical filters in the cascade) are given in Table 3-1.

TABLE 3-1. Bandwidth Shrinkage Factors and AC Noise Bandwidth Ratios for a Cascade of n, Low-Pass (0 to f_c) Filters.

Number of Filters in Cascade	Bandwidth Shrinkage Factor f_c'/f_c	AC Noise Bandwidth Ratio f_{cn}/f_c'
1	1.00	1.57
2	0.64	1.22
3	0.51	1.16
4	0.44	1.13
5	0.39	1.12

A single filter is seen to increase the ac noise by 57%, but two will only cause a 22% increase. Notice that the signal bandwidth that results from two filters in cascade will be only 64% of the corner frequency of the individual filters.

To eliminate the shrinkage in the signal bandwidth and to provide a steeper skirt, an Mth order Butterworth filter should be considered. We will discuss this type of filter in Chapter 5.

For a Butterworth filter, the signal bandwidth remains constant, independent of M, as shown in Figure 3-26. The ac noise bandwidth ratios for different orders of this filter are listed in Table 3-2.

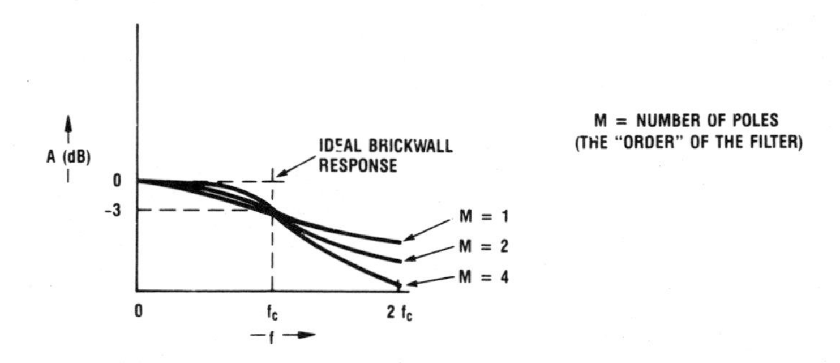

Fig. 3-26. The Butterworth Approximation to the Ideal Brickwall Response

TABLE 3-2. AC Noise Bandwidth Ratios for M^{th} Order Butterworth Low-Pass (0 to f_c) Filters

Number of Poles, M, in Filter	AC Noise Bandwidth Ratio f_{cn}/f_c' ($f_c = f_c'$)
1	1.57
2	1.11
3	1.05
4	1.03
5	1.02

Notice the more rapid improvements for increasing filter complexity; a two-pole (M = 2) Butterworth low-pass filter only increases the ac noise by 11%. (If transient performance is important in the application, this should be checked with the Butterworth filter in place. A different filter type may be beneficial in this case.)

In the case of filtering white noise, if a bandpass filter is used, and $f_{HI} \gg f_{LO}$, then the ac noise can be approximated by simply omitting the effects of f_{LO} and considering the bandpass filter as a low-pass filter with a corner frequency of f_{HI}. This results because we take the square root of the bandwidth. For example, a bandwidth extending from f to 10f has a square root given by

$$\sqrt{10f - f} = \sqrt{9f} = 3\sqrt{f}$$

If we consider this as simply a low-pass filter with a corner frequency of 10f (f_{HI} = 10f) we would have obtained

$$\sqrt{10f} = 3.16\sqrt{f}$$

which is too large by the following percentage error

$$\% \text{ error} = \frac{0.16 \times 100}{3} = 5.3\%$$

and, therefore, if the bandwidth were wider than one decade, the error would be less than 5%.

The frequency, f_{HI}, should be corrected for the ac noise bandwidth ratio depending on the type of filter that is used. The application bandwidth ($f_{HI} - f_{LO}$), when entirely in the white noise range (where $f_{LO} \gg$ 1/f corner frequency, which we will discuss in the next section) is corrected by multiplying by the ac noise bandwidth ratio. The square root of this effective bandwidth for white noise, $\sqrt{BW'_W}$, is given by

$$\sqrt{BW'_W} \cong \sqrt{\left(\frac{f_{cn}}{f_c'}\right) f_{HI}}$$

for $f_{HI} \gg f_{LO}$. The term (f_{cn}/f_c') is the ac noise bandwidth ratio (this depends on the filter used).

When the application bandwidth extends to lower frequencies, we must consider the effects of the increased ac noise that exists in the low frequency 1/f noise region.

Flicker (1/f) Noise. An additional ac noise mechanism exists at low frequencies (less than 100 Hz for a *good* bipolar op amp). Both the noise voltage and the noise current sources have a spectral density roughly proportional to 1/f, which is called *pink noise,* because of the higher noise energy content in the lower frequencies. (Lower visible frequencies are at the red, or *pink,* end of the visual spectrum.) This 1/f noise occurs in all conducting materials, therefore it is also associated with resistors.

Special steps can be taken in the fabrication of the op amp wafers to reduce the frequency where this 1/f ac noise starts increasing over the white noise level. This is the *1/f corner* frequency and *lowering the frequency where this 1/f corner occurs is all that can be done to reduce this ac noise component in the design and fabrication of an IC op amp.*

For ac coupled applications, this 1/f noise may be advantageously reduced by the low frequency attenuation of a bandpass or high-pass filter. In dc coupled applications, this 1/f noise rejection benefit cannot be obtained.

Our previous discussions have covered the cases where we limit the bandwidth of the application circuit to cover only the constant, or white noise higher frequency regions. The next question is how to handle ac noise in this low frequency 1/f region. We could be entirely within it, or only the lower range of our application bandwidth could extend into this region, as shown in Figure 3-27.

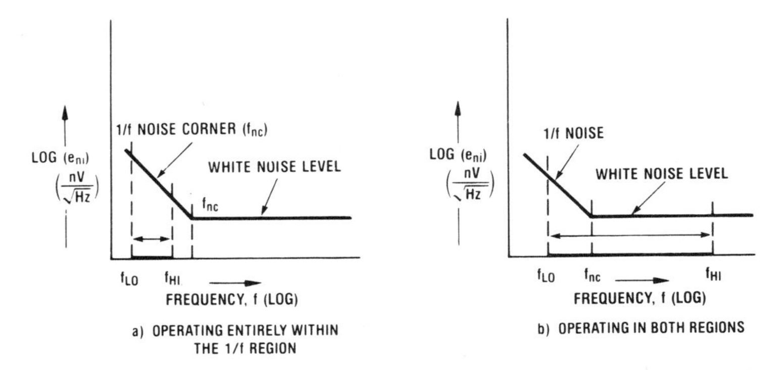

Fig. 3-27. Operating in the 1/f Region of an Op Amp Noise-Characteristic

We will first look at the case where our application bandwidth is entirely within the 1/f region, Figure 3-27a. If this bandwidth is established by a basic filter with skirt slopes of 20 dB/decade, it has been shown (Reference #4) that the square root of an effective bandwidth for this pink noise, $\sqrt{BW'_P}$, is given by

$$\sqrt{BW'_P} = \sqrt{f_{nc} \ln \left(\frac{f_{HI}}{f_{LO}}\right)}$$

The surprising thing about this equation is that the increased noise in this region depends on only the ratio of the band edges; the same noise increase exists over the bandwidth from 0.01 Hz to 1 Hz as exists from 0.1 Hz to 10 Hz. Additionally, in this 1/f region, this simple band-pass filter, rather curiously, can be shown to provide the same total noise as an ideal brickwall filter. Remember, the concept of noise bandwidth is limited to considerations of white noise where a constant noise spectral density exists.

As a numerical example, if we had an op amp with a 1/f noise corner, f_{nc}, of 100 Hz and a bandwidth of 0.1 Hz to 10 Hz, $\sqrt{BW'_P}$ would be given by

$$\sqrt{BW'_P} = \sqrt{f_{nc} \ln \left(\frac{f_{HI}}{f_{LO}}\right)}$$

$$\sqrt{BW'_P} = \sqrt{10^2 \ln \left(\frac{10}{0.1}\right)} = 21.5$$

If the 1/f noise was not present, the root of the effective bandwidth for white noise, $\sqrt{BW'_W}$, over this same bandwidth would provide a value of

$$\sqrt{BW'_W} \cong \sqrt{1.57\,(10)} = 4$$

So, this pink noise is causing a five-fold increase in the ac noise voltage. This indicates that *we can generally neglect the effects of the previously discussed white noise in this low frequency region.* For a bipolar op amp and $R_S' < R_n$ (where R_S' includes R_P, and $R_n = e_{ni}/i_{ni}$), the ac noise voltage spectral density, $\overline{e^2_{ni}}$, spec of the op amp replaces the equivalent $\overline{\text{ac input}}$ noise we have used in the white noise example and for $R_S' > R_n$; the ac noise current, $\overline{i^2_{ni}}\,(R_s')^2$ is used. (For a Bi-FET op amp, the ac noise voltage spec can typically be used for $R_S' < 10\ M\Omega$.)

Questions always arise concerning how low the frequency, f_{LO}, should be taken for ac noise calculations in the case of dc coupling. The equation for this 1/f region indicates that ac noise continues to be provided down to extremely low frequencies. *The problem is: at very low frequencies, it rapidly becomes impossible to distinguish between 1/f noise and dc drift effects.* For example, the frequency of 0.01 Hz has a period of 100 seconds, or 1.7 minutes. Observations of ac noise over such long time intervals will also include dc drift effects.

Notice that when the ratio of the band edges is relatively large, as in our previous example (100/1), there is only a 22% increase in $\sqrt{BW'_P}$ by extending this an extra decade to include 0.01 Hz, as a bandwidth from 0.01 Hz to 10 Hz would imply a $\sqrt{BW'_P}$ of

$$\sqrt{BW'_P} = \sqrt{f_{nc} \ln \left(\frac{f_{HI}}{f_{LO}}\right)}$$

or

$$\sqrt{BW'_P} = \sqrt{10^2 \ln \left(\frac{10}{0.01}\right)} = 26.3$$

and the percent increase over the previous example is given by

$$\Delta\sqrt{BW'_P} = \frac{26.3 - 21.5}{21.5} \times 100 = 22\%$$

and extending this through two more decades to 0.0001 Hz causes an increase of only 41% over the 0.01 Hz example.

If we use 0.01 Hz as a practical limit on the lower band edge for 1/f calculations for dc coupled amplifiers, we can find that a significant reduction in the upper band edge also has little effect on the total ac noise produced. For example, we have just found that the band 10 Hz to 0.01 Hz produced (for f_{cn} = 100 Hz) an effective root-bandwidth factor of 26.3. If we were to reduce the upper band edge to 0.1 Hz (a factor of 100:1) this would provide

$$\sqrt{BW'_P} = \sqrt{f_{nc} \ln \left(\frac{f_{HI}}{f_{LO}}\right)}$$

$$\sqrt{BW'_P} = \sqrt{10^2 \ln \left(\frac{0.1}{0.01}\right)} = 15.2$$

or only approximately a 42% reduction in noise.

In the dc coupled case, the op amp smoothly crosses over to where dc drift effects start to appear at frequencies less than approximately 1 Hz. If a lower frequency of 0.01 Hz is used in the noise calculation, it is impossible to insure that the output changes that exist over a 100-second time interval are all attributed to 1/f noise. DC drift effects will also appear in the measurement and increase the apparent ac noise.

When the bandwidth of an application circuit extends only partly into this 1/f region, as shown in Figure 3-27b, we can determine $\sqrt{BW'_{WP}}$ (the effective root-bandwidth in both the white, W, and pink, P, regions) as

$$\sqrt{BW'_{WP}} \cong \sqrt{\left(\frac{f_{cn}}{f_c'}\right) f_{HI} + f_{nc} \ln \left(\frac{f_{HI}}{f_{LO}}\right)}$$

where (f_{cn}/f_c') is the noise bandwidth ratio and f_{nc} is the 1/f noise corner frequency.

To summarize, we have three cases:

Case I:
The application is bandlimited to the white noise region by a bandpass filter $(f_{HI} - f_{LO})$ where $f_{LO} \gg f_{nc}$. For $f_{HI} \gg f_{LO}$, the bandwidth-corrected rms equivalent input noise, $e_{n\ in}$, is given by

$$e_{n\ in} \cong \sqrt{\overline{e^2_{n\ eq}}}\left[BW'_W\right]$$

$$e_{n\ in} \cong \sqrt{\overline{e^2_{n\ eq}}}\left[\sqrt{\left(\frac{f_{cn}}{f_c'}\right) f_{HI}}\right] \text{ volts (rms)}$$

Case II:
The application is bandlimited to the pink noise region by a bandpass filter $(f_{HI} - f_{LO})$ where $f_{nc} \gg f_{HI}$. The bandwidth-corrected rms equivalent input noise, $e_{n\ in}$, is given by either

$$e_{n\ in} = \sqrt{\overline{e^2_{ni}}}\ [\sqrt{BW'_p}]$$

$$e_{n\ in} = \sqrt{\overline{e^2_{ni}}}\left[\sqrt{f_{nc}\ \ln\left(\frac{f_{HI}}{f_{LO}}\right)}\right] \text{ volts (rms)}$$

(where $\overline{e^2_{ni}}$ is the op amp noise voltage spec) or for a bipolar op amp with $R_S' > R_n$,

$$e_{n\ in} = \sqrt{i^2_{ni}\,(R_S')^2}\ \ [\sqrt{BW'_p}]$$

or

$$e_{n\ in} = \sqrt{i^2_{ni}\,(R_{S'})^2}\left[\sqrt{f_{nc}\ \ln\left(\frac{f_{HI}}{f_{LO}}\right)}\right] \text{ volts (rms)}$$

(and $\overline{i^2_{ni}}$ is the op amp noise current spec).

Case III:
The application is bandlimited $(f_{HI} - f_{LO})$ and extends only partly into the 1/f region $(f_{LO} < f_{nc} < f_{HI})$. The bandwidth-corrected rms equivalent input noise, $e_{n\ in}$, is given by

$$e_{n\ in} = \sqrt{\overline{e^2_{n\ eq}}}\,[\sqrt{BW'_{wp}}]$$

or

$$e_{n\ in} = \sqrt{\overline{e^2_{ni\ eq}}}\left[\sqrt{\left(\frac{f_{cn}}{f_c'}\ f_{HI} + f_{nc}\ \ln\left(\frac{f_{HI}}{f_{LO}}\right)\right)}\right] \text{ volts (rms)}$$

The next question concerns the gain to use in translating this equivalent ac input noise voltage to the output of an application circuit. Just as the bandwidth to be used for ac noise was not straightforward and obvious, so it is with the ac gain. We will now consider the parameter, *ac noise gain*, and see why it is that this, too, has to be nonobvious and not the same as simply the *ac signal gain*.

AC Noise Gain. You may recall from Chapter 2 that a given β or feedback network doesn't completely specify the closed-loop voltage gain of an op amp application circuit because the signal can be entered at two different points to produce either noninverting or inverting closed-loop gain. This is again shown in Figure 3-28, and a large gain difference is obtained with R1 = R2: −1 or +2. The question now becomes: which is the correct voltage gain to use for ac noise?

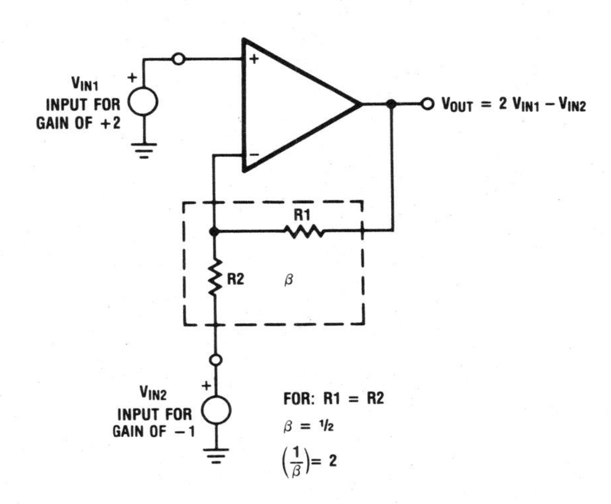

Fig. 3-28. Closed-Loop Gain is Not Always $(1/\beta)$

We have seen earlier in this chapter, when we considered the effects of the dc noise sources, that if the noise model is used properly, it will give the correct answer. Errors only exist if we jump to conclusions and assume that we already know the ac noise gain, and simply use the signal gain.

Notice, that to provide the smaller value of inverting gain, the input signal has to be applied remote from the op amp — at the outer end of R2. *AC noise is more intimate to the op amp and, therefore, gets special treatment; the largest value of gain possible!* (This result could have been anticipated by *Murphy's Law*, it also represents *the innate perversity of inanimate objects*).

A more subtle consequence of the reality of ac noise gain can be seen by considering the multiple-input summing amplifier of Figure 3-29. If all of the resistors are equal in value, the signal gain is -1 at any input. But ac noise, that can be considered as entering the (+) input, sees this as a higher gain circuit because all of the input resistors are effectively in parallel (when we use superposition) from the inverting input to ground. This resulting equivalent low valued input resistor, R/N, acts with R_F (R_F = R) to provide a feedback factor, β, of

$$\beta = \frac{R/N}{R + R/N} = \frac{1}{N + 1}$$

or an ac noise gain of $1/\beta$, which becomes

$$\text{ac noise gain} = \frac{1}{\beta} = N + 1$$

or, in general,

$$\text{ac noise gain} = 1 + \frac{R_F}{R_1} + \frac{R_F}{R_2} + \frac{R_F}{R_3} + \ldots + \frac{R_F}{RN}$$

[If any input had a larger signal gain (a smaller input resistor), the ac noise gain would also increase.]

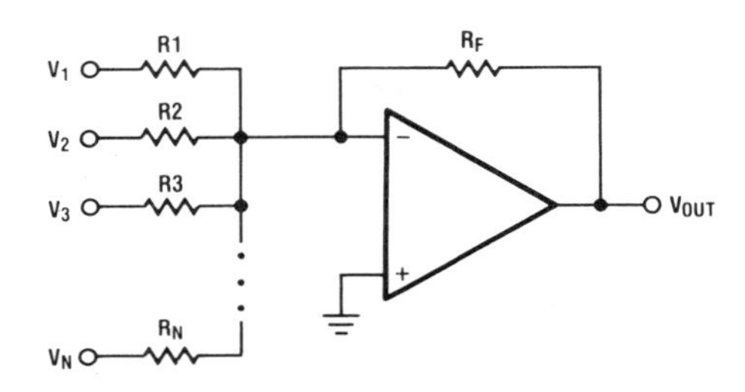

$$\text{AC NOISE GAIN} = 1 + \frac{R_F}{R1} + \frac{R_F}{R2} + \frac{R_F}{R3} + \ldots + \frac{R_F}{R_N}$$

Fig. 3-29. AC Noise-Gain is Larger than the Largest Signal-Gain

This same ac noise gain also multiplies the dc input offset voltage (a dc noise source) of this multiple-input application circuit and causes larger dc errors at the output than the lower value of the signal gain would indicate. (A capacitor, placed in series with each input resistor, would provide the same ac noise gain, but the dc noise gain would be reduced to 1.)

Because of the relatively large ac noise gain of multiple-input circuits, each low level signal should first be amplified by a separate amplifier before being combined or manipulated.

We can now multiply this ac noise gain by our noise-bandwidth-corrected equivalent input ac noise voltage to determine the output rms noise in an application circuit. (This would be read on a true-rms voltmeter: the common average responding voltmeter, that is calibrated for sinewaves, will be approximately −12% in error.)

There is one last consideration; the rms value of a noise voltage doesn't *nicely* convert into a peak-to-peak value that can be read from an oscilloscope. The statistical nature of ac noise creates a probability of finding peak noise amplitudes that are greater than a chosen multiple of the rms value of the noise. This relationship is listed in Table 3-3. A practical conversion is that the peak-to-peak voltage of an ac noise waveform is five times the rms value.

TABLE 3.3 Relating the Peak-to-Peak Value of an AC Noise Voltage to the rms Value

Peak-to-Peak Amplitude	Probability of Having a Larger Amplitude
2 × rms	32%
3 × rms	13%
4 × rms	4.6%
5 × rms	1.2%
6 × rms	0.3%
7 × rms	0.05%

"Popcorn" Noise. A peculiar additional ac noise mechanism is sometimes seen when an oscilloscope is used to observe the output ac noise voltage of an op amp. Digital signalling from outer space seems to be taking place. Pulses can exist with widths of many milliseconds where the constant baseline of random ac noise abruptly steps up by a large enough factor (say, 5:1) to trace out a rectangular pulse, with the baseline ac noise now occuring across the top of this pulse, as shown in Figure 3-30. This *abrupt type of noise is called popcorn noise* because of the sound it makes coming from a loudspeaker. Popcorn noise is caused by a defect that is dependent on the IC manufacturing technique.

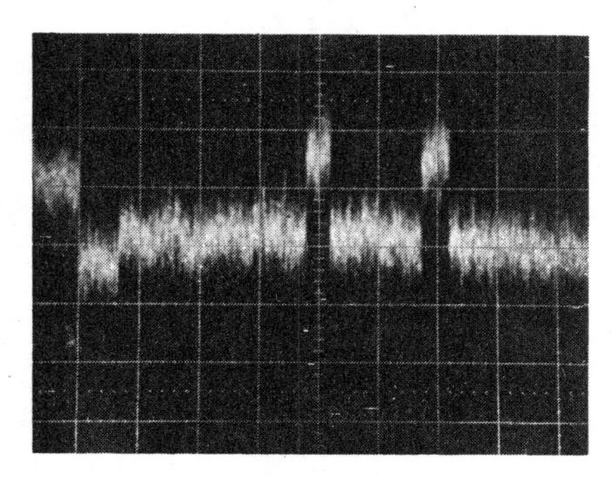

Fig. 3-30. The Abrupt Step-Type Nature of Popcorn Noise

This ac noise results from abrupt changes in the input current between the normal level and a second, quasistable level that differs by approximately 100 pA. Evidence of popcorn noise is closely monitored by the manufacturer. However, to guarantee that it is not present requires special, time consuming (and therefore expensive) testing.

With many of the basic considerations of op amps behind us, we will now take a closer look at the frequency stability, or the undesired oscillation problem of op amp application circuits.

CHAPTER IV

Frequency Stability, The Oscillation Problem

Linear circuit designers are often perceived as magicians who walk around with small-valued resistors and capacitors in their pockets, who can somehow tame the wildest oscillating circuit by just adding a few of these components in the *appropriate* places. One is never sure if the whole circuit will come *unglued* again sometime in the future, when the magician is gone. Therefore, many circuit design engineers fear the op amp and the unintentional oscillators that can be easily created with almost any of the application circuits.

This oscillation problem results because of the phase lag or the delay of the signal as it propagates through the op amp and then back through the feedback network, to be returned once again to the inverting or negative input of the op amp. Linear circuits use negative feedback, but when the phase lag of the fed-back signal becomes excessive (−180 degrees), *the sign of the feedback changes from negative to positive*. If the feedback is negative, the op amp is stabilized, providing predictable closed-loop performance. When the sign of the feedback changes to positive, it becomes regenerative. This simplifies the system because we no longer have to supply an input signal. The circuit takes care of this requirement all by itself. It "has its tail in its mouth"; it is oscillating!

4.1 STABILITY MARGINS, GAIN AND PHASE

Questions of stability involve the closed-loop or fed-back amplifier. To understand the stability problem requires that we once again consider the key equation for feedback control systems that was derived in Chapter 2, the feedback equation:

$$A_{CL} = \frac{A}{1 + A\beta}$$

Notice that anything that would make this denominator, $(1 + A\beta)$, approach zero would make the closed-loop gain approach infinity. This mathematical *blowing-up* of the equation for closed-loop gain *indicates that the circuit would oscillate*. Oscillation is not always bad; sometimes an oscillator is the intended function. Problems exist when *amplifiers oscillate and oscillators only amplify.*

Even digital systems can't handle division by zero; an error flag is usually set if this occurs. Undesired oscillations in an op amp circuit should therefore be thought of as the *linear, division-by-zero, error flag*. In either case, assistance is needed to get the electronic system operational again.

We can investigate the stability of a circuit by determining the conditions on $A\beta$ that make the denominator of the above equation equal to zero, or let's see what is implied when we write

$$(1 + A\beta) = 0$$

The open-loop gain, A, is a phasor. (A is just a dimensionless number at dc, but for any particular input signal frequency the open-loop gain has both magnitude and phase shift associated with it, due to the dominant pole and other internal causes of excess phase shift, and A is therefore a phasor.) The feedback network, β, can be simply a constant fraction, essentially independent of frequency (if only resistors are used in the β network), or it, too, can also be a phasor. Thus stating that

$$(1 + A\beta) = 0$$

or

$$A\beta = -1$$

means (because this is a phasor equation) that two conditions are simultaneously imposed: a magnitude condition,

$$|A\beta| = 1$$

and a phase or angle condition, the phase or angle associated with $(A\beta) = \pm 180°$. To meet these conditions, we must consider both A and β.

We can obtain an understanding of the stability problem by considering opening up the feedback loop and then passing a signal through the components of this disconnected feedback loop (just through the open-loop amplifier and then through the feedback network), as shown in Figure 4-1. This is difficult to actually do because we have to add some dc biasing circuitry to keep the op amp from saturating.

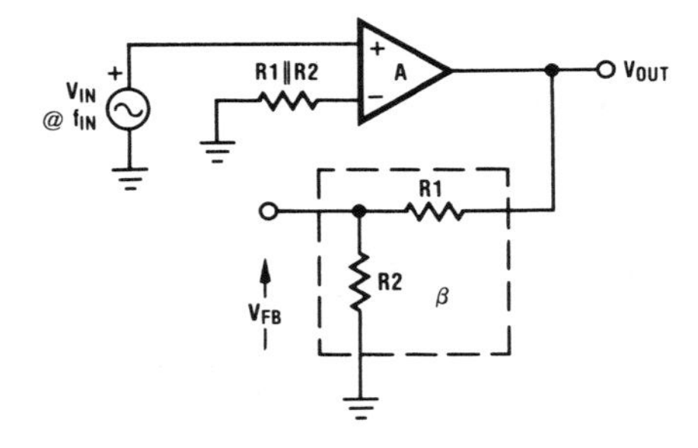

Fig. 4-1. Measuring the Signal Propagation through $A\beta$

To cause an undesired oscillation, the signal that comes out of the β network, V_{fb}, is required to have a magnitude at least as large as that of the input signal to the amplifier (this is the magnitude condition) and this output voltage waveform would also have to be shifted in phase until it became inverted when compared to the input voltage waveform (this is the 180° phase shift condition). If, in the above testing of the A and β networks, *a frequency could be found where the above two conditions could be simultaneously met,* then using these A and β networks for a feedback amplifier *would result in an oscillating circuit.* The frequency of oscillation would be essentially the particular frequency that was found to simultaneously satisfy both the magnitude and phase conditions that are implied in

$$A\beta = -1$$

If you were more fortunate in your A and β selections, you may find that just as you met one condition, say that the magnitude equaled 1, the phase shift (usually phase lag) may not be as large as 180°. For example, if it was only 120°, you would have some *cushion* or *margin* in the design; a *phase margin* of 180° minus 120° or 60°.

Similarly, if you continued to increase the input frequency, you may find that by the time you actually did have 180° phase lag, the amplitude of the signal out of the β network, V_{fb} was not as large as V_{IN}. For example, if V_{fb} was only one-quarter as large, you would also have another *cushion* or *margin* in the design; a *gain margin,* where

$$\text{gain margin (dB)} = 20 \log\left(\frac{V_{IN}}{V_{fb}}\right) = 20 \log\left(\frac{V_{IN}}{1/4\ V_{IN}}\right)$$

$$= 20\ (0.60)$$

$$= 12\text{dB}$$

The definitions of these stability margins are shown graphically in Figure 4-2, where for simplicity we have assumed $\beta = 1$ (as in the voltage follower) so the plot of $A\beta$ for this example is the same as the plot of A, the open-loop gain of the op amp.

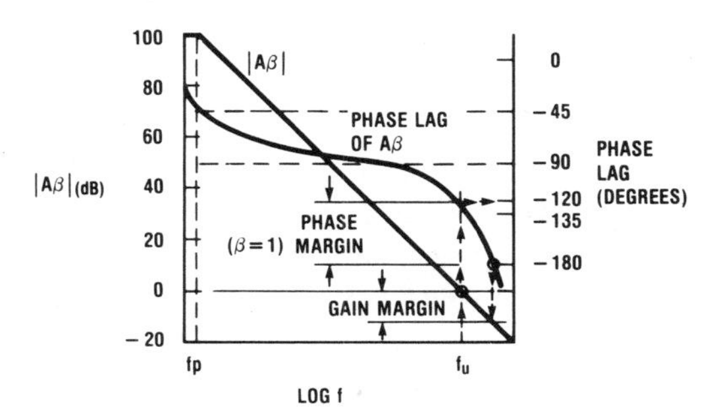

Fig. 4-2. Graphical Display of Stability Margins

To determine the phase margin, we first find the frequency where $|A\beta| = 1$. In this example, this occurs at a frequency that is equal to f_u, the frequency where the open-loop gain of the op amp equals unity. The amount of phase lag of $A\beta$ is determined at this same frequency from the other curve (phase lag of $A\beta$) and phase margin is the number of degrees of phase shift that remain before the total phase lag would equal 180°. A typical phase margin for a practical amplifier is 60°.

Gain margin is found by first locating the frequency where the phase lag of $A\beta$ is equal to 180 degrees. The magnitude response at this frequency is then found from the other curve ($|A\beta|$). Gain margin is the magnitude (in dB) that the *amplitude could be increased* at this frequency to equal unity gain, 0 dB. A typical gain margin for a practical amplifier is 12 dB.

For more complex feedback networks, the actual feedback factor that exists may not be obvious. As an example of this, Figure 4-3a, shows may Rs and Cs around an op amp and the question now is: What is β? This can most easily be handled by redrawing the circuit, as shown in Figure 4-3b, where to simplify the analysis, the two input voltages, V1 and V2, have been replaced by short circuits to ground. The value of β can now more easily be seen as the transfer function of this passive circuit. This trick, of redrawing the circuit to emphasize β, is useful in complex op amp application circuits.

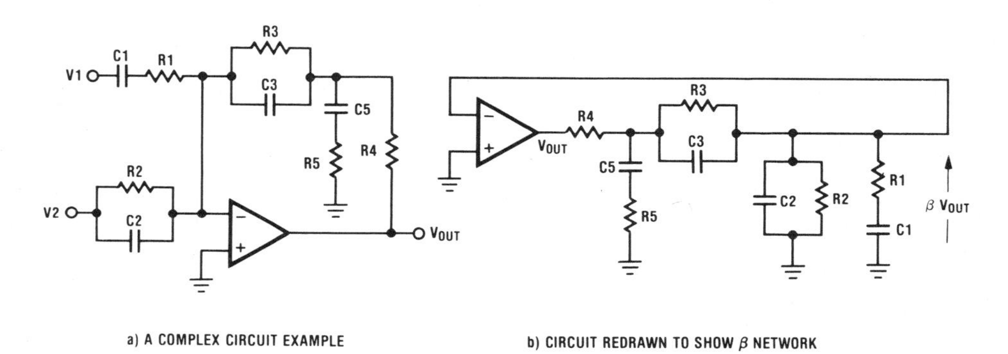

a) A COMPLEX CIRCUIT EXAMPLE

b) CIRCUIT REDRAWN TO SHOW β NETWORK

Fig. 4-3. Redrawing a Circuit to Make β More Obvious

4.2 POLES AND ZEROS

Whenever stability is discussed, *poles* and *zeros* often enter the conversation. These are powerful analytical tools and graphical representations that aid the designer of control systems. We will now introduce the concept of poles and zeros so those who may not be familiar with these terms can obtain an understanding of the basic way poles and zeros are used to help the designer. This will serve as an introduction to further study, if and when a more in-depth appreciation for these design aids is needed.

Some Background Material

Poles and zeros are used to describe the characteristics of transfer functions, the mathematical expressions for V_{OUT}/V_{IN} (or other output-input relationships), and also impedances or admittances. Our interest is in transfer functions. As we will soon see, *poles* are associated with a transfer function *going to infinity*, "blowing up," and *zeros* are associated with a transfer function *going to zero*. Before we see what this all means, we will review the concepts of reactance, impedance, and complex numbers.

Reactances and Impedance Diagrams. When we are dealing with a single sinusoidal excitation frequency, the reactance of a capacitor, X_C, can be written as

$$X_C = \frac{1}{j\omega C} \text{ or } X_C = -j\left(\frac{1}{\omega C}\right)$$

where the "j" accounts for the phase shift between the voltage across and the current flow through a capacitor. Similarly, the reactance of an indicator, X_L, is given by

$$X_L = j\omega L$$

Impedance diagrams can be drawn, as shown in Figure 4-4, to allow adding the individual phasors that are associated with an R, an L, or a C. A resistance phasor is represented by a scaled length along the + x axis, an inductive phasor is represented by a scaled length along the + j axis, and a capacitive phasor is represented by a scaled length along the -j axis. For example, the impedance diagram of a series-connected RC network, Figure 4-5, shows the graphical construction for the phasor addition, Z, that has a magnitude, |Z|, given by

$$|Z| = \left|\left[R + \left(\frac{-j}{\omega_C}\right)\right]\right| = \sqrt{R^2 + \left(\frac{1}{\omega_C}\right)^2}$$

and the phase angle of Z, known as the *argument* of Z or Arg (Z), is given by

$$\text{Arg }(Z) = -\Theta = -\tan^{-1}\frac{|X_c|}{R}$$

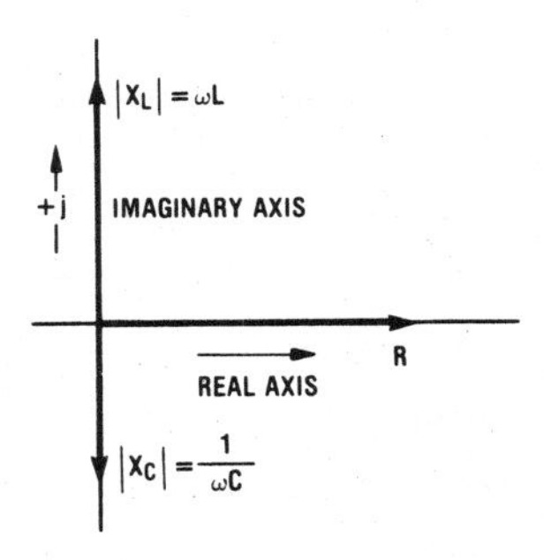

Fig. 4-4. Locating the Impedances of an R, L, and C on an Impedance Diagram

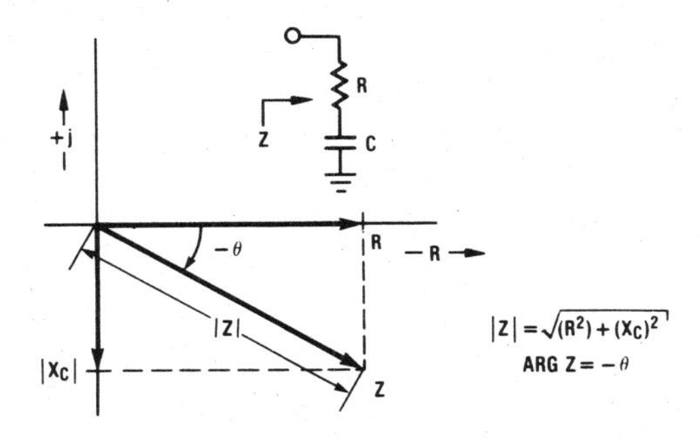

Fig. 4-5. The Impedance, Z, of a Series RC Circuit

Complex Frequencies and Complex Numbers. In the study of control systems, the response to more general excitation waveforms than simply sinusoids is very useful. This has popularized the use of *complex frequency,* s, where s is a complex number. (This is not a new concept and was used in 1900 by the early circuit theorist, Heavyside.)

In general,

$$s = \sigma + j\omega$$

Unfortunately, this causes much confusion; because in this equation, if $\sigma = 0$, *the imaginary part,* ω, of this general complex frequency variable *represents the real* or *"usual" sinusoidal frequency* that we associate with a sinewave signal generator. This frequency is expressed as radians per second (with the dimensions of 1/seconds), and therefore ω is called *radian frequency,* as opposed to the usual use of frequency, f, that has the dimensions of cycles per second, or Hertz (Hz) where $\omega = 2\pi f$ (1 cycle = 360° = 2π radians).

Because s represents the general complex frequency variable; and complex numbers have both a real and an imaginary part, we get into strange sounding things like *real and imaginary frequencies.* The only thing either real or imaginary about these frequencies is that they are associated with this complex frequency variable. *Nothing metaphysical is intended about these complex frequency components.*

The strangeness of the complex number historically was associated with the *unimaginable, and therefore imaginary,* "j" itself (where $j = \sqrt{-1}$) because this did not seem as real to the early mathematicians of commerce as the more familiar integers. When multiplication by $+j$ was later considered as a rotation by +90° [and therefore $(j)^2$ by 180°, or -1], this got the number system off of a line and spread it out all over the complex number plane (this is the *s-plane,* in the case of the complex frequency variable).

The use of complex frequency allows a convenient way to mathematically express many useful input excitation waveforms. Some basic complex frequency waveforms of the form

$$V(t) = H(\exp \sigma t) \cos \omega t$$

where the constant, H, is usually separately noted on the graph of the s-plane (as we will soon see) are shown in Figure 4-6. For example, Figure 4-6d shows the usual sinusoidal frequency that results when $\sigma = 0$. Figure 4-6a represents the decaying exponential waveform that results when $\omega = 0$ and $s = -\sigma$. (If σ were positive as shown in Figure 4-6e, we would have an increasing exponential waveform in the time domain that is associated with an oscillatory system.) Figure 4-6b combines a negative real component with an imaginary part to provide what is known as a *damped sinewave.*

The complex frequency variable, s, is the same as the Laplace transform variable, s. Transforms of voltages and currents are, therefore, also functions of this same s. Using this transform convention, we show, in Table 4-1 how the impedance, Z (s), for an R, L, and C are derived from the following three basic Laplace transforms (where we are neglecting the initial conditions):

$$\mathcal{L}\, i(t) = I(s) \qquad \mathcal{L}\, \frac{di(t)}{dt} = sI(s) \qquad \mathcal{L} \int i(t)\, dt = \frac{I}{s} I(s)$$

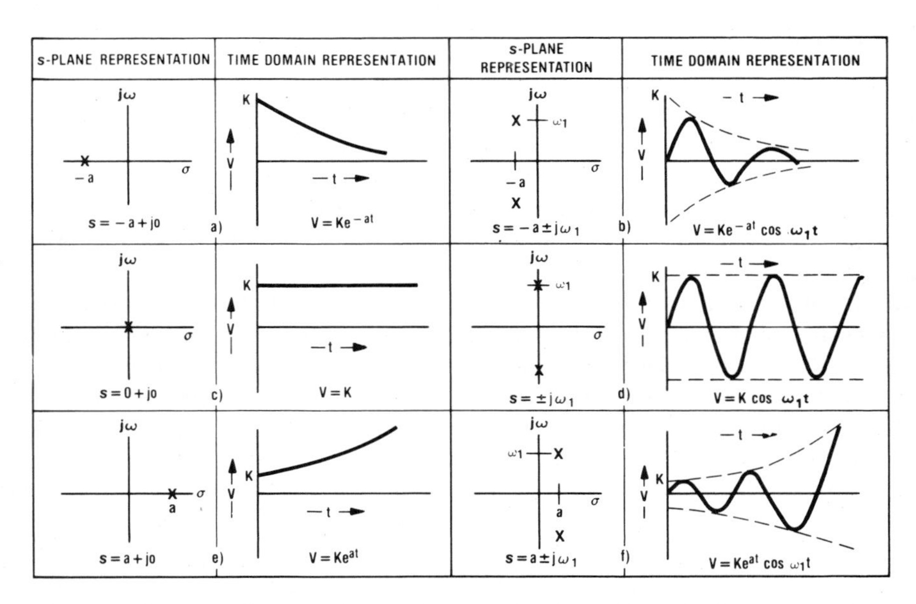

Fig. 4-6. Complex Frequencies: s-plane Versus Time Domain Representations

TABLE 4-1. Complex Impedances

Component	v = f (i)	Transform Equation	Impedance Z (s)
R	$v = Ri$	$V = RI$	R
L	$v = L\frac{di}{dt}$	$V = LsI$	sL
C	$v = \frac{1}{C}\int idt$	$V = \frac{1}{sC}I$	$\frac{1}{sC}$

(If you feel comfortable writing the impedance of a capacitor as "1/sC", you can omit this background material. This is only intended to refresh your memory and justify our use of s in the last column of Table 4-1, Impedance Z (s), because we will use this form for impedance in our discussions of poles and zeros.)

When we restrict our interest to only sinusoids, we can substitute

$$s = j\omega$$

in these expressions for impedance to obtain the previous reactance equations for the circuit elements.

Transfer Functions

When the stability of an op amp is considered, the expression for the closed-loop voltage gain is evaluated. This is also called a transfer function because it expresses the relation V_{OUT}/V_{IN} (how V_{IN} is "transferred" through the op amp application circuit to produce V_{OUT}). Before we consider the op amp, we will consider the transfer functions of a few passive circuits and see how the frequency dependence of these transfer functions can be evaluated by making use of the locations of the poles and zeros of these transfer functions on the complex frequency plane.

An RC Low-Pass Filter. With this background, we can now write an expression for the gain or transfer function for an RC low-pass filter. It is sometimes confusing that the *transfer function* and the *input signal* (or excitation) can both be expressed as *functions of the complex frequency variable s*. Our interest now will be in *evaluating transfer functions*, so we will *not make use of specific excitation frequencies*, but we will be interested in *how a transfer function can be evaluated as a function of frequency.*

This low-pass filter transfer function, as shown in Figure 4-7, becomes

$$\frac{V_{OUT}(s)}{V_{IN}(s)} = \frac{\left(\frac{1}{sC}\right)}{\left(R + \frac{1}{sC}\right)} = \frac{1}{RC} \frac{1}{\left(s + \frac{1}{RC}\right)}$$

$$\frac{V_{OUT}(s)}{V_{IN}(s)} = H \frac{1}{\left(s + \frac{1}{RC}\right)}$$

where H equals 1/RC, the constant in this example.

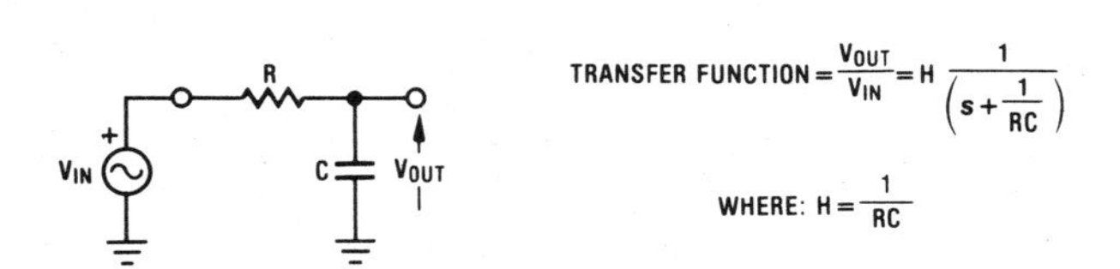

Fig. 4-7. The RC Low-Pass Filter Transfer-Function

This transfer function will blow up (go to infinity) for a value of s that would make the denominator have a value of zero. We can find this value of s by equating the denominator to zero,

$$\left(s + \frac{1}{RC}\right) = 0$$

This implies that if

$$s = -\frac{1}{RC}$$

we would find that the transfer function blows up. This transfer function therefore is said to have a "pole" that is located at $s = -1/RC$ on the complex frequency plane, the place where the transfer function blows up. (Conversely, a "zero" is a place where the transfer function has a value of zero.)

When we restrict our interest to the sinusoidal steady-state, the only complex frequencies we will consider are $s = j\omega$, so we will only be graphically evaluating transfer functions that have been plotted on the complex frequency plane at points (sinusoidal steady-state frequencies) along the $+j\omega$ axis. This graphical evaluation technique will allow us to determine the frequency dependence of a transfer function after we plot the poles and zeros of the transfer function.

When we *plot* a transfer function (or any function of s) on the complex frequency plane (the s-plane), only the points of interest — those complex frequencies where the function either has a value of zero (the *zeros*) or infinity (the *poles*) — are plotted. [Notice that this is quite different from the continuous plots or curves that result with functions of real variables, such as $y = f(x)$.]

A function of the complex frequency variable has unique values at these specific *singularities* or points of singularity; it is either equal to zero or infinity. For all other values of s, the transfer function can be evaluated by substituting any desired s, such as $s_1 = \sigma_1 + j\omega_1$, in place of the general complex frequency variable, s, in the original transfer function.

These poles and zeros have been called the *lifeblood* of a transfer function. Without these points of interest, there would only be a flat "desert" due to the scaling factor that is represented as H. But, when the evaluation frequency ($s = j\omega$) gets near a pole, the magnitude of the transfer function becomes very large (remember, at the pole it is equal to infinity!) and when near a zero, the magnitude of the transfer function tends to be small, near zero. This graphical *feeling* about networks is useful to the designers of control systems.

If we *plot* our low-pass filter transfer function on the s-plane, we obtain the rather strange looking result for a plot, Figure 4-8, where all we see is the "X" that represents the pole location. (We do not have any "explicit" zeros in this transfer function; but, if they were to exist, they would be plotted as a "0". A zero does exist for this function at $s = \infty$, so we have a *zero at infinity*, but our plot isn't large enough to locate this point, ∞.)

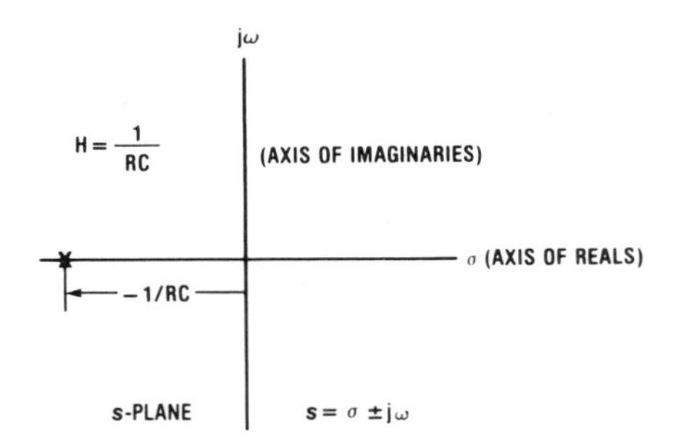

Fig. 4-8. The s-plane "Plot" of an RC Low-Pass Filter

The question may arise, *where is infinity on the s-plane?* The best way to think about the complex plane is that it is a small, apparently flat, plane on the surface of an infinitely large sphere. With this in mind, infinity is then located on the reverse side of the sphere, "half-way" around. We can therefore get to infinity on the s-plane by wandering off in any direction away from the origin.

To see how we will make use of an s-plane plot, we will determine the value of the transfer function of the RC low-pass filter for various sinusoidal steady-state frequencies. For such a limited input signal, the complex frequency variable becomes $s = j\omega$. By substituting $s = j\omega$ in our previous transfer function and then using various values of ω (called ω_i), we can find the *frequency response* of this RC filter. This transfer function therefore becomes, for our first value of ω, ω_1,

$$\frac{V_{OUT}(j\omega_1)}{V_{IN}(j\omega_1)} = H \frac{1}{\left(j\omega_1 + \frac{1}{RC}\right)}$$

The evaluation of this denominator can be interpreted on our complex plane as the graphical addition of the two phasors, $j\omega_1$ and $+1/RC$, as shown in Figure 4-9. So, if we have an accurately scaled plot, we can simply measure the length of this vector sum and also measure its phase angle. Now we can find the value of our transfer function from

$$\left|\frac{V_{OUT}(j\omega_1)}{V_{IN}(j\omega_1)}\right| = H \frac{1}{\left|\left(j\omega_1 + \frac{1}{RC}\right)\right|}$$

where the magnitude of the denominator was determined graphically and

$$\text{Arg}\left[\frac{V_{OUT}(j\omega_1)}{V_{IN}(j\omega_1)}\right] = -\text{Arg}\left(j\omega_1 + \frac{1}{RC}\right)$$

is also determined graphically. (The minus sign for the angle that is associated with the transfer function results because we change the sign of the angle as we move the measured angle that we obtain for the denominator up into the numerator.)

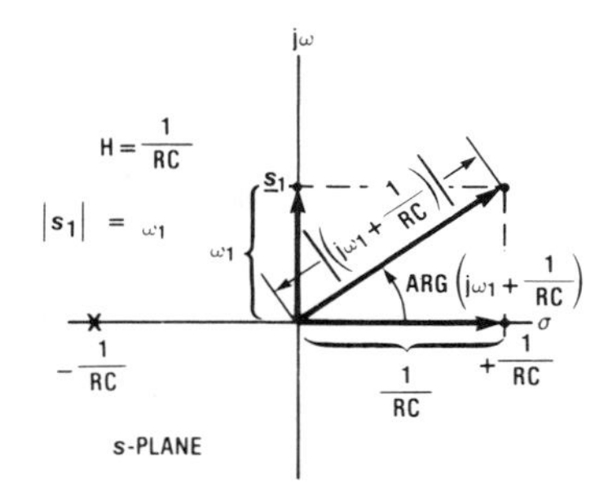

Fig. 4-9. Adding Phasors on the Complex Plane to Evaluate a Transfer Function

Notice that although we had previously located and plotted the pole, we had to make use of a new point, + 1/RC. This extra plotting labor can be saved if we do our phasor addition slightly differently so as to make use of the already plotted pole location, -1/RC. Figure 4-10 shows this change; we notice that the phasor sum is now represented by the line that connects the pole location to the particular s_1 on the $j\omega$ axis.

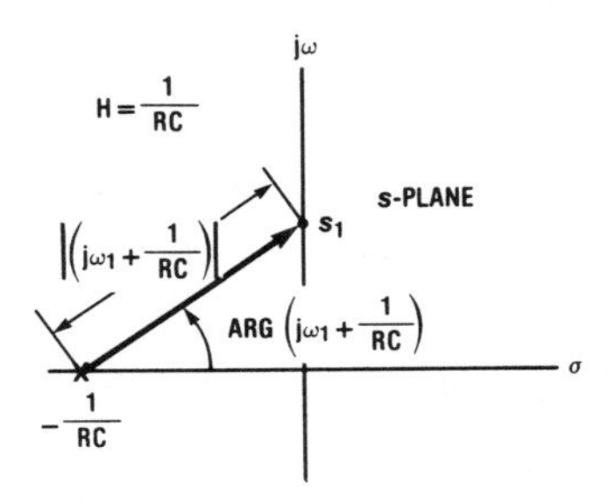

Fig. 4-10. Making Use of the Pole Location

By starting at dc ($s = j\omega = 0$), we find that the RC low-pass filter has a transfer function magnitude that is given by

$$\left|\frac{V_{OUT}\,(jo)}{V_{IN}\,(jo)}\right| = \frac{H}{\text{(length of the vector from the pole location to } \omega = 0)}$$

$$= \frac{H}{\frac{1}{RC}}$$

but

$$H = \frac{1}{RC}$$

so

$$\left|\frac{V_{OUT}\,(jo)}{V_{IN}\,(jo)}\right| = 1$$

and the phase angle is given by

$$\text{Arg}\left[\frac{V_{OUT}\,(jo)}{V_{IN}\,(jo)}\right] = 0°$$

This looks good, so far, because this is the proper transfer function for an RC low-pass filter with a dc input voltage.

We could continue taking steps along the $+j\omega$ axis and for each s_i measuring both the angle and the magnitude of the phasor sum that is indicated in the denominator. We could then calculate the magnitude of the transfer function for each of these frequencies. Let's just take one more value of s_i in this example; the particularly interesting one where

$$|s| = |j\omega| = \frac{1}{RC}$$

as shown in Figure 4-11. We can see that both phasor lengths are equal to 1/RC, so the magnitude of the phasor sum is given by

$$\sqrt{\left(\frac{1}{RC}\right)^2 + \left(\frac{1}{RC}\right)^2} = \sqrt{\frac{2}{(RC)^2}} = \frac{1.414}{RC}$$

which gives the magnitude of the transfer function a value of

$$\left|\frac{V_{OUT}\left(j\frac{1}{RC}\right)}{V_{IN}\left(j\frac{1}{RC}\right)}\right| = \frac{H}{\frac{1.414}{RC}} = \frac{\frac{1}{RC}}{\frac{1.414}{RC}}$$

$$\left|\frac{V_{OUT}\left(j\frac{1}{RC}\right)}{V_{IN}\left(j\frac{1}{RC}\right)}\right| = 0.707$$

and

$$\operatorname{Arg}\left[\frac{V_{OUT}\left(j\frac{1}{RC}\right)}{V_{IN}\left(j\frac{1}{RC}\right)}\right] = -45°$$

(We have just made a proper evaluation of the transfer function of a low-pass filter at the corner frequency.)

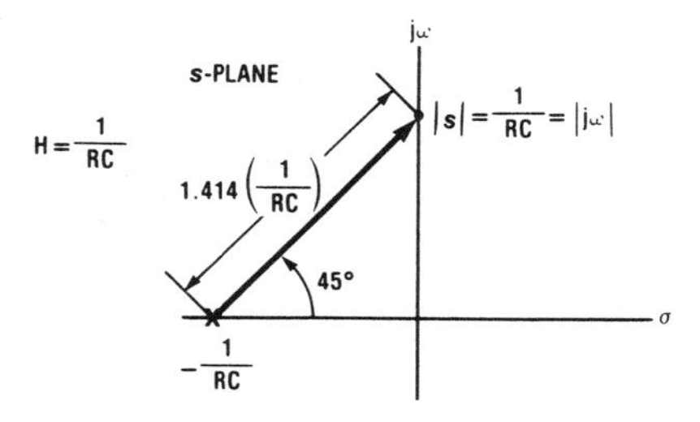

Fig. 4-11. Evaluating the Transfer Function of a Low-Pass Filter at the Corner Frequency

As the evaluation frequency increases, notice that the magnitude of the transfer function decreases (because we are dividing H by a larger and larger denominator) and the phase angle approaches -90°. This is the way the sinusoidal steady-state response of a transfer function is graphically evaluated by making use of constructions for each evaluation frequency in the s-plane. As an example, we show four values for ω_i in the construction of Figure 4-12. The vector lengths from the pole (at -1/RC) to each ω_i was scaled off this drawing and these lengths are indicated in Figure 4-12a. The chart of Figure 4-12b shows the calculation of $|V_{OUT}/V_{IN}|$ (the theoretical values for dB are shown for comparison) and these resulting data are plotted in Figure 4-12c, the low-pass transfer characteristic as a function of frequency.

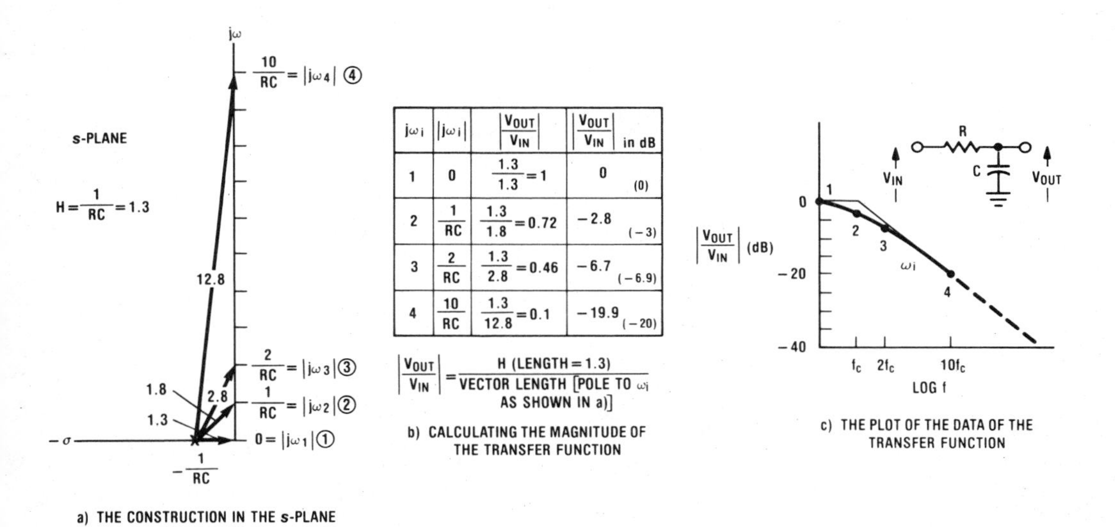

$j\omega_i$	$\lvert j\omega_i \rvert$	$\left\lvert \frac{V_{OUT}}{V_{IN}} \right\rvert$	$\left\lvert \frac{V_{OUT}}{V_{IN}} \right\rvert$ in dB
1	0	$\frac{1.3}{1.3} = 1$	0 (0)
2	$\frac{1}{RC}$	$\frac{1.3}{1.8} = 0.72$	−2.8 (−3)
3	$\frac{2}{RC}$	$\frac{1.3}{2.8} = 0.46$	−6.7 (−6.9)
4	$\frac{10}{RC}$	$\frac{1.3}{12.8} = 0.1$	−19.9 (−20)

Fig. 4-12. Graphically Determining the Magnitude of a Transfer Function

An RC High-Pass Filter. As an example of a transfer function with a zero, we can use an RC high-pass filter, Figure 4-13.

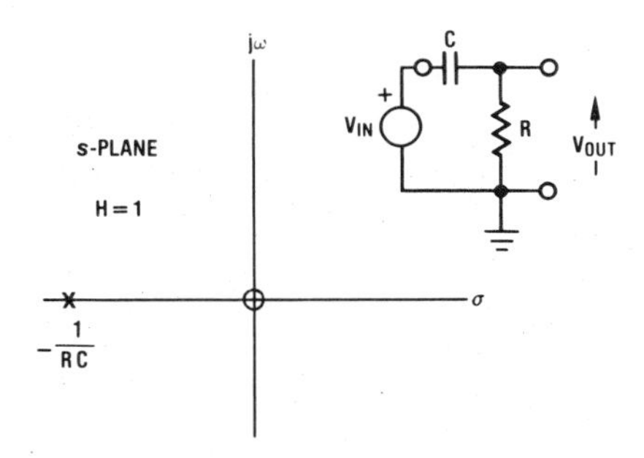

Fig. 4-13. The s-plane Plot of the Transfer Function of a High-Pass RC Filter

This high-pass transfer function is

$$\frac{V_{OUT}(s)}{V_{IN}(s)} = \frac{s}{\left(s + \frac{1}{RC}\right)}$$

and a zero exists at zero frequency. (This means that the magnitude of the transfer function is equal to zero at dc. This results because of the dc blocking, series position of the capacitor). The s-plane plot of this high-pass transfer function is also shown in this figure. (The location of the pole is the same as in the previous low-pass example.)

The only differences when we have both poles and zeros are that, in general, the magnitude is given by,

$$\left|\frac{V_{OUT}(s)}{V_{IN}(s)}\right| = H \frac{\Pi \text{ (magnitudes of the phasors from each zero to } s_i)}{\Pi \text{ (magnitudes of the phasors from each pole to } s_i)}$$

where: Π means the product of factors

and s_i $(= j\omega_i)$ is the value of the complex frequency at which the transfer function is currently being evaluated

and the phase angle is given by,

$$\text{Arg}\left[\frac{V_{OUT}(s)}{V_{IN}(s)}\right] = \Sigma \text{ (angles from zeros to } s_i) - \Sigma \text{ (angles from poles to } s_i)$$

where: Σ means a summation of terms.

A Useful Frequency Compensation Network. A very useful application of this theory can be made by considering the circuit of Figure 4-14. This RC network is often used to frequency compensate op amp application circuits. We will now see why it is so useful.

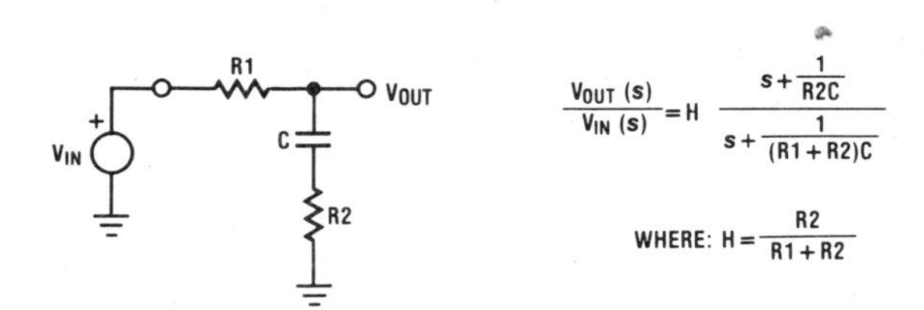

Fig. 4-14. Adding a Resistor in Series with the Capacitor Provides a Zero

The transfer function for this circuit is

$$\frac{V_{OUT}(s)}{V_{IN}(s)} = \frac{R2 + \frac{1}{sC}}{(R1 + R2) + \frac{1}{sC}}$$

or, after simplifying,

$$\frac{V_{OUT}(s)}{V_{IN}(s)} = \left(\frac{R2}{R1 + R2}\right)\left\{\frac{\left(s + \frac{1}{R2C}\right)}{\left[s + \frac{1}{(R1 + R2)\,C}\right]}\right\}$$

Note that we again have both a pole and a zero.

Instead of adding just a pole (a capacitor to ground), the small valued resistor, placed in series with the capacitor, provides this latest transfer function. We also get a zero. This zero reduces the limiting phase lag at high frequencies to nearly zero.

This transfer function is useful to keep in mind when attempting to frequency stabilize a complex op amp application circuit: for example, one that involves an added transistor amplifier in the overall loop as shown in Figure 4-15. (The signal inversion of this added transistor amplifier requires that we feedback to the noninverting input of the op amp in order to obtain an overall negative feedback loop.)

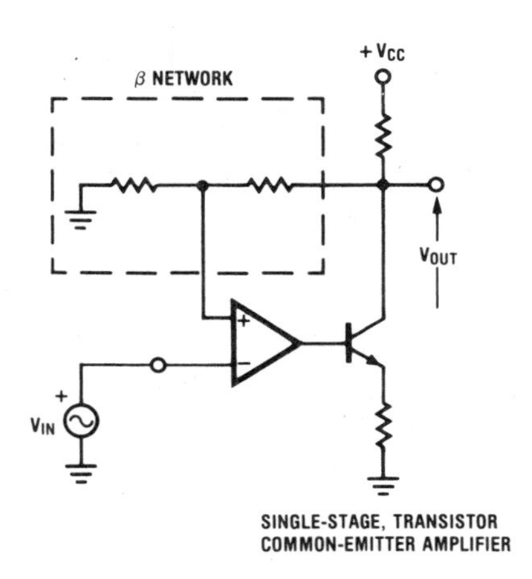

Fig. 4-15. Adding a Single-Transistor Amplifier Stage at the Output of an Op Amp

This extra voltage gain stage reduces both the gain margin and the phase margin and makes the frequency compensation problem more difficult. This use of a single-transistor voltage gain stage is very useful because a high breakdown-voltage transistor can increase the magnitude of the output voltage swing that can be obtained in the overall amplifier.

As we have seen earlier in this section, an added frequency compensation capacitor to ground across the collector load resistor of this added amplifying stage would introduce another pole and additional phase lag that would approach −90° at high frequencies. Because the system already has one dominate pole (from the op amp), adding an additional pole (and the extra phase lag that is associated with this pole) will usually cause frequency instability (oscillations) to occur.

This RC network to ground does cause a loss in signal magnitude because of the gain-limiting resistive attenuator that exists at high frequencies, but the high frequency phase lag is nearly zero. The gain and phase characteristics of this network are compared with the RC low-pass filter that would result without the added resistor in Figure 4-16.

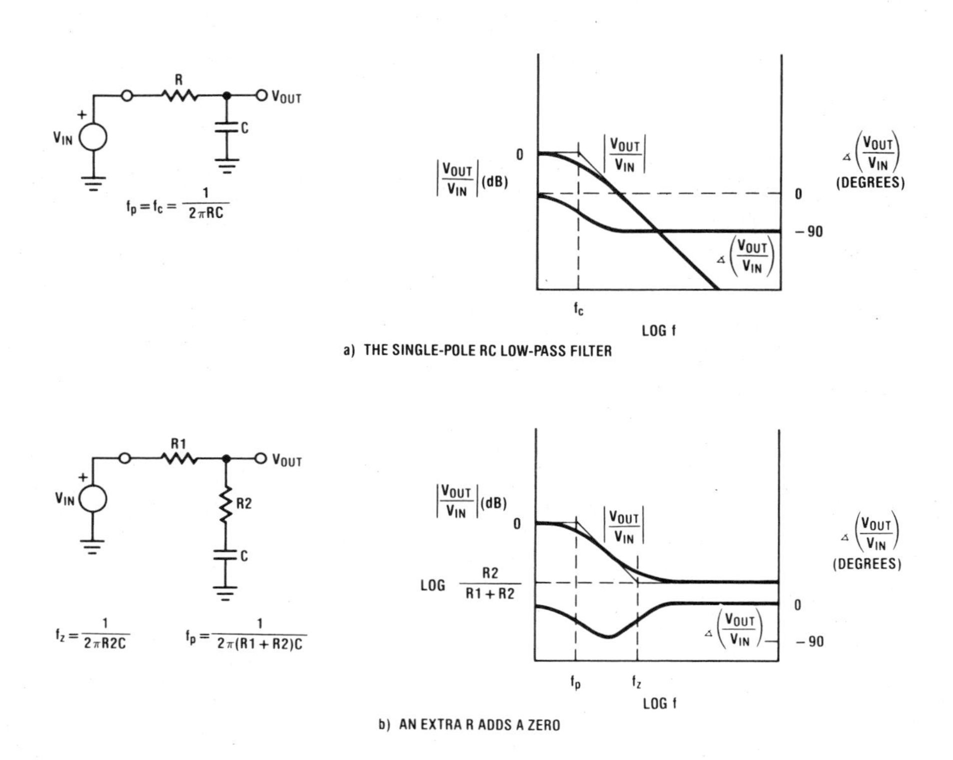

Fig. 4-16. Gain and Phase Characteristics of RC Low-Pass Transfer Functions

Using One Op Amp and Two Rs and Two Cs. By using a maximum of only two resistors and two capacitors, a wide range of op amp transfer functions is possible. A few of these are shown in Table 4-2 for reference. To avoid clutter, the R and C values are not indicated on each circuit drawing, but R_i and C_i are the input components and R_f and C_f are the feedback components.

The general transfer functions of these circuits can be anticipated by remembering that the impedance of a capacitor (the capacitive reactance) falls as the operating frequency is increased. Further, because these are all inverting amplifiers, we can think of the output voltage as resulting from the *input current* flowing through the *feedback impedance* (the network that ties from V_{OUT} to the inverting input). The *input impedance* (the network that ties from V_{IN} to the inverting input) determines the *input current* that is taken from the V_{IN} source. Therefore, the closed-loop voltage gain will increase as the input frequency is increased if the magnitude of the feedback impedance *increases* with increasing frequency or if the magnitude of the input impedance *decreases* with increasing frequency. For example, the first circuit shown in Table 4-2 has a capacitor, C_f, in parallel with the feedback resistor R_f. At dc this capacitor has no effect and the voltage gain is therefore established by the ratio of the two resistors ($A_v = -R_f/R_i$). As the input frequency is increased, the reactance of C_f decreases; therefore, the feedback impedance decreases, and the gain falls, which is indicated by the pole in the

TABLE 4-2. Some Useful Op Amp Transfer Functions

CIRCUIT	TRANSFER FUNCTION, $V_{OUT}(s)/V_{IN}(s)$
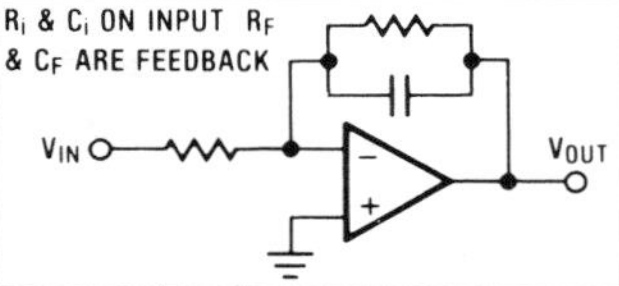 R_i & C_i ON INPUT R_F & C_F ARE FEEDBACK; V_{IN}; V_{OUT}	$-\frac{R_f}{R_i}\left(\frac{1}{1+sR_fC_f}\right)$
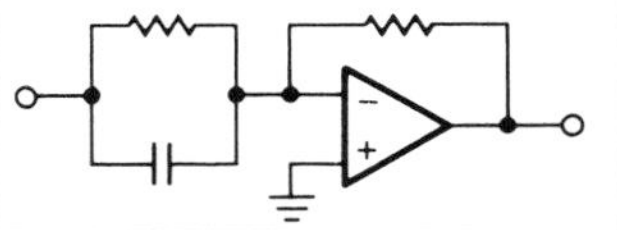	$-\frac{R_F}{R_i}(1+sR_iC_i)$
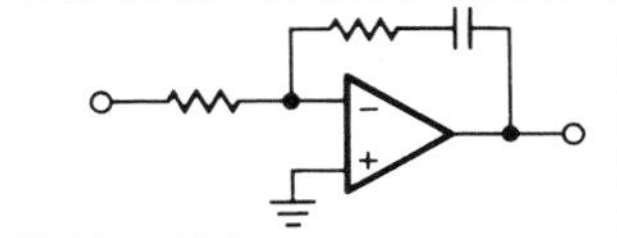	$-\frac{1+sR_FC_F}{sR_iC_F}$
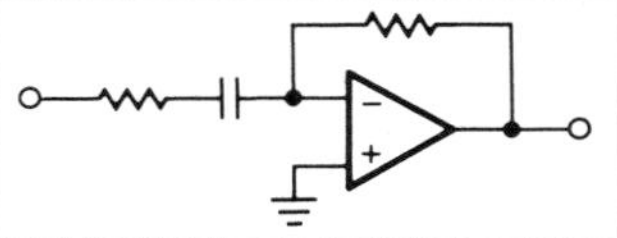	$-\frac{sR_fC_i}{1+sR_iC_i}$
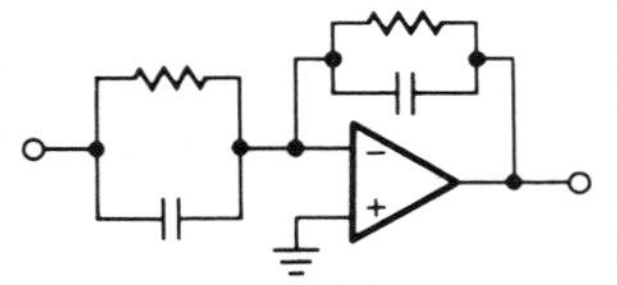	$\frac{-R_F}{R_i}\left(\frac{1+sR_iC_i}{1+sR_FC_F}\right)$
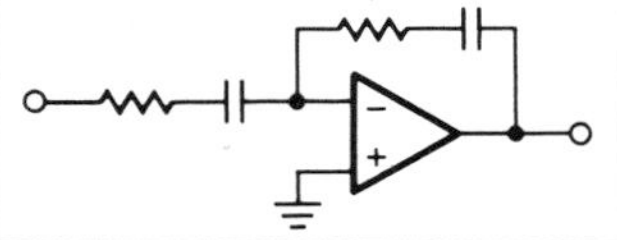	$-\frac{C_i}{C_F}\left(\frac{1+sR_FC_F}{1+sR_iC_i}\right)$
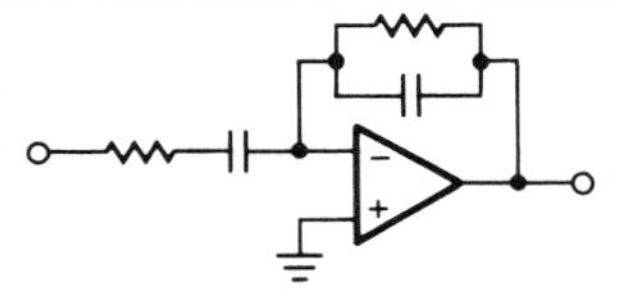	$-\left[\frac{sR_FC_i}{(1+sR_iC_i)(1+sR_FC_F)}\right]$
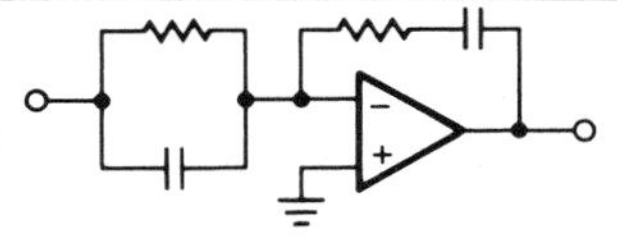	$-\frac{1}{sR_iC_F}[(1+sR_iC_i)(1+sR_FC_F)]$

transfer function shown. Notice that the second circuit example in this table has a capacitor, C_i, shunting R_i. This causes the input impedance to decrease as the input frequency is increased. Therefore, the voltage gain increases (indicated by the zero in the transfer function).

If both impedances are frequency dependent, the resulting change in gain with frequency can be more interesting. Gain can change at a faster rate with the last two circuit examples of Table 4-2 where two poles or two zeros result.

These networks are useful in providing a relatively simple way to obtain frequency-dependent voltage gain in an op amp application circuit. As we will see later in this chapter, the first circuit example in Table 4-2 is also recommended to improve the stability of op amp application circuits.

Obtaining Complex Poles. Passive RC networks provide poles and zeros that lie along the negative real axis of the s-plane. If we have an LC network, we can get pole and zero locations that are not restricted to this negative real axis. For example, if we consider the idealized LC bandpass filter, which has no resistive losses indicated, Figure 4-17, we can solve for a slightly

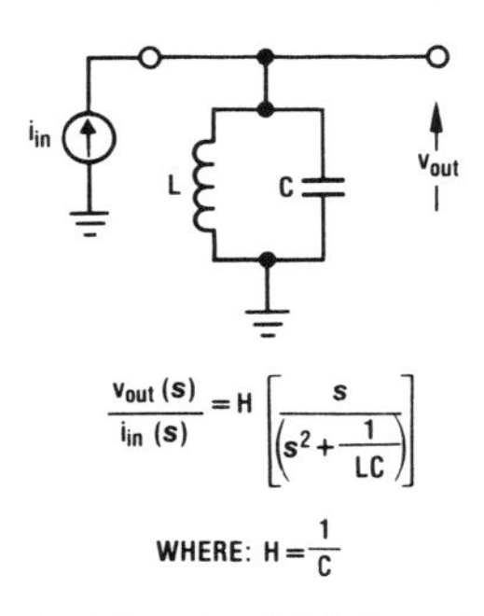

Fig. 4-17. An Idealized LC Bandpass-Filter

different transfer function, *the transfer impedance*. This transfer impedance, which is an output voltage in response to an input current, is given by

$$\frac{V_{OUT}(s)}{I_{IN}(s)} = \frac{(sL)\left(\frac{1}{sC}\right)}{\left(sL + \frac{1}{sC}\right)}$$

which, after simplifying, becomes

$$\frac{V_{OUT}(s)}{I_{IN}(s)} = \frac{1}{C}\,\frac{s}{\left(s^2 + \frac{1}{LC}\right)}$$

This has a zero at zero, because the L is a short circuit to dc, and the pole locations can be found by equating the denominator to zero,

$$\left(s^2 + \frac{1}{LC}\right) = 0$$

or

$$s^2 = -\left(\frac{1}{LC}\right)$$

$$s = \pm\sqrt{-\left(\frac{1}{LC}\right)}$$

and

$$s = \pm j\sqrt{\frac{1}{LC}}$$

These poles now lie on the jω axis, as shown in Figure 4-18. If relatively small losses, indicating high Q, were included, these poles would shift slightly to the left, off of the jω axis.

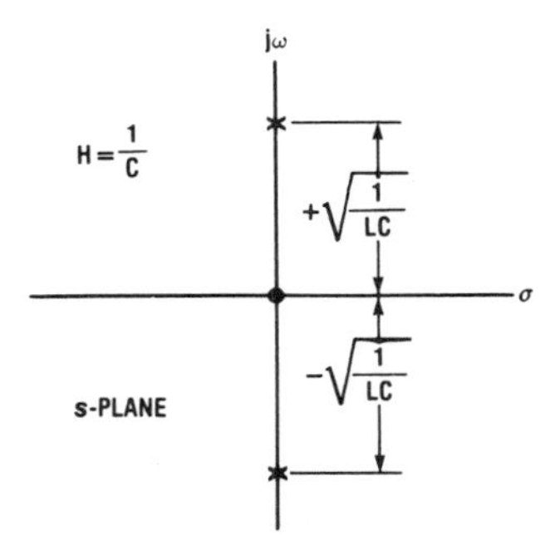

Fig. 4-18. The s-plane Plot of an Idealized LC Bandpass-Filter

If the input excitation current to this idealized LC network had a frequency that was equal to

$$s = +j\sqrt{\frac{1}{LC}}$$

(or $\omega = \sqrt{\frac{1}{LC}}$, the resonant frequency of the LC network), the output voltage would be infinite. Here we have a pole located where we can get to it. In fact, our sinusoidal evaluation frequency goes right over the infinite top, even though our evaluation frequency is restricted to the sinusoidal steady-state, $s = j\omega$. We can even end up "sitting in a zero" where we will find the transfer function, and therefore the output voltage, both going to zero.

In neglecting all loss, we have an infinite Q, so this transfer function provides an idealized bandpass filter response. Although it is not too useful, because it has *no* bandwidth, but infinite Q can't be achieved with real passive components. If infinite Q were attempted with active components, an oscillator would result.

For a practical bandpass circuit, the Q is finite and the poles lie to the left of the jω axis. Now when we make our frequency evaluation by moving our test frequency up the jω axis, we cross away from the peak of this "high mountain." (The pole, which has an elevation of infinity, now lies to the left of our path.) Therefore, we get a bandpass characteristic that does have some bandwidth. When the pole is located farther to the left of the jω axis, we will get a smaller resonant rise in the frequency response of the transfer function. This is the low Q case.

The strange thing is that op amps, with Rs, Cs, and feedback, can also produce poles and zeros that are not restricted to the negative real axis. If these poles should lie near the jω axis, but still be located on the left-hand side *(the left half plane)*, then the op amp is useful to replace inductors in bandpass filters. *This is the basis of RC active filters*, which we will consider in the next chapter. Unfortunately, in the case of a filter application, if these poles should move into the right half plane, *an undesired oscillator results*, because the right half plane is associated with waveforms that increase with time.

Now we will see how it happens that feedback can have such interesting effects on the locations of the poles of the closed-loop response of an op amp application circuit.

4.3 ROOT LOCUS

Predicting the movement, or *locus*, of the poles, or *roots*, of the closed-loop response that result with various amounts of feedback is what *root locus* is all about. Dr. Walter Evans is credited with this pioneering work. It was the subject of his Ph.D. thesis at the University of California, Los Angeles. He also devised a *spirule*, a plastic measuring device that could rapidly take care of the graphical measurements and the calculations that are involved in the root locus design techniques. These calculations are the multiplication and division of vector lengths and the summation of angles as we used in the graphical evaluation of transfer functions.

Assuming a Single-Pole Op Amp

To see why poles move, we can start with a rather simplified example, where the open-loop gain of the op amp is assumed to have only a single pole, ω_P, and therefore can be expressed as

$$A(s) = \frac{A_0}{\left(1 + \dfrac{s}{\omega_P}\right)}$$

where A_0 is the dc value of the open-loop gain.

For further simplicity, we will use a feedback network, β, that is made up of only resistors so there are no poles or zeros associated with β; it has a constant value, β_0, that is independent of frequency. This feedback factor, β_0, will be the only parameter we change to make the poles move. The rest of the parameters are fixed by the op amp that is used.

We return, again, to the expression for closed-loop gain from our basic feedback equation:

$$A_{CL} = \frac{A}{1 + A\beta}$$

So, for the assumption of a single-pole amplifier, this becomes

$$A_{CL}(s) = \frac{A_o}{\left(1 + \frac{s}{\omega_P}\right)} \; \frac{1}{1 + \frac{A_o}{\left(1 + \frac{s}{\omega_P}\right)} \beta_o}$$

or

$$A_{CL}(s) = \frac{A_o}{\left[(1 + A_o\beta_o) + \frac{s}{\omega_P}\right]}$$

and the root of this denominator can be found by equating the denominator to zero,

$$\left[(1 + A_o\beta_o) + \frac{s}{\omega_P}\right] = 0$$

$$s = -(1 + A_o\beta_o)\,\omega_P$$

This is plotted in Figure 4-19 for various values of β_o. For the open-loop case, $\beta_o = 0$, we find the pole is at $-\omega_P$. As we use larger amounts of feedback (as β_o approaches 1) this pole is moved out along the "root locus" to a higher frequency or the pole is *broadbanded*. (This pole migration is another way of *seeing* the increase in the small-signal bandwidth that results with increasing amounts of negative feedback.)

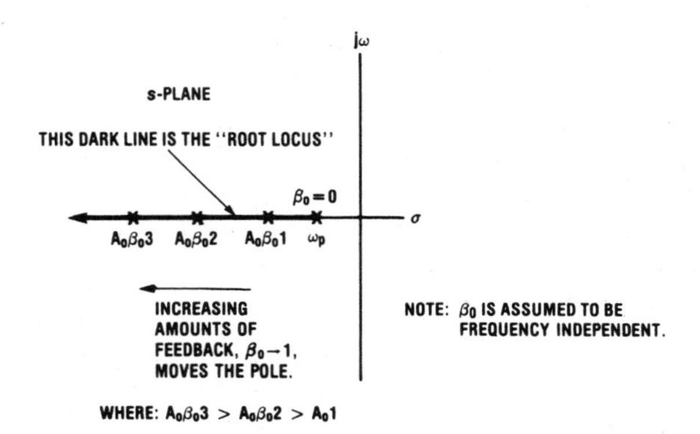

Fig. 4-19. The Root Locus for a Single-Pole Op Amp

This pole, although it moves as the value of β_o is changed, never leaves the negative real axis, so there is no stability problem for any β_o that is used with a single-pole system. With no excess phase, there is never more than a -90° phase lag possible, and 180° is needed for instability.

The technique for graphically finding the root locus, as opposed to the analytical approach, involves finding all of the complex frequency locations (the locus) that produce a "vector" (really a phasor) angle of 180°, where this vector originates on the open-loop pole, ω_P, and terminates on the complex test frequency, s_i. The testing of three complex frequencies s_1, s_2, and s_3 to determine if any of these lie on the root locus is shown in Figure 4-20, where the angles produced by these three different test frequencies are indicated. Only one of these, s_3 of Figure 4-20c, produces the required 180°. It can easily be seen that any complex frequency that lies on the negative real axis and is positioned to the left of ω_P (that is, higher in frequency) will be on the root locus; so the root locus is found to extend to the left of ω_P.

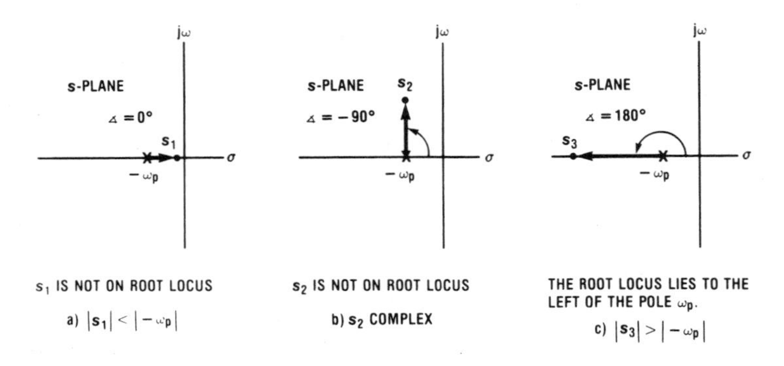

Fig. 4-20. Graphically Locating the Root Locus

With a Two-Pole Op Amp

We can now complicate the analysis somewhat by assuming an op amp with a two-pole open-loop response, or

$$A(s) = \frac{A_o}{1 + a_1 s + a_2 s^2}$$

where the second-order denominator has not been factored to indicate the locations of these open-loop poles. (We know they must both lie on the negative real axis because they result from RC networks within the amplifier.) We will assume that these poles are located simply at ω_{P1} and ω_{P2}, for now. Again, we use a constant β network, β_o, and return to the feedback equation,

$$A_{CL}(s) = \frac{A(s)}{1 + A(s)\beta_o}$$

which now becomes

$$A_{CL}(s) = \frac{A_o}{1 + a_1 s + a_2 s^2}\left[\frac{1}{1 + \left(\dfrac{A_o\beta_o}{1 + a_1 s + a_2 s^2}\right)}\right]$$

or

$$A_{CL}(s) = \frac{A_o}{(1 + A_o\beta_o) + a_1 s + a_2 s^2}$$

To find the roots of this denominator, we equate it to zero, or

$$a_2 s^2 + a_1 s + (1 + A_o\beta_o) = 0$$

We solve this quadratic in s by making use of the standard solution of a quadratic equation to find

$$s = \frac{-a_1 \pm \sqrt{(a_1)^2 - 4(1 + A_o\beta_o) a_2}}{2 a_2}$$

When the expression inside the square root (the radical) is negative (that is, for a large enough β_o), the roots will move off of the negative real axis and become complex.

These poles will move into symmetrical (called *complex conjugate*) positions as shown in Figure 4-21, because of the ± sign in front of the now imaginary term. This locus is still all within the left half plane, so we also have no oscillation problems with this ideal two-pole system. Again, with no excess phase lag the ultimate limit of exactly −180° phase shift occurs at such a high frequency that the magnitude is severely attenuated. This provides a large gain margin.

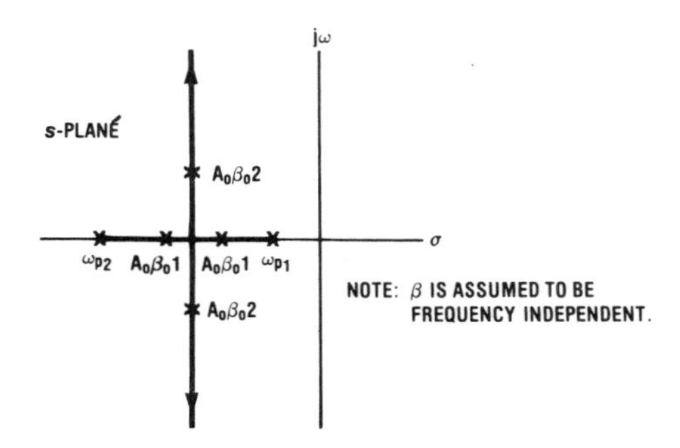

Fig. 4-21. The Root Locus for a Two-Pole Op Amp

The graphical search for this root locus involves both of the pole locations, ω_{P1} and ω_{P2}, of the open-loop response, as shown in Figure 4-22. The angles from the pole locations are shown as negative values. Angles from zeros, if available, would have positive values. In Figure 4-22c, it is seen that the root locus extends between the poles on the real axis before it becomes complex, because the angle in this region is 180° to ω_{P1} and 0° to ω_{P2}. The locus breaks away from the negative real axis at a point midway between the poles ω_{P1} and ω_{P2}.

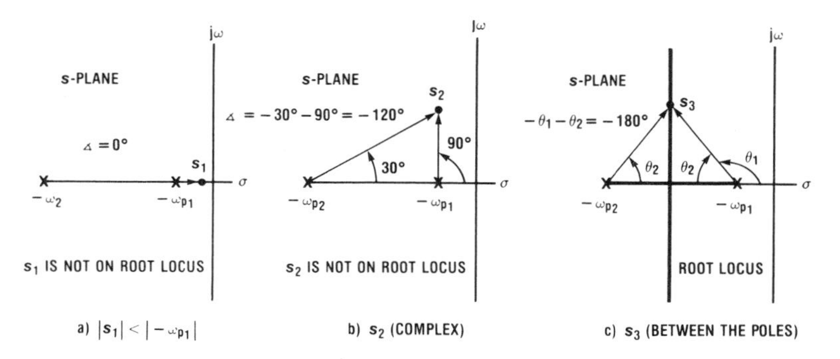

Fig. 4-22. Graphically Locating the Root Locus for a Two-Pole Op Amp

For a Real Op Amp

A real op amp has one dominate pole (located very close to the origin, dc), a second or non-dominate pole, and much excess phase shift that exists because of the presence of many higher frequency poles. This excess phase shift causes this root locus to bend to the right after the poles become complex, Figure 4-23.

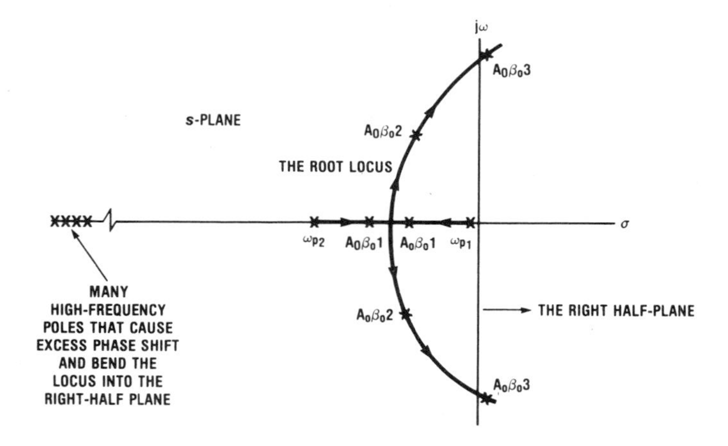

Fig. 4-23. The Root Locus of a Decompensated Op Amp

If the internal frequency compensation capacitor was not designed to be large enough in value to handle 100% feedback (this is the case for the "decompensated" op amps), with the larger amount of feedback indicated at $A_0\beta_03$ the poles will move into the right half plane. This feedback condition will therefore cause oscillations, but the system will be stable if less feedback is used, as shown at $A_0\beta_02$, where the closed-loop gain is 5 or greater, as is required for the decompensated op amps.

The reason for the bending of the root locus toward the right half plane can be seen in Figure 4-24a where, for clarity, only one high frequency pole has been included. (Poles at higher frequencies will each contribute less of an angle to the root locus.) The point, s_1, is included for reference on both constructions shown in this figure to indicate where the root locus previously passed, before the angle contributions from the high frequency pole of Figure 4-24a and the pole and zero of Figure 4-24b were included.

In Figure 4-24b, the added zero is at a lower frequency than the high frequency pole, so this zero contributes a larger angle (of the opposite sign) so that the root locus is actually bent away from the right half plane; therefore, more feedback could be allowed. This indicates the value of zeros to steer the root locus away from an oscillatory condition, or at least away from a poor transient response, in the closed-loop response and allows the use of larger desensitivity factors.

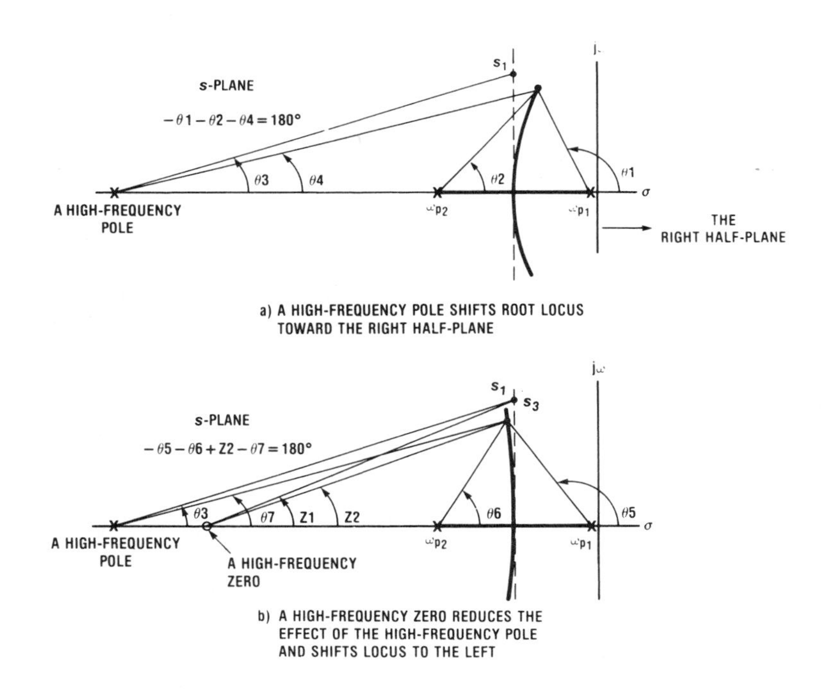

Fig. 4-24. Bending of the Root Locus by High-Frequency Poles and Zeros

There is certainly more rigor associated with root locus than we have presented here. There are also many rules which simplify the plotting of the actual locus. The intent of this discussion is to provide an introduction to this concept and to indicate the basic ideas that are involved.

The power of root locus is that it offers a graphical way to get a *feel* for a feedback control system and to rapidly *see* the effects of adding extra *compensating networks.* These networks contribute poles and zeros at various locations to the transfer function. These added poles and zeros are used to *steer* the locus toward a more favorable closed-loop response. The goal is to allow using large amounts of feedback for improved desensitivity and yet still have an adequate closed-loop response, a good step response without oscillating or excessive ringing.

Final Pole Locations Determine Frequency Response

If we assume that the closed-loop response of an op amp is dominated by a pair of complex poles (this is a useful simplification of the actual case), then the Q of these poles controls the closed-loop step response. The determination of the Q of this dominate pair of complex poles is shown in Figure 4-25a and the resulting low-pass filter response of the closed-loop op amp is shown in Figure 4-25b.

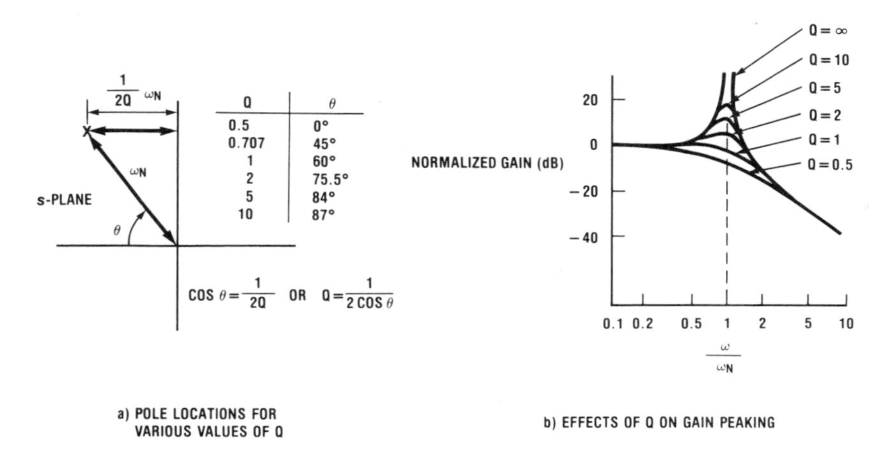

Fig. 4-25. The Positions of the Poles Determines the Q of a Low-Pass Filter

If the Q of these poles is greater than 0.5 (the *critically damped case)* they will lie off of the negative real axis and are called *underdamped.* This will cause ringing in the step response of a closed-loop amplifier. For greater Q ($Q \geq 0.707$), there will be peaking in the frequency response as well. For smaller Q ($Q < 0.5$), these poles will separate on the negative real axis: the *overdamped* case.

Different circuit applications can tolerate different transient performance and therefore different Qs. For no ringing, Q has to be less than 0.5, but an audio application could probably allow a Q equal to 1.

Adequate closed-loop frequency stability is usually indicated by an overshoot of less than 10% in the small-signal output voltage response to step input voltages. (These abrupt inputs can most easily be supplied by a squarewave signal source.) The phase margin can be estimated by measurements of the percent overshoot in the step response as shown in the following table. (Note that overshoot tests are more useful than noting the resonant rise, the peaking, in the high-frequency closed-loop magnitude response.)

For the internally compensated op amps, there is usually no access to the internal frequency compensating capacitor. This on-chip capacitor is designed to provide an acceptable

Phase Margin (degrees)	Overshoot in Step Response (%)	Peaking in Magnitude Response (dB)
75	0.0	—
70	1.4	—
65	4.7	0.0
60	8.8	0.3
55	13.3	0.8

phase margin for the worst case, noninverting voltage follower. The stability of application circuits that use internally compensated op amps can be improved, when necessary, by adding a small-valued capacitor across the feedback resistor from the output to the inverting input. This capacitor will also restrict the high frequency response of the application circuit, a system benefit if this bandwidth is not needed. (It will also eliminate the *feedback pole*. This concept is discussed later in this chapter.)

4.4 AN INTRODUCTION TO BODE PLOT ANALYSIS AND OTHER TECHNIQUES

A Bode plot of an RC low-pass filter is shown in Figure 4-26. Because the voltage transfer function (V_{OUT}/V_{IN}) of this network produces a lagging phase shift, it is called a *lag network*. Lag networks reduce the phase margin of a fed back op amp and are therefore not desired as feedback networks. (Notice that a lagging phase shift is associated with a pole in the left half of the s-plane.)

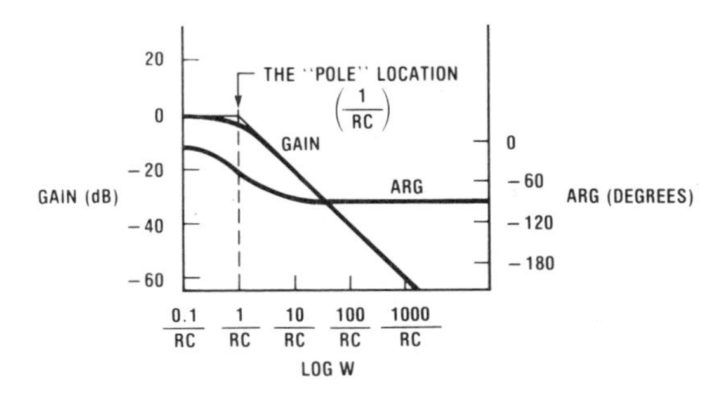

Fig. 4-26. A Bode Plot of a Low-Pass Filter (A Lag Network)

A high-pass RC network, that results with the R and C components interchanged (a zero in the s-plane), produces a leading phase shift in the resulting voltage transfer function. A *lead network* (any network that provides a leading phase shift), used to establish the feedback factor of a fed back op amp, increases the phase margin and therefore aids stability.

An alternative technique for evaluating closed-loop stability, the *Bode stability analysis*, involves graphically combining the Bode plots of A and β to form the $A\beta$ product. (These factors multiply, so the logs of the magnitudes can be added and the phase angles can also be added.) The gain margin and the phase margin of this final $A\beta$ plot provide the stability indicators. Additional frequency compensation networks can then be graphically included to evaluate the improvements that may be gained. (This is made easier by the use of standard response templates.) Historically, the root locus technique was developed as an improvement on this earlier Bode technique, but both are still being used today.

A special analog computer was also built and sold a number of years ago that consisted of a large sheet of uniform resistance paper (Western Union *Teledeltos* paper) that served as an analog of the s-plane. Moveable metallic probes would enter current at each pole location or remove current at each zero location from this resistance paper. A high input impedance voltmeter could then be moved up the "$j\omega$ axis"—for example, to obtain a voltage analog of the frequency response of a transfer function. This special s-plane analog computer was also useful for analyzing complex multiple-loop control systems. Digital computer solutions are now replacing these earlier stability analysis techniques for the design of analog control systems.

Op amp application circuits can usually be stabilized by a general knowledge of the stability problem and by spending some time in the lab doing pulse or squarewave testing to evaluate the results of a few stabilization techniques. We will now consider some of these practical factors.

4.5 IF IT OSCILLATES, THE FREQUENCY INDICATES WHY

The important thing to remember about op amp applications is that oscillations can usually be brought under control; *have faith!* Oscillations might be expected the first time a new op amp circuit idea is tried out. This serves to indicate that things are properly connected, very much alive and ready to be tamed.

Before the feedback loop is suspected, first insure that other areas have been considered. For example, it is important that the ground loops are under control. That is, insure that there is no coupling from output to input because of a common impedance in the grounding of the circuit components. Grounding problems can be avoided by a careful layout plan prior to building the circuit. If improper grounding is expected, the ground point of suspected components can be temporarily relocated in a circuit breadboard and the effects on the stability of the amplifier can be observed.

The power supplies for the op amp should also be properly bypassed; locate the bypass capacitors close to the op amp and use an adequate capacitor value. Many times, disc-ceramic bypass capacitors, of values 0.1 μF to 0.01 μF, will create high frequency (a few MHz) resonant parasitic LC networks in combination with the parasitic inductance of long power supply leads. Use of 1 μF to 10 μF tantalum bypass capacitors will prevent this by greatly reducing this resonant frequency to a value where there is poor Q in the parasitic inductors that are created by the power supply leads. A 3Ω to 10Ω resistor in series with a disc-ceramic capacitor can also be used to spoil the Q of a parasitic resonant circuit. Sometimes these RC networks have to be used when aluminum, instead of tantalum, electrolytics are used. This involves the use of two components and doesn't provide the local energy storage benefits of a larger valued bypass capacitor.

If the instability is due to the feedback loop, then temporarily taking a larger closed-loop gain (say, by a factor of 4 to 10) should increase the stability of the circuit. If this increase in closed-loop gain does not at least decrease the frequency of oscillation, then the instability problem is not due to the negative feedback loop. Look elsewhere to find the cause.

In chasing down feedback stability problems, it is very important to insure that the fundamental component of an existing oscillation is in the vicinity of the unity-gain frequency of the open-loop op amp. When this is the case, you most likely have a feedback-loop stability problem.

As obvious as it sounds, be sure that the frequency of the "instability" is not 60 Hz or 120 Hz. This comes from *hum* pick up from the power supply or stray power line coupling to exposed components. Other strange valued "oscillation frequencies" have been traced to local radio stations or even radiations from the guy on the next bench. (Is the circuit *quiet* during lunch time or after 5 p.m.?)

Integrated circuits have been found to produce 50 MHz to 200 MHz signals as a result of a local instability within the output stage. This is more common with audio amplifiers and voltage regulators, but should be kept in mind when working with high frequency op amps.

A final possibility is parasitic coupling between the circuit components, especially from those components at the output to those components at the noninverting input. This creates parasitic positive feedback. To reduce this problem, it is important to keep the leads from the op amp inputs as short as possible by locating the components of the β network close to the input of the op amp. Also, keep the output components well away from the input components. High frequency op amps also require the use of careful breadboarding techniques. Use a wide copper sheet of an unetched PC board as a ground plane and don't use universal breadboards that don't have a good ground plane. Move some of the components and see if the output voltage waveform cleans up or is significantly affected, or try using grounded, conductive shields between components that are suspected of *talking to each other.*

Remember, it's most likely NOT the feedback loop if the oscillations are not near the unity-gain frequency of the op amp. Also, if you simultaneously have multiple problems, you have to learn to leave in place any fixes that seem to help clean up the output voltage waveform. When the amplifier is finally tamed, you can then see which of the fixes are really important.

Many times op amp users unknowingly aggravate the stability of their op amp application circuit. We will now consider one problem that is often overlooked; the effects of a capacitance load at the output of the op amp.

4.6 EFFECTS OF CAPACITANCE LOADING AT THE OUTPUT

One of the worst things about capacitance loads at the output of an op amp is that *they can exist when there is no apparent capacitor.* This usually happens when an op amp is used to supply a signal through a long length of coaxial cable that is serving as a shielded line to a remote load.

Coaxial Cables Can Be Capacitors

A coaxial cable *looks* like a capacitance (and not its characteristic impedance: *a resistance*) if the length of the cable is less than one-fortieth of the wavelength in the cable at the frequency of interest, f. This length, l, is given by

$$l \le \frac{1}{40}\left(\frac{Kc}{f}\right) \text{ meters}$$

where K is a propagation constant that is sometimes called the *velocity factor* (0.66 for coaxial cable) and c is the velocity of light (3.00×10^8 m/s).

As an example, if f = 10 kHz,

$$l \le \frac{1}{40}\left[\frac{(0.66)\,(3 \times 10^8)}{10^4}\right] = 495 \text{ meters (1624 feet)}$$

or, cables less than this length will appear capacitive at the rate of 95 pF/meter (29 pF/foot) for RG-58A/U, a commonly used coaxial cable.

Smaller values of capacitance per foot can be obtained with shielded wire rather than as compared with the low characteristic impedance rf cables. Also, good results can generally be obtained with shielded two-wire cable for the input leads to an op amp. When considering capacitive loading on the output of an op amp, remember that the capacitance of a no. 22 insulated wire in a bundle of no. 22 wires can be 40 pF/ft.

One of the IC op amps, the LF356, can drive large values of capacitance loads; up to 0.01 μF, even in the inverting, unity-gain configuration. Most op amps have difficulty for one of two reasons: feedback-loop instability or local output stage instability. Let's consider the reason for feedback-loop instability.

Load Capacitance Causes Phase Lag

Op amps have an open-loop output resistance, R_o, that can be modeled as shown in Figure 4-27. The presence of C_{LOAD} provides an R_o C_{LOAD} low-pass filter that introduces phase lag in the output voltage. This increased phase shift reduces the phase margin of a feedback circuit.

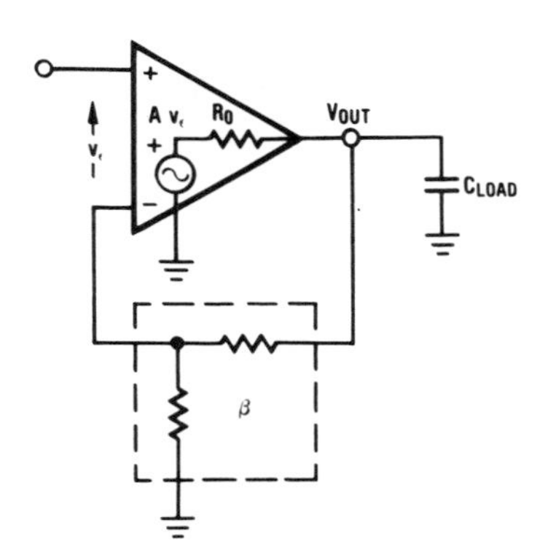

Fig. 4-27. A Model to Evaluate the Effects of Capacitive Loads

The phase lag, ϕ, of an RC low-pass filter is given by

$$\phi = \tan^{-1}\left(\frac{f}{f_c}\right) \text{ degrees}$$

The surprising thing about phase lag is that a significant phase shift can be provided at frequencies much lower than the corner frequency, where the magnitude response of the circuit is still quite close to unity. This characteristic can be seen by looking at the following data.

f/f_p	Response Magnitude	Phase Lag (in degrees)
0.01	1.0000	0.6
0.03	0.9996	1.7
0.10	0.9950	5.7
0.30	0.9580	16.7
0.50	0.8900	26.6
1.00	0.7070	45.0

As a typical example, with $R_o = 100\ \Omega$ and $C_{LOAD} = 1000$ pF, we would find

$$f_c = \frac{1}{2\pi R_o C_{LOAD}} = \frac{1}{2\pi (10^2)(10^3)(10^{-12})}$$

$$f_c = 1.6 \text{ MHz}$$

At the critical frequency, f_u, of the op amp (assumed 1 MHz) this would provide a phase lag of

$$\phi = \tan^{-1}\left(\frac{1 \text{ MHz}}{1.6 \text{ MHz}}\right)$$

$$\phi = 32 \text{ degrees}$$

This is a significant reduction in phase margin and the transient response will therefore suffer greatly, or the amplifier may be unstable when this large capacitive load is in place.

Output Stage Instability

If the loading capacitor at the output of the op amp causes an instability at a frequency considerably in excess of f_u, then the output stage is undergoing a localized instability. This can be cured by the use of a 1 to 20 μF local power supply bypass capacitor or by introducing losses by using a resistor of 3 to 10Ω in series with a disc-ceramic bypass capacitor to reduce the Q *(Q spoilers)* of the VHF parasitic tank circuit that exists in the power supply or maybe the output line. These RC Q-spoilers should be used directly from the power supply pins or the output pin to ground.

Isolating a Load Capacitance

Large values of capacitance can be isolated from the amplifier by making use of the circuit shown in Figure 4-28a, where the LF356 is now driving a load capacitance of 0.5 μF! Notice that a small-valued capacitor, 20 pF, is used to feedback directly from the output of the op amp to the inverting input and a small-valued resistor (10Ω, although larger values may be required for other op amps) isolates C_{LOAD} from the output of the op amp. In general, the values to use for the isolating resistor and the feedback capacitor are determined experimentally while observing the small-signal output voltage response to an input squarewave.

The main feedback (at dc and low frequencies) is still taken from the voltage that exists across R_L, so the 10Ω resistor is *in-the-loop* as shown in Figure 4-28b and does not significantly degrade the closed-loop output impedance. The 20 pF capacitor can be thought of as directly coupling the output voltage back to the inverting input (feeding forward) to prevent the appearance of the phase-lagging signal that would otherwise result at this node at high frequencies because of C_{LOAD}. *This trick, the small-valued direct feedback capacitor, is often useful for many other applications and should be kept in mind when stabilizing application circuits.*

Another parasitic capacitor lurks at the inverting input of the op amp. We will now consider the effects of the *feedback pole* that it produces.

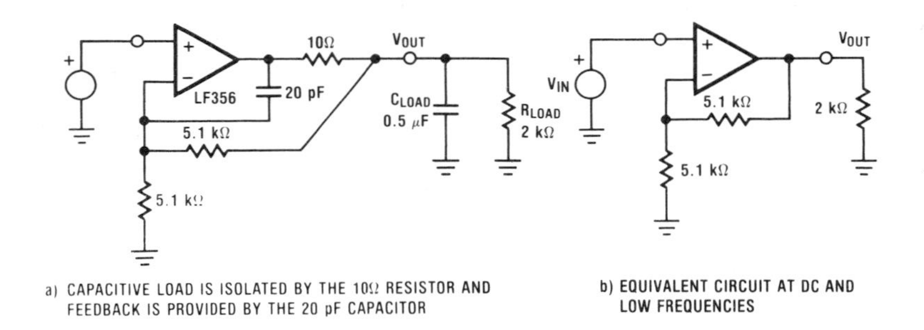

a) CAPACITIVE LOAD IS ISOLATED BY THE 10Ω RESISTOR AND FEEDBACK IS PROVIDED BY THE 20 pF CAPACITOR

b) EQUIVALENT CIRCUIT AT DC AND LOW FREQUENCIES

Fig. 4-28. Isolating a Large Load Capacitance

4.7 THE EFFECT OF THE FEEDBACK POLE

A *feedback pole* exists with any op amp, but the low values of feedback resistors (≅10 kΩ) that are typically used with bipolar op amps and the relatively low bandwidths of these op amps has prevented many designers from ever seeing this problem. Therefore, the feedback pole is often considered to be unique to the Bi-FET op amps simply because larger resistor values are typically used.

We have just seen how an output loading capacitance can introduce phase lag because of the open-loop output resistance of an op amp. Another similar RC low-pass phase lag circuit exists when a feedback network, that is composed of only resistors, is tied to the inverting input of the op amp. A stray capacitance of approximately 3 to 5 pF to ground *typically* exists at this input and is the culprit that creates the *feedback pole*.

This circuit is shown in Figure 4-29. It is the parallel combination of the feedback resistors that must be considered. For example, if these resistors were R1 = 100 kΩ and R2 = 1 kΩ, the resulting corner frequency, f_{fp}, would be

$$f_{fp} = \frac{1}{2\pi R_{eq} C_{IN}}$$

$$f_{fp} = \frac{1}{2\pi (10^3)(3 \times 10^{-12})}$$

$$f_{fp} = 53 \text{ MHz}$$

and this would create only a small phase lag at the much lower frequency (f_u) of the op amp.

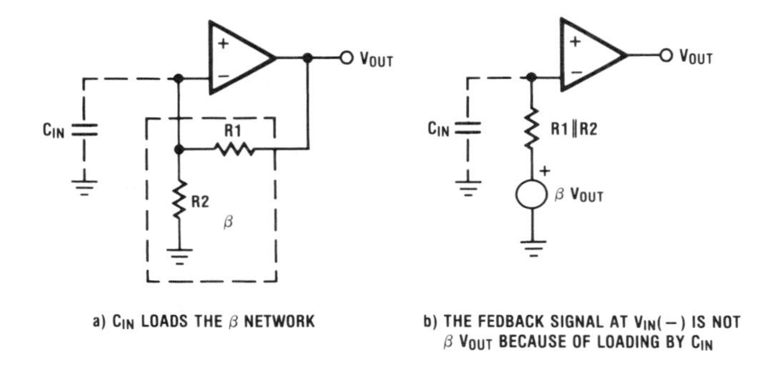

a) C_{IN} LOADS THE β NETWORK

b) THE FEDBACK SIGNAL AT $V_{IN}(-)$ IS NOT β V_{OUT} BECAUSE OF LOADING BY C_{IN}

Fig. 4-29. The Feedback Pole

If we had a Bi-FET op amp, and therefore used larger resistance values, the equivalent resistance of the β network would be increased and the larger f_u of these op amps would also aggravate this problem. (The results of the previous section also apply here to determine the detrimental phase lag of the feedback pole.)

One solution to this problem is to add a capacitor, C_f, across R1 such that

$$R2\ C_{IN} \leq R1\ C_f$$

The value that is used for C_f can be made larger (as the above equation indicates) to reduce the bandwidth, if desired. It can also be made smaller, without causing instability. The condition of equality provides a good starting value to use. A value somewhere between 2 pF and 50 pF is typical. To regain some response-time advantages, an additional small-valued capacitor can be used across the input resistor. Proper component selection is best made experimentally, while observing the squarewave response of the op amp.

It is recommended that a small-valued capacitor always be used across the feedback resistor. This eliminates stability problems, prevents peaking of the response, and limits the bandwidth of the circuit, reducing not only noise, but the susceptibility to interference. Changes in stray capacitances between breadboards and production PC boards are also less likely to cause stability problems.

4.8 SOME PRACTICAL TRICKS

The real world of electronic circuits is plagued with many surprises and many parasitics. High frequency circuit performance depends on the lead lengths of circuit components. Long leads to resistors and capacitors not only add the inductance of these leads in series with the components; excessive lead lengths also tend to create stray pickup and coupling problems.

An *ideal ground plane* is also often assumed to exist, which is as nebulous as the *frictionless plane* of the physicist or the *ideal marriage* that is discussed in the humanities. The simple sounding act of obtaining a valid voltage measurement in a circuit is complicated by the capacitive loading that is introduced by the measuring device. We will now consider how these problems complicate the design and evaluation of a feedback amplifier.

Taming an Oscillating Amplifier

When a new op amp circuit idea is first breadboarded, it can happen that it "wakes up" oscillating violently. The series RC damper or Q spoiler (10Ω to 10 kΩ and 50 to 10,000 pF), mentioned earlier in this chapter, can be used from the summing point to ground, and sometimes from the output of the op amp to ground, to bring an application circuit under control while the cause of the instability is found. These RC dampers can then be momentarily disconnected while other compensation circuits are tested. If the damper components are carefully chosen—for example, use as small a value for C as possible to reduce the low frequency noise gain—these dampers can also be useful compensating elements.

Problems with Measuring A, the Open-Loop Gain

From a theoretical standpoint, if we can mathematically describe both A and β, we can analytically determine the stability margins of a feedback amplifier. The β network is generally made up of passive components that we select and therefore a mathematical description of β can be determined by calculating the transfer function of this β network. We then need to describe or find a way to measure A, the open-loop voltage gain of the op amp.

It is impossible to hold the output voltage of a high gain open-loop op amp within its linear operating range while we obtain measurements without the use of feedback. The problems of measuring the open-loop voltage gain and phase shift of an op amp therefore relate to: (1) insuring that the feedback network that is used to keep the op amp within its linear range is not affecting the ac measurements and, (2) measuring the small output voltages that result at high frequencies. The input test signal must be kept small to prevent slew-rate problems.

If it is desired to determine the magnitude and phase shift of A over a broad range of frequencies, many different circuit layouts with different values and types of bypass capacitors must often be used. To determine stability, measurements can be restricted to a relatively narrow band of frequencies that are near the unity-gain frequency of the op amp. The problem with this approach is that there are then no low frequency data points available to compare these high frequency measurements with. Large errors can result from improper breadboarding techniques or the use of bypass capacitors that have either large-enough parasitic series R or L associated with their construction to affect the measurements.

It would certainly seem that some other measurement technique should be considered.

Dr. R. D. Middlebrook of Cal Tech has suggested a solution (class notes from EE114) to this measurement problem as shown in Figure 4-30. Notice, in Figure 4-30b, that the test signal is transformer coupled and no bypass capacitors are used to open the loop. The voltage at the inverting input of the op amp, V_{IN}, and the voltage out of the feedback network, $A\beta\ V_{IN}$, are free to take on values as required by the op amp. The ratio of these resulting voltages provides the magnitude of $A\beta$. (The phase shift of $A\beta$ is measured as the phase of the $A\beta\ V_{IN}$ signal using the V_{IN} signal as the phase reference.) The test voltage at the secondary of the transformer, V_{TEST}, should be adjusted in magnitude to provide a small signal level at the $A\beta\ V_{IN}$ measurement point. We are most interested in test frequencies that provide $|A\beta| \cong 1$, so this implies that the magnitudes of V_{IN} and $A\beta\ V_{IN}$ are not very far apart. This circuit approach is recommended because it solves the many practical problems of directly measuring A.

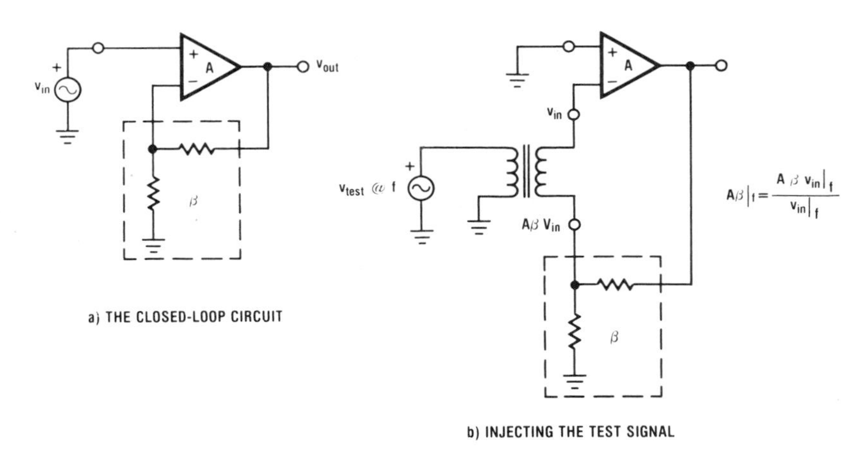

Fig. 4-30. Middlebrook's Technique to Directly Measure $A\beta$

In most of the application circuits, stability margins are more rapidly inferred by square-wave tests on the closed-loop amplifier. This testing technique can be modified to simultaneously include dynamic testing, as suggested by R. A. Pease, to uncover some unusual stability problems that are often observed with op amps. We will now consider this method of stability testing.

Dynamic Stability Testing

The open-loop gain and phase characteristics of most op amps depend on ambient temperature, output current magnitude (and direction), the output voltage level, and the stray capacitances that are associated with the IC socket and/or the PC board layout. For these reasons, guaranteeing the stability of an application circuit (and especially a tricky one) requires that a number of tests be made.

To insure that an application circuit has no hidden problems, dynamic stability testing with the output voltage swept over the complete dynamic range should be done, as shown in Figure 4-31. During this testing R_L, C_L, a sample of op amps, and the ambient temperature should all be systematically varied over their expected ranges to determine the effects on stability. This squarewave testing during dynamic V_{OUT} sweeping can also be used to evaluate various frequency compensation alternatives.

With all of the basics of the op amp behind us, let's now take a look at some of the ways op amps can be used to solve circuit problems for the system designers.

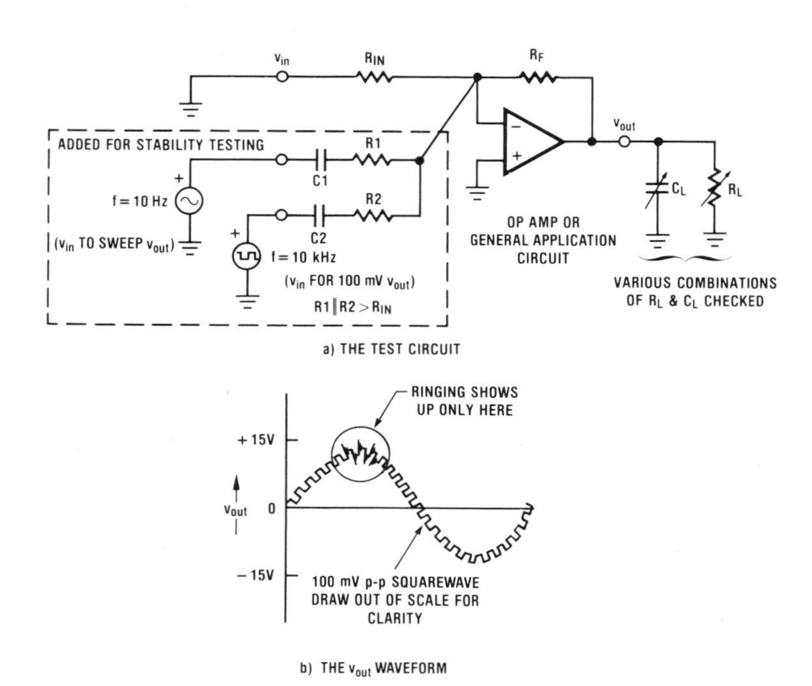

Fig. 4-31. Sweeping V_{OUT} While Testing Stability

CHAPTER V

Some of the Key Op Amp Application Circuits

There is no end to the possible ways of using an op amp. Many cleverly designed circuits have been published and many more probably exist as trade secrets. A wide range of op amp circuits have also been used to solve some very specialized application problems and therefore have little general use.

If you study a large number of op amp application circuits you can get a feel for the way the game is played. Linear application circuits are most useful to stimulate thinking; it rarely happens that you can use a circuit just as it is shown. The trick in linear design is to have a big bag of op amp application tricks, keep it stocked by *devouring* all of the application circuits you can and then, starting with these, come up with the unusual circuit that is needed for the job at hand. There is still room for original contributions because the electronic art is very dynamic. Continual changes in the costs that are associated with a design is the way of life. A design approach that was once considered costly and therefore, not widely used, will appear in tomorrow's electronic toy.

The purpose of this chapter is to present a collection of fundamental circuits (as a background) and to add a few that are not as common, to stimulate your thinking. In all the circuits presented there are many changes that can be made. Changing the power supply voltages that are used or the dc biasing levels, including input current matching resistors or not, varying the quality of the external components, changing the operating bandwidths, and using different op amp types. The proper selection of all of these components comes from the particular needs of your application circuit. A final application circuit usually results that uses bits and pieces of other circuits so it can accomplish something quite different, because this is what designing with op amps is all about. (We have omitted some application circuits because of the wide variety of specialty linear ICs that are available today.)

In the schematics that are given, a voltage source is used to represent a signal source with a zero source resistance. When a signal source is loaded by an input resistor that connects to the virtual ground of an op amp, the value of this input resistor should be reduced by an amount equal to any source resistance that may exist with the actual signal source in your application. If the input signal is provided by the output of a preceding fed back op amp, then the source resistance in this case can be assumed to be equal to zero ohms.

In looking at a new application circuit, keep in mind three simplifying assumptions:

1. $V_{IN}(-)$ follows $V_{IN}(+)$, therefore find the $V_{IN}(+)$ voltage first
2. No current enters the inputs to the op amp

3. Look for the feedback path:
   a) None
   b) Positive
   c) Negative
   d) Both

An additional aid to quick application circuit analysis, that also saves fumbling with your calculator, is to develop the ability to mentally impedance scale an RC network so you can estimate high frequency roll-offs or the low frequency capabilities of capacitor coupling. The key is to remember, as the mental point of reference, that 1 pF and 1 kΩ corner at 160 MHz. (160 is easier to work with than the actual value of 159 and introduces a very small error.) Raising either the capacitance or the resistance values *lowers* the corner frequency and vice versa. For example, it is now easy to mentally calculate that 100 pF and 10 kΩ corner at 160 kHz. (Sometimes a scratch pad is handy, but remember to round-off R and C values if you only want an estimate.) For lower frequency ranges, a useful reference is that 1 μF and 1 kΩ provide a 160 Hz corner frequency.

5.1 ±15 V_{DC} POWER SUPPLIES VERSUS A SINGLE +5 V_{DC} SUPPLY

Electronic systems can be built for less cost if only a single power supply voltage is used. It is definitely a design benefit for linear circuits to have both polarities of supply voltages available and many sophisticated linear system designers will insist on this. A wide range of linear systems have been built, however, using the single + 5 V_{DC} supply voltage standard of the digital world. This has been made possible by the appearance of linear ICs that will work within this low voltage power supply constraint. The power supply noise contamination that results from the high-speed switching of digital circuits should be isolated from the linear circuits. A separate power supply voltage regulator IC is recommended for the linear circuitry.

In considering IC op amps for a single 5V supply application, the biasing circuitry must be operational, the output voltage swing loss not too great, and the input common-mode voltage range should, ideally, include ground. Some of the linear ICs that will work with a single 5 V_{DC} power supply are listed in Table 5-1.

The largest benefits of single supply linear circuits are that they can easily operate off of the single battery of an automobile and they are also capable of operating within a digital system. The new electronic control systems for automobiles represent a large market for single supply linear circuits. Many low cost industrial control systems have also followed this trend. To add to the low voltage ICs, both analog-to-digital converters and digital-to-analog converters are available that operate from a single 5V supply.

In some single supply linear systems, a *pseudoground* (a voltage level of + $V_{CC}/2$) is used. This is harder to keep at a low impedance and free of noise than the more conventional *analog ground*. Additionally, the restricted voltage swings allowed at the inputs and obtained at the outputs of the op amps do not usually allow the signal to cover the complete 5V range. This, in addition to the already low value of supply voltage, greatly reduces the signal-to-noise ratio that can be obtained in these systems. There are also not many linear ICs that will operate with a 5V supply and it now appears that by the time the linear ICs respond to this 5V requirement, the digital ICs may be changing to a new 2 or 3V power supply voltage.

TABLE 5-1. Linear ICs That Operate With a $+5\ V_{DC}$ Power Supply

Part #	Description
LM3900	Quad Current-Differencing Amp
LM324	Quad Single-Supply Op Amp
LM358	Dual Single-Supply Op Amp
LM308	Single Op Amp
LM312	Single Op Amp
LM10	Single Op Amp and Voltage Reference
LM11	Single Op Amp
LM4250	Programmable Op Amp
LM346	Quad Programmable Op Amp
LM392	Mixed Dual: One Op Amp and One Comparator
LM339	Quad Single-Supply Comparator
LM393	Dual Single-Supply Comparator
LM311	Single Comparator
LM319	Dual Comparator
LM331	Voltage-to-Frequency Converter
LM555	Single Timer
LM556	Dual Timer
LM322	Precision Timer
LM3080	Operational g_m Amplifier
LM13600	Dual Operational g_m Amplifier with Buffers
LM103	1.8 to 4.2V Low Current Zeners
LM313	1.2V Voltage Reference
LM336-2.5	2.5V Voltage Reference
LM385-1.2 (2.5)	Low Current 1.2V and 2.5V Voltage References
LM334	3-Terminal Adjustable Current Source
LM3909	LED Flasher/Oscillator
LM3914	Dot/Bar Display Driver

Better dynamic range can be obtained by operating linear circuits with a single 12-volt or a 15-volt power supply. The higher operating voltage range of the older metal gate CMOS logic circuits (15 V_{DC}) allows CMOS logic to also be used in these high voltage single supply systems. (Metal gate CMOS logic circuits also operate at speeds slower than T^2L or Schottky logic. This is beneficial because it reduces the digital noise and therefore makes CMOS a recommended logic family to be used with, or near, linear circuits).

5.2 WORKING WITH STANDARD RESISTOR VALUES

Problems are generally encountered when attempting to select the best fit resistor values for an application circuit out of the standard resistor values that are provided. In some designs we have to select only two resistors (as in a voltage divider), but sometimes we need a

TABLE 5-2. Resistor Ratios Obtainable with 5% Resistor Values

Standard 5% Multipliers	All Possible Resistor Ratios Referenced to: (Neglecting Powers of Ten)																						
	1.1	1.2	1.3	1.5	1.6	1.8	2.0	2.2	2.4	2.7	3.0	3.3	3.6	3.9	4.3	4.7	5.1	5.6	6.2	6.8	7.5	8.2	9.1
1.0	.909	.833	.769	.667	.625	.556	.500	.455	.417	.370	.333	.303	.278	.256	.233	.213	.196	.179	.161	.147	.133	.122	.110
1.1	1.0	.917	.846	.733	.688	.611	.550	.500	.458	.407	.367	.333	.306	.282	.256	.234	.216	.196	.177	.162	.147	.134	.121
1.2	1.09	1.0	.923	.800	.750	.667	.600	.545	.500	.444	.400	.364	.333	.308	.279	.255	.235	.214	.194	.176	.160	.146	.132
1.3	1.18	1.08	1.0	.867	.813	.722	.650	.591	.542	.481	.433	.394	.361	.333	.302	.277	.255	.232	.210	.191	.173	.159	.143
1.5	1.36	1.25	1.15	1.0	.938	.833	.750	.682	.625	.556	.500	.455	.417	.385	.349	.319	.294	.268	.242	.221	.200	.183	.165
1.6	1.45	1.33	1.23	1.07	1.0	.889	.800	.727	.667	.593	.533	.485	.444	.410	.372	.340	.314	.286	.258	.235	.213	.195	.176
1.8	1.64	1.50	1.38	1.20	1.13	1.0	.900	.818	.750	.667	.600	.545	.500	.462	.419	.383	.353	.321	.290	.265	.240	.220	.198
2.0	1.82	1.67	1.54	1.33	1.25	1.11	1.0	.909	.833	.741	.667	.606	.556	.513	.465	.426	.392	.357	.323	.294	.267	.244	.220
2.2	2.00	1.83	1.69	1.47	1.38	1.22	1.10	1.0	.917	.815	.733	.667	.611	.564	.512	.468	.431	.393	.355	.324	.293	.268	.242
2.4	2.18	2.00	1.85	1.60	1.50	1.33	1.20	1.09	1.0	.889	.800	.727	.667	.615	.558	.511	.471	.429	.387	.353	.320	.293	.264
2.7	2.45	2.25	2.08	1.80	1.69	1.50	1.35	1.23	1.13	1.0	.900	.818	.750	.692	.628	.574	.529	.482	.435	.397	.360	.329	.297
3.0	2.73	2.50	2.31	2.00	1.88	1.67	1.50	1.36	1.25	1.11	1.0	.909	.833	.769	.698	.638	.588	.536	.484	.441	.400	.366	.330
3.3	3.00	2.75	2.54	2.20	2.06	1.83	1.65	1.50	1.38	1.22	1.10	1.0	.917	.846	.767	.702	.647	.589	.532	.485	.440	.402	.363
3.6	3.27	3.00	2.77	2.40	2.25	2.00	1.80	1.64	1.50	1.33	1.20	1.09	1.0	.923	.837	.766	.706	.643	.581	.529	.480	.439	.396
3.9	3.55	3.25	3.00	2.60	2.44	2.17	1.95	1.77	1.63	1.44	1.30	1.18	1.08	1.0	.907	.830	.765	.696	.629	.574	.520	.476	.429
4.3	3.91	3.58	3.31	2.87	2.69	2.39	2.15	1.95	1.79	1.59	1.43	1.30	1.19	1.10	1.0	.915	.843	.768	.694	.632	.573	.524	.473
4.7	4.27	3.92	3.62	3.13	2.94	2.61	2.35	2.14	1.96	1.74	1.57	1.42	1.31	1.21	1.09	1.0	.922	.839	.758	.691	.627	.573	.516
5.1	4.64	4.25	3.92	3.40	3.19	2.83	2.55	2.32	2.13	1.89	1.70	1.55	1.42	1.31	1.19	1.09	1.0	.911	.823	.750	.680	.622	.560
5.6	5.09	4.67	4.31	3.73	3.50	3.11	2.80	2.55	2.33	2.07	1.87	1.70	1.56	1.44	1.30	1.19	1.10	1.0	.903	.824	.747	.683	.615
6.2	5.64	5.17	4.77	4.13	3.88	3.44	3.10	2.82	2.58	2.30	2.07	1.88	1.72	1.59	1.44	1.32	1.22	1.11	1.0	.912	.827	.756	.681
6.8	6.18	5.67	5.23	4.53	4.25	3.78	3.40	3.09	2.83	2.52	2.27	2.06	1.89	1.74	1.58	1.45	1.33	1.21	1.10	1.0	.907	.829	.747
7.5	6.82	6.25	5.77	5.00	4.69	4.17	3.75	3.41	3.13	2.78	2.50	2.27	2.08	1.92	1.74	1.60	1.47	1.34	1.21	1.10	1.0	.915	.824
8.2	7.45	6.83	6.31	5.47	5.13	4.56	4.10	3.73	3.42	3.04	2.73	2.48	2.28	2.10	1.91	1.74	1.61	1.46	1.32	1.21	1.09	1.0	.901
9.1	8.27	7.58	7.00	6.07	5.69	5.06	4.55	4.14	3.79	3.37	3.03	2.76	2.53	2.33	2.12	1.94	1.78	1.63	1.47	1.34	1.21	1.11	1.0

few more. If we neglect the powers of ten, there are twenty-four 5% multipliers and ninety-six 1% multipliers.

To simplify the best 5% resistor selection problem, we have tabulated all of the resistor ratios that are possible with the 5% multipliers in Table 5-2. Each column in this table uses the successive standard 5% multipliers as the normalizing factor and these resulting columns show all of the resistor ratios that can be obtained when a set of resistors is normalized to each one of these standard 5% column multipliers. (This chart might have been clearer if each of these columns started at 1.0 and then ran down, vertically, to the highest ratio for that column. Unfortunately, this would have made a much larger chart.)

To make use of this table, simply normalize all resistor values to the smallest value in the group of the related resistors you need. Glance down each column and find that column that most closely approximates the single ratio or the multiple resistor ratios you need. Each desired resistor ratio, plus the column heading multiplier, supplies the set of resistor multipliers to use. Finally, choose the appropriate decimal scaling factor for each of these resistor multipliers to correspond to the general impedance level you want for the resistor network. This completes the selection of the best fit set of 5% resistors to optimumly center the distributions of these resistor ratios in the production of a circuit.

A problem would exist if we attempted to make a similar tabulation for the 1% resistor multipliers. This chart would be 96 × 96 and would contain 9216 ratios. Selecting 1% resistors is best handled by choosing the closest 1% resistor values to match the calculated values of the resistors you need. If this doesn't work out as well as you'd like, try scaling all of the resistor values by multiplying by various constant factors to see if this produces a better fit. In general, the large number of 1% multipliers that is available should ease this selection problem.

An example of the use of Table 5-2 to select the best fit 5% resistor values for a simple voltage divider is indicated in Figure 5-1 where we have arbitrarily chosen a divider that will supply 0.34V from an existing 2.5V reference voltage. As indicated on this figure, we need a resistor ratio of 635 (neglecting the powers of ten). A study of Table 5-2 indicates that the standard 5% multipliers of 3 and 4.7 provide a ratio of 638 (neglecting powers of ten) and 4.3 and 6.8 provide 632. (These are the best fit ratios to 635 and they are equally good, unless you would prefer a particular trend in the error distribution.) If we used the first multipliers, the resistor values would be $R1 = 30\ k\Omega$ and $R2 = 4.7\ k\Omega$ and a calculation for the center value of the resulting tap voltage V_T, using these resistors is

$$V_T = 2.5\left(\frac{4.7\ k\Omega}{34.7\ k\Omega}\right) = 0.339V$$

For comparison, the other multipliers of 4.3 and 6.8 would provide $R1 = 43\ k\Omega$ and $R2 = 6.8\ k\Omega$ and V_T would then center around a value that is given by

$$V_T = 2.5\left(\frac{6.8\ k\Omega}{49.8\ k\Omega}\right) = 0.341V$$

Other powers of ten could be used to scale the source resistance of V_T. Multiplying or dividing each resistor by the same scaling factor to achieve a desired source resistance and then choosing the closest standard resistor values can also be used to exchange a greater ratio error for a desired source resistance.

Fig. 5-1. Choosing Best-Fit 5% Resistor Values for a Voltage Divider

5.3 SOME MISCELLANEOUS CIRCUITS

Grouped into this section are many application circuits that were selected because of their novelty, their illustration of a basic way op amps solve problems, or their inability to fit into one of the other subsections.

Current to Voltage Converter

This deceptively simple looking circuit, Figure 5-2, can be used to measure a current that is flowing into or out of ground (or any dc biasing level that the $V_{IN}(+)$ input is tied to) while the measured node is held at ground. This circuit concept is widely used in the design of electronic test equipment. An important thing to keep in mind is the effects of the dc error sources. For example, a 741 with a 500 nA max I_B spec would represent a 10% error for an input current of 5 μA, but many op amps are available with a smaller I_B spec.

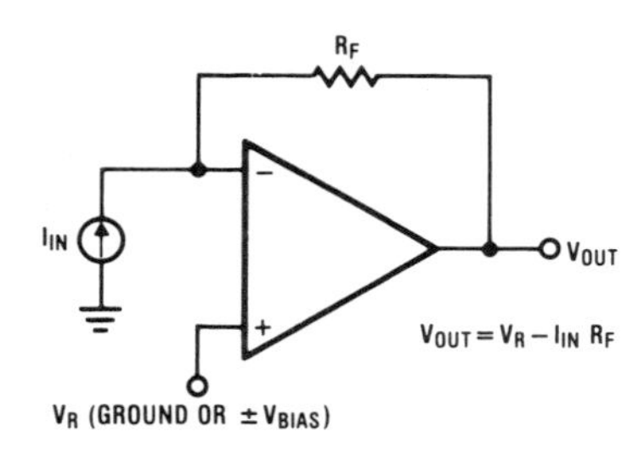

Fig. 5-2. The Current-to-Voltage Converter

To show how op amp applications can build on a basic idea, we can derive a circuit that provides a logical output voltage signal that indicates the direction of current flow in a *grounded* resistor, R_G, as shown in Figure 5-3. There is a positive feedback loop around the comparator used in this circuit (200 kΩ and 5.1 kΩ) to provide hysteresis and eliminate any electrical *chatter* or oscillation for values of I_R near zero.

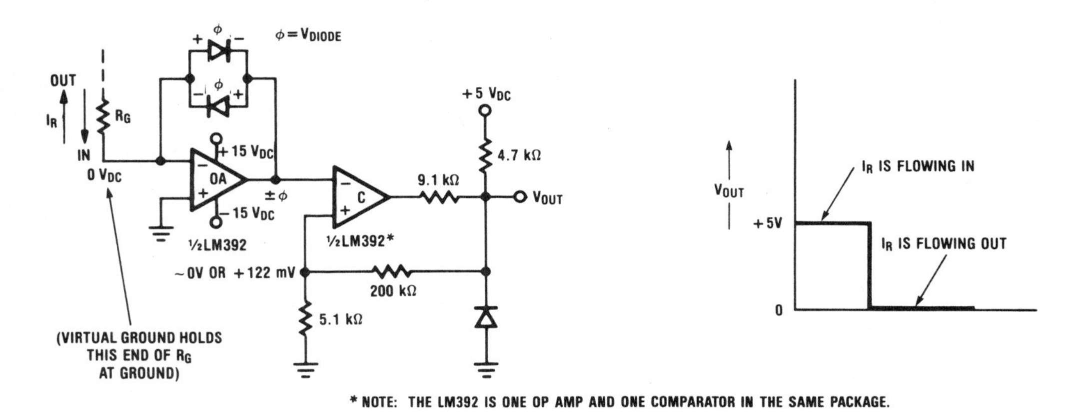

Fig. 5-3. Generating a Logic Signal That Indicates Current Direction

Measuring Junction Capacitance

An adaptation of this summing-node measuring concept can be used to measure the junction capacitance of a semiconductor diode for various dc values of reverse bias voltage, Figure 5-4. Additionally, the ac signal across the diode can be kept small and the op amp isolates the ac voltmeter that is used to read V_{OUT}.

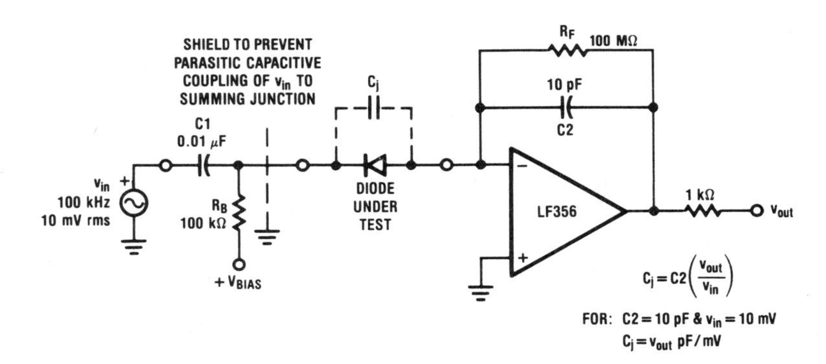

Fig. 5-4. Measuring the Junction Capacitance of a Diode

A High Input Impedance Differential Voltmeter

The circuit shown in Figure 5-5a uses two voltage followers to present a high input impedance to the sources of the dc input voltages V1 and V2. Notice that the inverting inputs track these respective voltages so that the full difference voltage appears across R. The resulting current will now flow through the floating meter and *the meter resistance, R_m, will not cause an accuracy problem; it's in the loop.* This idea can be used for ac differential input voltages as shown in Figure 5-5b.

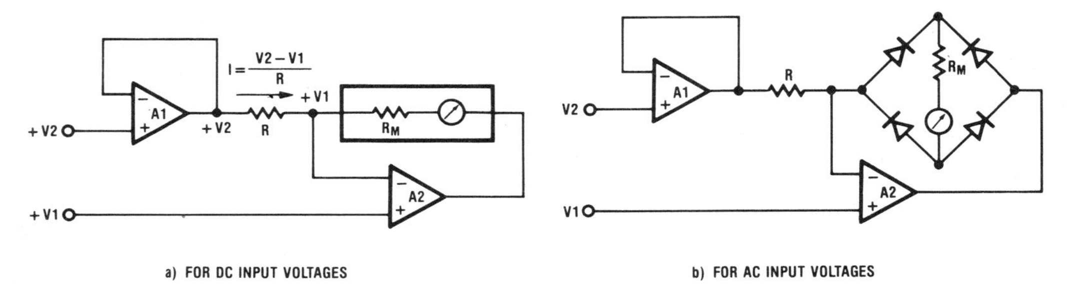

Fig. 5-5. A High Input-Impedance Voltmeter

Operating Simultaneously With Two Inputs

An op amp can operate with two simultaneous inputs, as shown in Figure 5-6a. This circuit is evaluated by making use of superposition. Consider each input voltage acting separately, with the other input temporarily shorted to ground. The V2 input is a convenient way to introduce a correcting offset voltage into an inverting gain circuit with no disturbance on the V1 input.

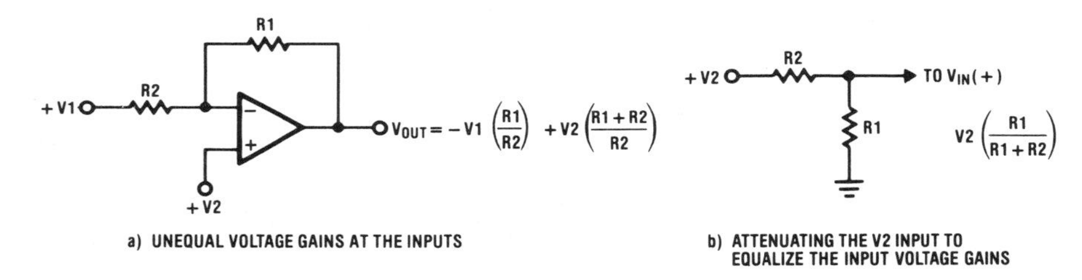

Fig. 5-6. Operating With Two Inputs

It is interesting to notice that the current flow (even the direction of the current flow) from the V1 input signal source will also depend on the magnitude of V2. Two examples will illustrate this. First consider that both resistors are equal and have a value of 10 kΩ. V1 is equal to 1V and V2 is equal to 0V. A current of *0.1 mA will therefore flow out of the V1 source* for this case and V_{OUT} will be equal to -1V.

Now, if V1 is kept at the same voltage, and V2 is raised to 2V, *0.1 mA will enter the V1 source (although V1 did not change)* and the output now will be 3V, as it should be. These inputs are independent in their effects on V_{OUT}, *even though they interact* with each other as indicated in the above example.

If the V2 input signal were attenuated by similar resistors, R2 and R1 as shown in Figure 5-6b, then the expression for V_{OUT} becomes

$$V_{OUT} = -V1 \left(\frac{R1}{R2}\right) + V2 \left(\frac{R1}{R1 + R2}\right) \left(\frac{R1 + R2}{R2}\right)$$

or

$$V_{OUT} = \frac{R1}{R2}(V2 - V1)$$

This circuit now becomes a differential amplifier with a large difference in the values for the input impedance at each input.

The circuit of Figure 5-6a is a useful application circuit to provide both output voltage polarities from a single polarity output voltage digital-to-analog converter (DAC), as shown in Figure 5-7. This circuit also doubles the output voltage span of the DAC.

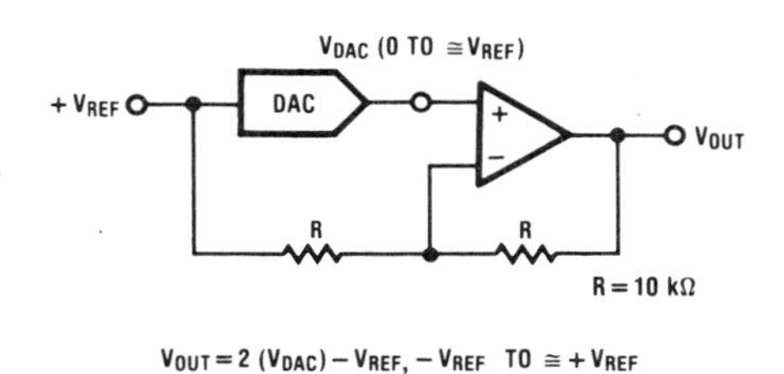

$V_{OUT} = 2\,(V_{DAC}) - V_{REF},\ -V_{REF}$ TO $\cong +V_{REF}$

Fig. 5-7. Making Use of Simultaneous Inputs to an Op Amp

A very useful application of this two-input op amp is to simultaneously provide voltage gain for an input signal and also to accomplish a dc level shift in the resulting output signal. A dc reference voltage is used at one input to provide the desired dc level shifting and the input signal can be amplified and dc shifted, if this input is applied to the noninverting input, or the input signal can be inverted, amplified, and dc level shifted if this input is applied to the inverting input.

Assume we have a signal that ranges from + 1V to + 3V (a 2-volt span) and we want to provide a noninverted output signal that ranges from -4V to + 10V (a 14-volt span). This requires a noninverting voltage gain of 7 as shown in Figure 5-8a. We next determine the dc voltage that is required at the inverting input as shown in Figure 5-8b. Generally this unusual dc voltage is supplied from a reference voltage that is available in the system by making use of the Thevenin equivalent circuit as shown in Figure 5-8c.

As an example of simultaneously providing signal inversion, voltage gain, and dc level shifting, we could require that an input signal that ranges from -2V to + 0.5V (a 2.5-volt span) provide an inverted and dc level shifted output voltage that ranges from 0V to + 5V (a 5-volt span). This requires an inverting gain of 2 as shown in Figure 5-9a. We next calculate the required dc input voltage that is needed at the noninverting input as shown in Figure 5-9b. This voltage can be supplied from a system reference voltage as shown in Figure 5-9c. (Note that we are again matching the resistance seen at each input.)

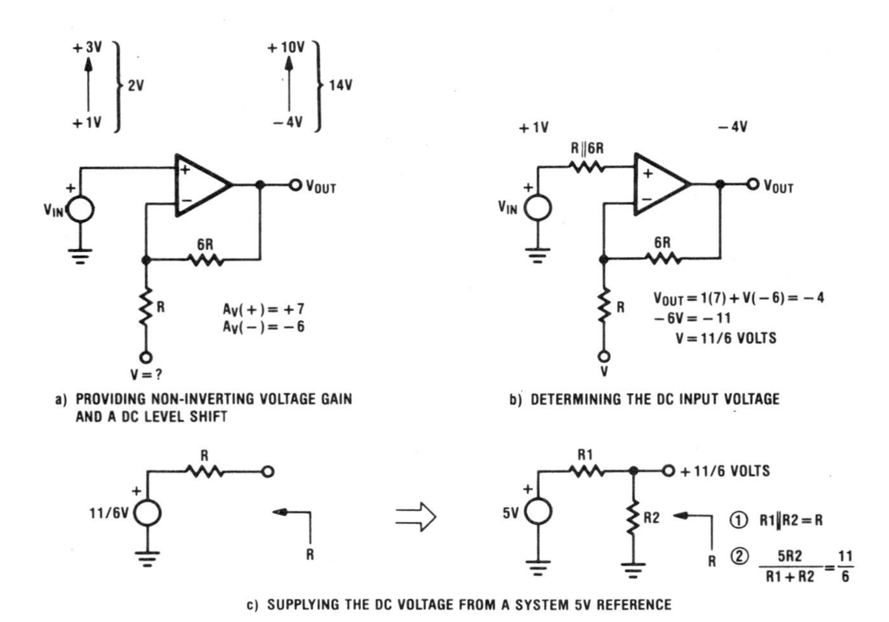

Fig. 5-8. Amplifying and DC Shifting an Input Signal

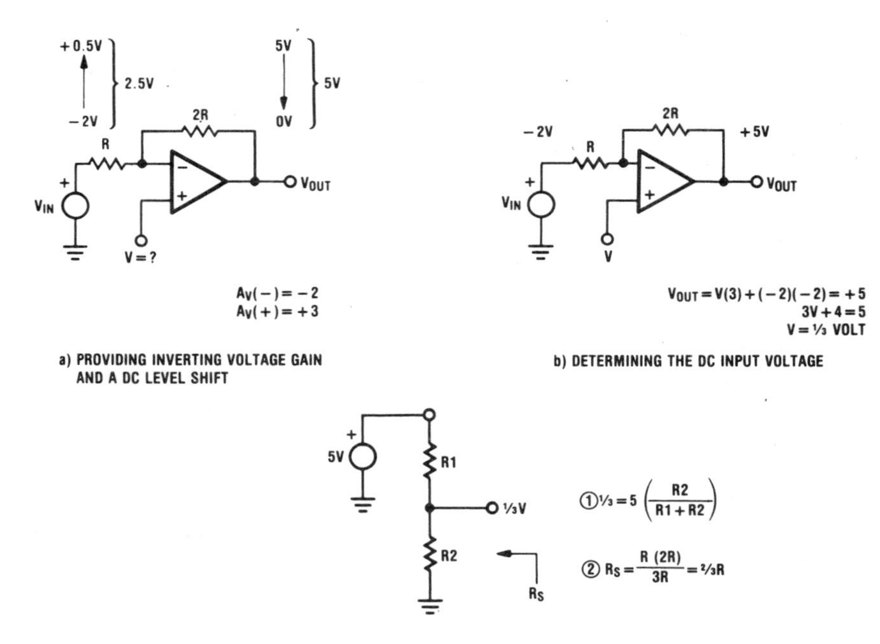

Fig. 5-9. Amplifying, Inverting, and DC Level Shifting an Input Signal

Operating a Decompensated Op Amp at Unity Gain

Several op amp products are available that are not sufficiently internally frequency compensated to operate at unity gain. These are called *decompensated* op amps and are usually limited to a minimum closed-loop gain of 5. The application circuits of Figure 5-10 show how these higher frequency amplifiers can be used with a signal gain of ± 1 (or any gain from ± 1 to ± 5) without instability problems.

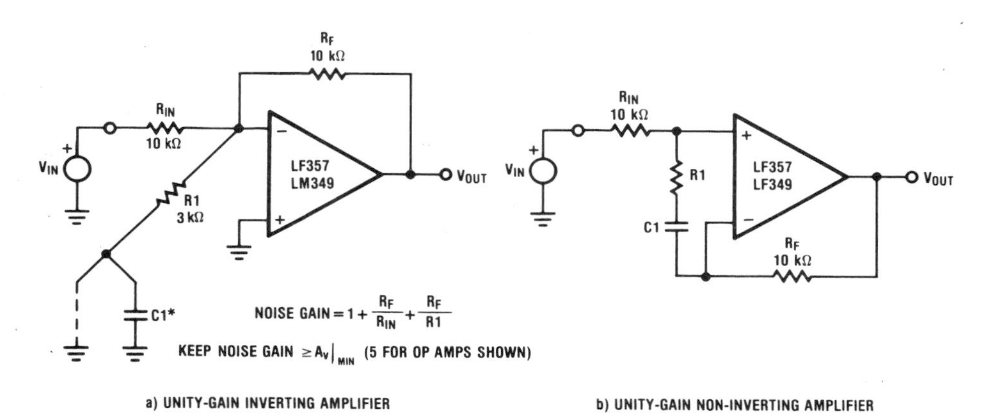

Fig. 5-10. Operating a Decompensated Op Amp at Unity-Gain

The trick here is to raise the noise gain by adding an additional resistor at the input summing junction. Two possibilities exist: increase both the dc and ac noise gain with R1 returned to ground or raise only the ac noise gain by adding a capacitor in series with this resistor. The benefit of adding the capacitor is that there will then be no increase in the offset voltage at the output (a dc noise effect).

Unfortunately, the ac noise at the output will be increased by the larger value of ac noise gain (5) that results. (Some low frequency noise reduction benefits can be obtained by using a relatively small-valued capacitor. This will cause the ac noise gain to increase to the minimum value of gain only at high frequencies, where stability is important, just before reaching the unity-gain frequency.)

This circuit trick, that allows the use of a decompensated op amp, can increase the slew rate and power bandwidth beyond what can be achieved with a unity-gain compensated op amp. This improvement factor is almost equal to the minimum gain of the decompensated op amp in the family. As an example, the decompensated LF356, the LF357 (A_V min = 5), has a slew rate of 50 V/μsec and a unity-gain frequency of 20 MHz: the LF356; 12 V/μsec and 5 MHz. The decompensated LM348 quad, the LM349, has 2 V/μsec and 4 MHz: the LM348; 0.5 V/μsec and 1 MHz. This can increase performance in those applications where the minimum gain specified for the decompensated op amp is larger than desired. For large-signal applications, the resulting increase in noise can usually be tolerated, but the increase in noise that exists with this circuit trick should be kept in mind.

An extension of this technique (to increase bandwidth and slew rate) is to consider an externally compensated op amp. This can provide improved performance, especially at high values of closed-loop gain.

Neutralizing the Input Pole

Stray capacitance at the input leads of an op amp causes both a feedback pole and also an *input pole.* In Figure 5-11, the feedback pole is taken care of by adding C2. Notice that R_S, in combination with the common-mode input capacitance at the noninverting input, C_{IN}, will also cause an input pole that can reduce high frequencies if R_S is large. This can be broadbanded by introducing a neutralizing capacitor, C_N, that is placed from the output of the op amp to the noninverting input. This careful use of positive feedback can remove the effects of the input pole, but the value used for C_N will require a careful adjustment to prevent ringing or oscillations.

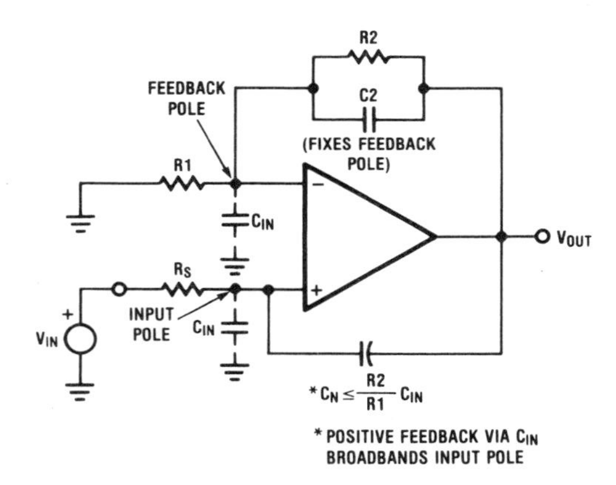

Fig. 5-11. Neutralizing the Input Pole

An alternative procedure is to omit C2, keep $R1||R2 = R_S$, and use $C_N \cong 0.5$ pF to prevent oscillations owing to the feedback pole. This *balanced poles* condition ($R_S C_{IN} = R1||R2\ C_{IN}$) tends to cancel the C_{IN}s at both inputs because the ac signal roll off becomes a common mode input and will provide a broadband response.

A Few Multi-Input, Noninverting, Summing Circuits

The standard inverting summing circuit has already been presented in Chapters 2 and 3. A noninverting summing circuit is shown in Figure 5-12. Notice that each of the input signals is attenuated by a factor of one-half (use superposition) directly at the noninverting input. (The inputs do interact with this noninverting summing circuit and this specific drawback led to the early popularity of the inverting op amp for audio summing and mixing applications.) The gain from the noninverting input to the output is + 2, so the sum of the input signals will properly appear at the output. (The feedback resistor can be increased in value if it is desired to provide gain to the sum of the input voltages.)

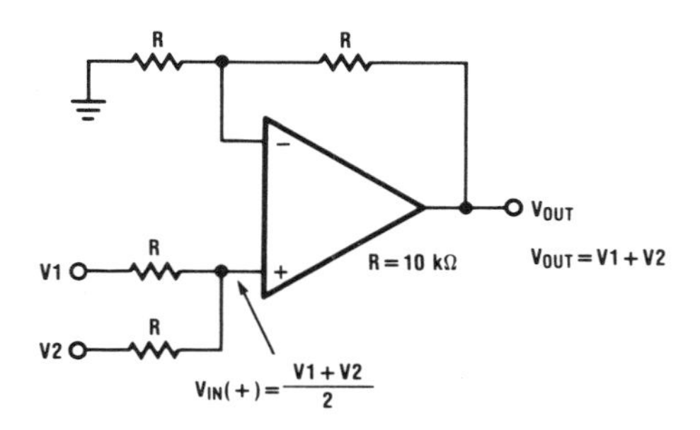

Fig. 5-12. A Two-Input Noninverting Adder

A three-input summing circuit is shown in Figure 5-13. The signal at the noninverting input now becomes one-third of the sum of the input voltages. When this is multiplied by the gain of + 3, the output voltage becomes equal to the sum of the three input voltages. (Any unused inputs of noninverting summing circuits must be grounded to maintain the proper values of signal gain for the other inputs.)

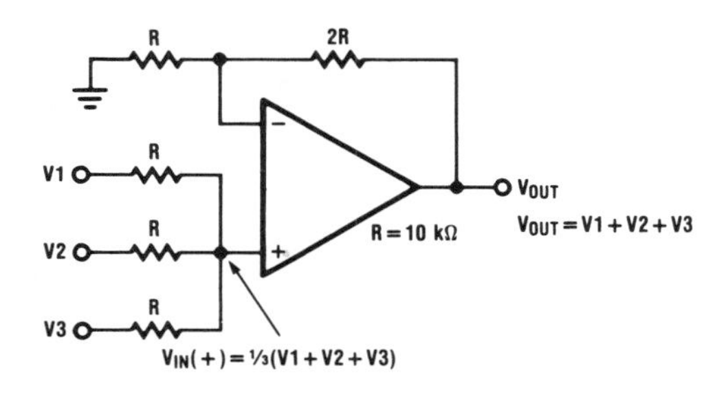

Fig. 5-13. A Three-Input Noninverting Summing Circuit

A general summing circuit is shown in Figure 5-14. Any number of inputs can be used on either the inverting or noninverting sides as long as the Ks are properly kept track of.

An example of an unbalanced, multi-input summing circuit that can be derived from the previous circuit is shown in Figure 5-15. Notice that the gain to the V1 input is -3 in order to keep the sum of the Ks equal for both the positive and the negative inputs.

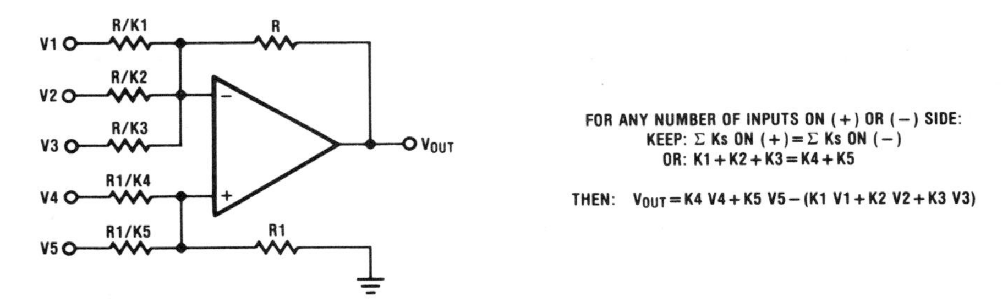

Fig. 5-14. A Generalized Summing Circuit

A Differential Input, Differential Output Amplifier

The two-op amp circuit shown in Figure 5-16 can be used to provide a high input impedance amplifier that provides a differential output voltage from a low source impedance.

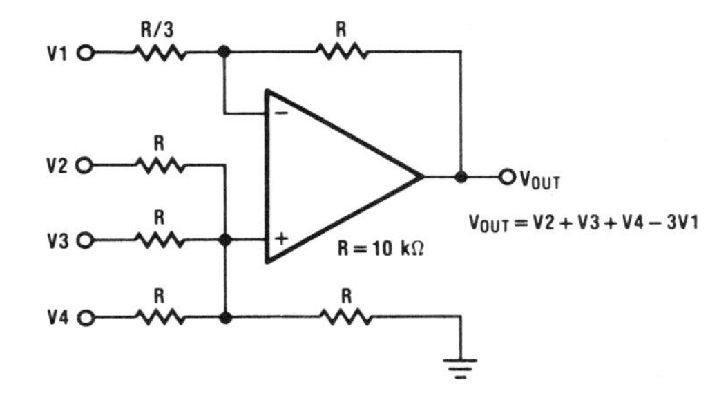

Fig. 5-15. An Unbalanced Multi-Input Summing Circuit

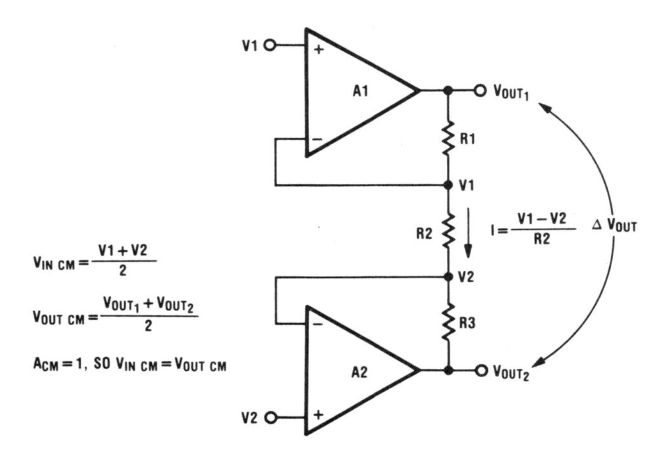

Fig. 5-16. A Differential-Input, Differential-Output Amplifier

This circuit has a common-mode voltage gain of 1 and the differential voltage gain can be determined from the figure by noticing that expressions for V_{OUT1} and V_{OUT2} can be written as

$$V_{OUT1} = V1 + (V1 - V2)\left(\frac{R1}{R2}\right)$$

and

$$V_{OUT2} = V2 - (V1 - V2)\left(\frac{R3}{R2}\right)$$

The differential-mode gain, A_D, is defined to be

$$A_D = \frac{\Delta V_{OUT}}{\Delta V_{IN}} = \frac{V_{OUT1} - V_{OUT2}}{V1 - V2}$$

so

$$A_D = \frac{\left[V1 + (V1 + V2)\left(\frac{R1}{R2}\right)\right] - \left[V2 - (V1 - V2)\left(\frac{R3}{R2}\right)\right]}{(V1 - V2)}$$

collecting terms

$$A_D = \frac{V1\left(1 + \frac{R1}{R2} + \frac{R3}{R2}\right) - V2\left(1 + \frac{R1}{R2} + \frac{R3}{R2}\right)}{(V1 - V2)}$$

or

$$A_D = 1 + \frac{R1}{R2} + \frac{R3}{R2} = \frac{R1 + R2 + R3}{R2} = \frac{\Sigma R_S}{R2}$$

If we let $R1 = R3 = NR2$

$$A_D = \frac{NR2 + R2 + NR2}{R2} = 2N + 1$$

The gain of this circuit can be easily altered by changing the value of only R2. None of the resistors ties to or loads the input leads, and a high input impedance therefore results at both of the differential input leads. These desirable features have made this circuit popular as the first stage of a two-op amp instrumentation amplifier. A second stage is added that accepts the differential voltage out of this first stage and then converts it to a single-ended output voltage, as shown in Figure 5-17. (It is expected that this application circuit will eventually be replaced by special instrumentation amplifier ICs.)

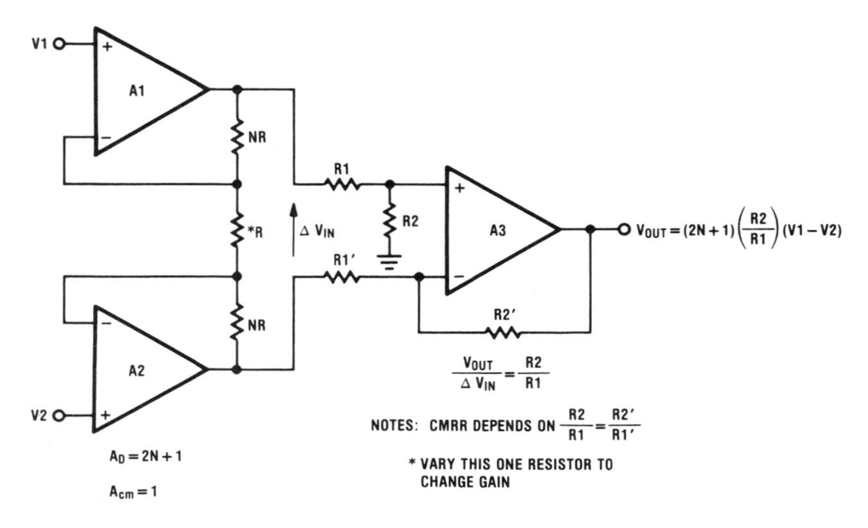

Fig. 5-17. A Two-Op Amp Instrumentation Amplifier

Single-Amplifier, Maximum Input Voltage Selector

A circuit, sometimes called an upper selector, that selects the maximum of several input voltages, Figure 5-18, appears much like a diode-logic OR gate. A lower selector circuit would result if all the diodes were reversed and the diode biasing voltage source were changed to $+V_{CC}$ ($+15\ V_{DC}$). The trick here is to bias two diodes with the same value of dc current flow so as to track out both their dc voltage drops and their temperature changes. (Matched diodes in the same package should be used to obtain the best performance.) For operating at high speeds, 1N914s or 1N4148s are recommended, but better accuracy can be obtained at lower speeds with planar diodes like 1N484s or 1N457s.

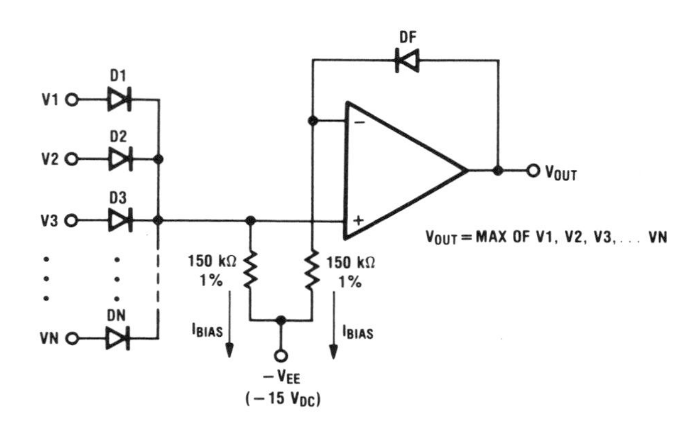

Fig. 5-18. A Single-Op Amp Max-V_{IN} Selector

This is a useful circuit for relatively large valued input signals where the small differences in diode drops will be less of an error source. Note that the current flow through the two 150 kΩ biasing resistors will be the same and only one of the input diodes will, ideally, be conducting. (If more than one of the input voltages is at the **exact** same value, more than one of the input diodes will conduct and errors will be introduced. For example, a dc error voltage of 18 mV would be introduced if two input diodes were to equally conduct.)

Computer-Controlled Window Comparator

A window comparator circuit easily adaptable to digital computer control is shown in Figure 5-19. In this application circuit, use of an op amp would undesirably increase the response time and diode logic would be needed at the outputs. For faster operation, comparators can be used because only logic outputs are needed and comparators have a faster response time than op amps.

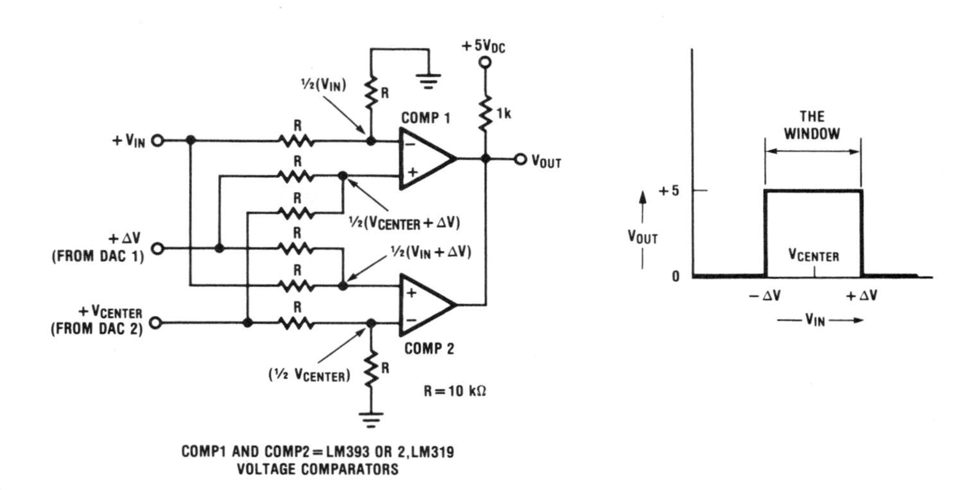

Fig. 5-19. A Computer-Controlled Window Comparator

The upper comparator, COMP1, will pull V_{OUT} to a LOW state for $V_{IN} > (V_{CENTER} + \Delta V)$. Comparator COMP2 will pull V_{OUT} to a LOW state for $V_{IN} < (V_{CENTER} - \Delta V)$. Therefore, the only time that the wired-OR output of these two comparators will be in the HIGH state is when V_{IN} is within the window, $(V_{CENTER} - \Delta V)$ to $(V_{CENTER} + \Delta V)$.

The positioning of this window is easily controlled by a computer. Both the center voltage, V_{CENTER}, and the width of the window, ΔV, are separately and independently established by two control voltages that can be provided by digital-to-analog converters (shown as DAC1 and DAC2 on this figure).

A Tri-State Window Comparator. If an application requires three different voltage levels out that depend on the value of the input voltage, the circuit of Figure 5-20 can be used. (In linear design, this is often a more useful three-state choice than the digital systems designers use.) This strange looking bridge circuit controls the output voltage swing limits of the op amp via the zener diode clamps and also provides for the flat spot in the transfer characteristic that is shown on the figure near the magnitude of the bias voltage, $|-V_B|$.

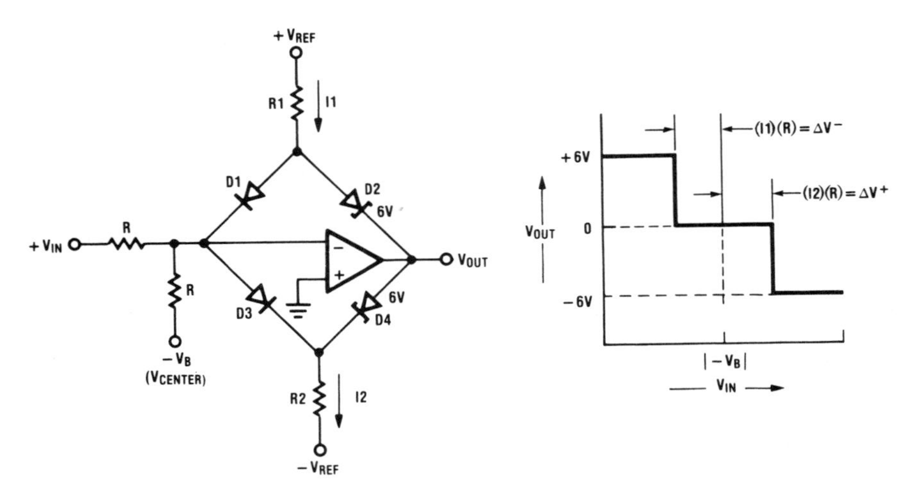

Fig. 5-20. A Tri-State Window Comparator

To obtain an understanding of the operation of this circuit, it is best to start with the particular input voltage, V_{IN}, where $V_{IN} = |-V_B|$. For this condition, the voltage at the inverting input of the op amp is zero and both currents, I1 and I2, split and all four of the diodes are conducting. The output voltage of the op amp, V_{OUT}, is therefore at zero volts.

If we now consider the sequence of events as V_{IN} increases, the key to the operation of this circuit can be easily seen. An increase in V_{IN} would tend to slightly raise the voltage at the inverting input of the op amp and cause V_{OUT} to slightly drop until the diode D3 absorbed the extra current that is now available at the summing junction because of the increase in V_{IN}. This teetering or unbalancing of the diode bridge continues until I2 is not large enough to sink the surplus current that V_{IN} is pouring into the summing junction of the op amp. To absorb the next increase in current, the output voltage of the op amp starts moving negative in response to the now, slightly increasing voltage at the summing node. (This results when I2 can no longer absorb the surplus current.) Because small negative changes in V_{OUT} have no effect on the conditions at the summing node, V_{OUT} rapidly, and somewhat desperately, continues to swing negative in response to the increasing voltage at the inverting input. V_{OUT} is on its way to negative-swing saturation.

When V_{OUT} reaches -6.6V, the Zener diode, D4, enters breakdown and conducts the excess current that is being forced into the summing node by the increasing V_{IN} signal source. This prevents any further change in V_{OUT} and keeps the op amp out of saturation. This transition in V_{OUT} from 0V to -6.6V therefore occurs rather abruptly when I2 needs help in absorbing the excess current that V_{IN} enters into the summing node, or when

$$\frac{V_{IN}}{R} \geq \frac{V_B}{R} + I2$$

or

$$V_{IN} \geq V_B + I2\ R = V_B + \Delta V^+$$

A similar condition exists for the lower limit and this deviation from $|-V_B|$, ΔV^-, becomes I1 R, as indicated in the transfer function plot that is shown on the figure.

Some circuit additions can be made to allow control by a computer via DACs as shown in Figure 5-21. Also, a slight amount of positive feedback (hysteresis) can be added, Figure 5-21b, to prevent indecision (chattering) when near the V_{OUT} switching points.

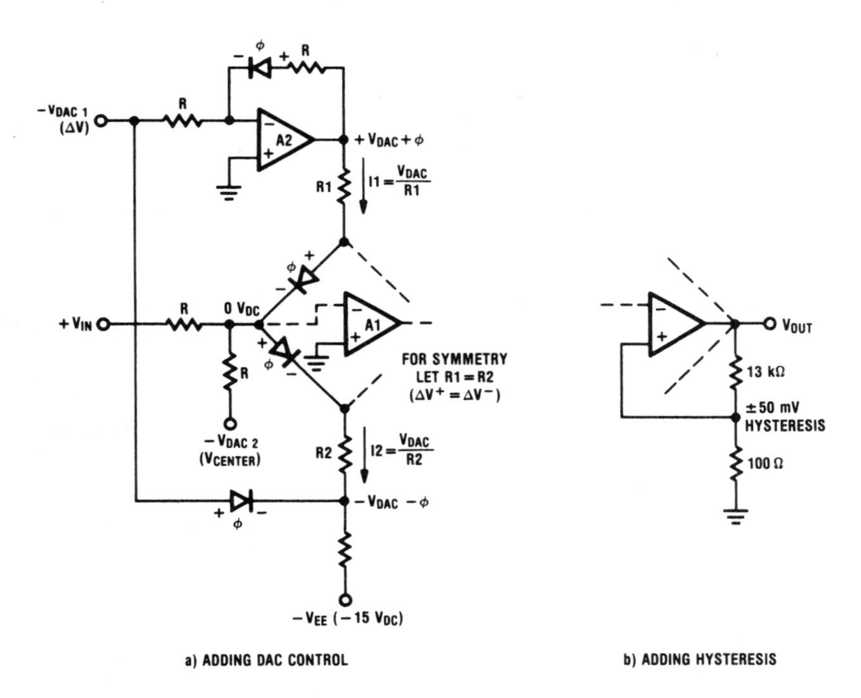

Fig. 5-21. Some Possible Modifications of the Tri-State Window Comparator

Rate Limiter

In some applications, especially when driving inductors, it may be necessary to limit the maximum allowed time rate-of-change of a voltage waveform. This can be accomplished with the circuit shown in Figure 5-22. This diode bridge, makes a useful **current gate.** (For highest precision, these diodes can be obtained as a matched set, but this degree of matching is not required in this application.)

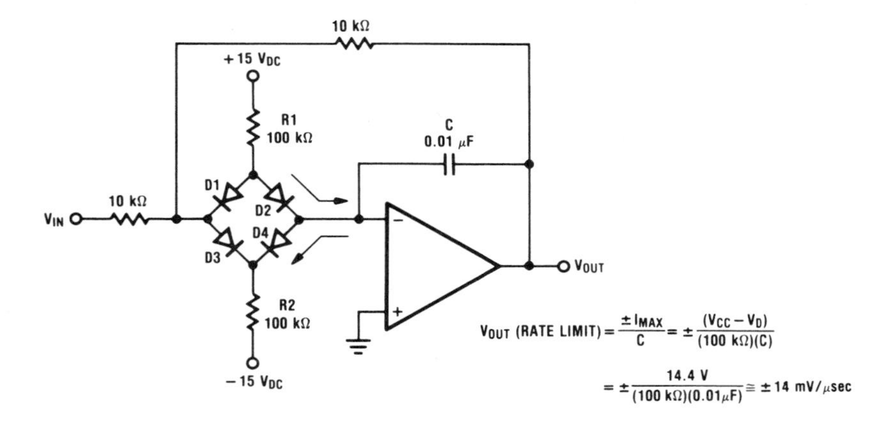

Fig. 5-22. A Rate Limiter

The output voltage provided from the integrator circuit has the ability to ramp up or down at a maximum rate that is determined by the current available through the current gate. The current gate isolates the input voltage from the input to the integrator. With a large positive-step increase in the input voltage, D1 will be OFF and D3 will carry the current demanded by R2. The diode D4 will also be OFF, and the current flow through R1 will forward-bias D2 and cause the output voltage of the integrator to ramp in a negative direction. A rate limit is therefore established for this negative ramp voltage by R1 and C. This rate is independent of the time rate-of-change of the input signal.

When V_{OUT} approaches the steady state, the current gate comes into balance and the current flow through R1 will eventually continue through R2. The action of the overall feedback loop will keep the input current to the integrator at zero to stop any further changes in V_{OUT}.

Similarly, with a large negative-step increase in the input voltage, the input current to the integrator is provided by R2. We will see other uses for this diode bridge, or current gate, when we discuss **bounding circuits** later in this chapter.

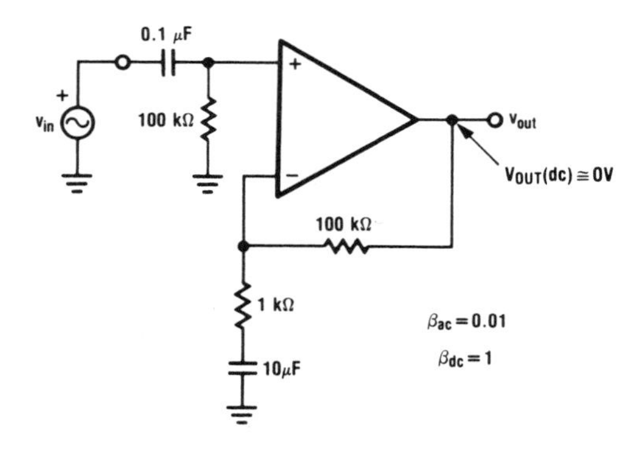

Fig. 5-23. An AC-Coupled Amplifier

AC-Coupled Amplifiers

For an ac-coupled amplifier, the feedback network can be dc isolated with a capacitor to provide ac gain, yet keep the dc gain at unity. This improves the predictability of the dc output voltage of the op amp.

A common way to achieve this is shown in Figure 5-23. This circuit typically requires a large valued capacitor in the feedback network. (Owing to the small dc voltage across it, a large sized capacitor is not necessary.)

Getting the Best of Two Op Amps

In Chapter 1 we saw how the vacuum tube op amp used an ac-coupled auxiliary amplifier with a mechanical chopper to reduce the dc offset voltage and increase the dc gain of the main op amp. A modern version of this idea is shown in Figure 5-24, where there are now two IC op amps connected in parallel at their inputs.

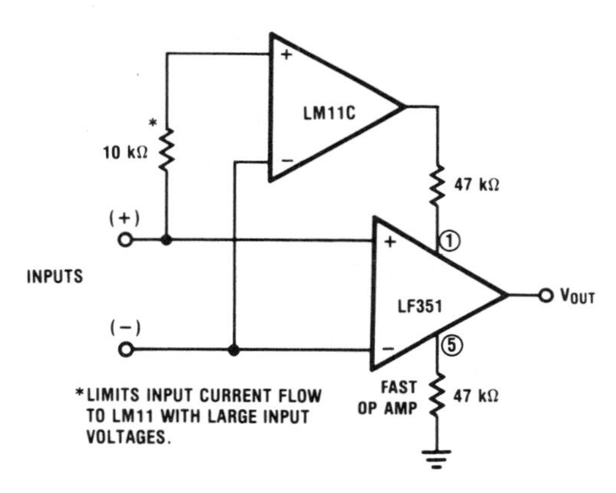

Fig. 5-24. Using a Low-V_{OS} Slow-Op Amp to V_{OS} Null a Fast Op Amp

The LM11 op amp has very precise dc characteristics, low offset voltage and also very low input bias current. The LF351 is a much higher frequency op amp. This combination of the two op amps therefore obtains the benefits of each in one composite op amp.

The LM11 is automatically adjusting the offset voltage of the LF351, via the balance pins of the LF351, in response to the differential input voltage. As in the case of the chopper-stabilized, vacuum tube op amp, this increases the dc gain of this combination to in excess of 140 dB. This circuit benefit should also be kept in mind, in case high dc gain is ever needed.

To see what is happening, the equivalent circuit of the LF351 is shown in Figure 5-25. The output voltage of the LM11 is seen to introduce a voltage across the 100Ω resistor in the input balancing (or offset voltage adjust) circuitry of the LF351. The 100Ω resistor on the opposite side is permanently unbalanced because of the 47 kΩ resistor that is inputting current from ground. If the output voltage of the LM11 were 0V (and all of the resistors were well-matched), there would be no offset correction taking place. As the output voltage of the LM11 swings positive or negative, it can provide a dynamic adjustment of the dc balance of the LF351 op amp.

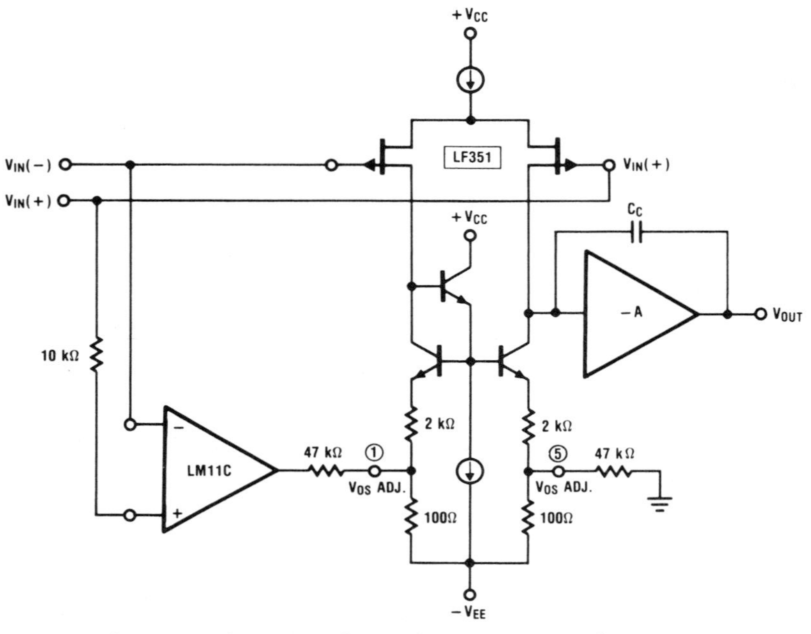

Fig. 5-25. The Internal Details of the V_{OS} Nulling Circuit

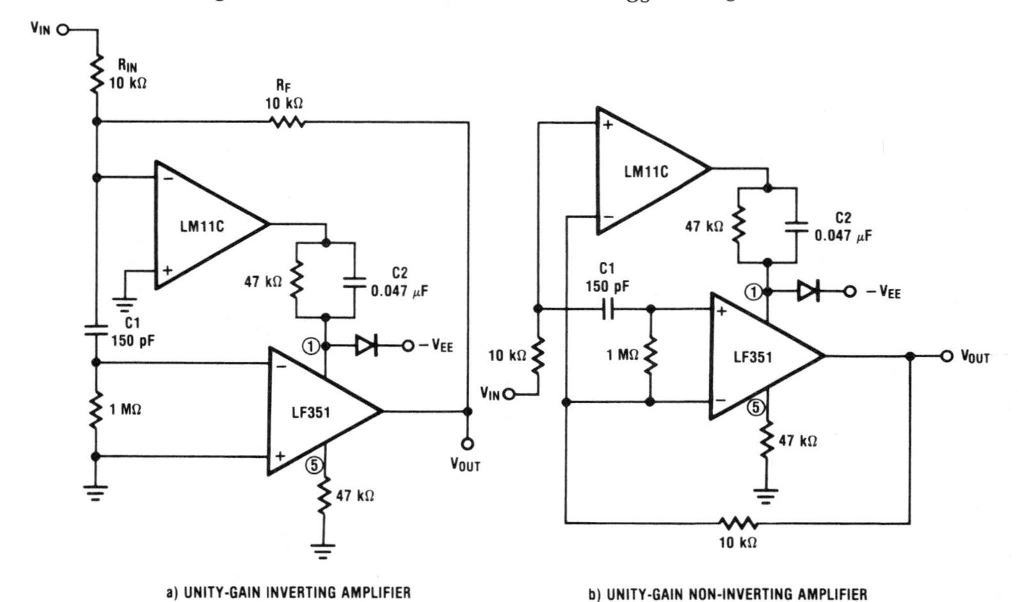

Fig. 5-26. Isolating the Input Current of the Bi-FET Op Amp

The dc input bias current of the LM11 is smaller than that of the LF351 and does not increase as much at high temperatures, so it may be desirable to remove the dc loading of the Bi-FET from the input of the overall op amp. This is done by using a capacitor to dc isolate the Bi-FET, yet still obtain the high frequency benefits of this fast op amp, as shown in Figure 5-26. The value of this capacitor, C1, should be kept small, because, for a large input voltage overload, this capacitor will pick up a charge as the summing node of the LM11 moves away from ground. Removing this charge will then create a recovery time problem.

5.4 CURRENT SINKS, SOURCES, AND PUMPS

A current source is theoretically expected to be capable of either supplying or absorbing current, depending upon the algebraic sign that is associated with the magnitude of the current source. Many practical current sources provide a current that flows in only one direction. Thus, three circuit capabilities exist: only supply current, only absorb current, or be able to do both. In an attempt to catalog these circuits and provide a systematic grouping that will aid future reference, we will introduce some new definitions, as shown in Figure 5-27.

Those circuits that can *only absorb current*, Figure 5-27a, will be called *current sinks*. If, instead, the circuit *supplies a current*, Figure 5-27b, it will be called a *current source*, and if the circuit can *do either*, Figure 5-27c, it will be called a *current pump*.

The main function of all of these circuits is to provide an output current that is essentially independent of the node voltage that results at the output terminal. This is useful if a load requires a current drive, for use in circuit biasing, for operating with high voltages, and also for providing a linear voltage sweep if a load capacitor is used.

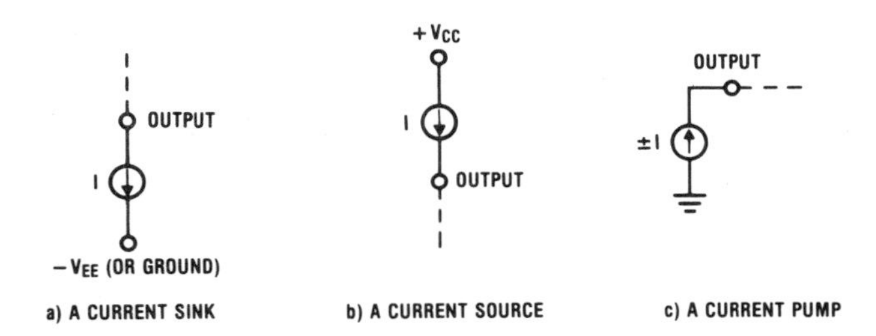

Fig. 5-27. Some New Definitions for the General "Current Source"

How well this constant current feature is achieved can be specified by a parameter called *output resistance*. An ideal current source would have an infinite output resistance. The current that is supplied would then be completely independent of the output voltage.

Practical circuits that provide constant currents are also limited both by a minimum voltage that must be provided across the output terminals to keep the circuitry properly biased and a maximum, or *compliance*, voltage limit. If, in the application, the output terminal of the constant current circuit remains within this voltage range, the proper current will be supplied.

Now, lets see how op amps can be used to perform this useful circuit function.

Current Sinks

A single supply current sink is shown in Figure 5-28. The input common-mode voltage range of the op amp shown includes ground, therefore V_{BIAS} can range down to small values. If the offset voltage becomes a problem; the LM358A or LM324A, with V_{OS} of 2 mV max, can be used. The advantage of using a small value for V_{BIAS} is that the output lead (the collector of the transistor) can then operate to lower values of output voltage.

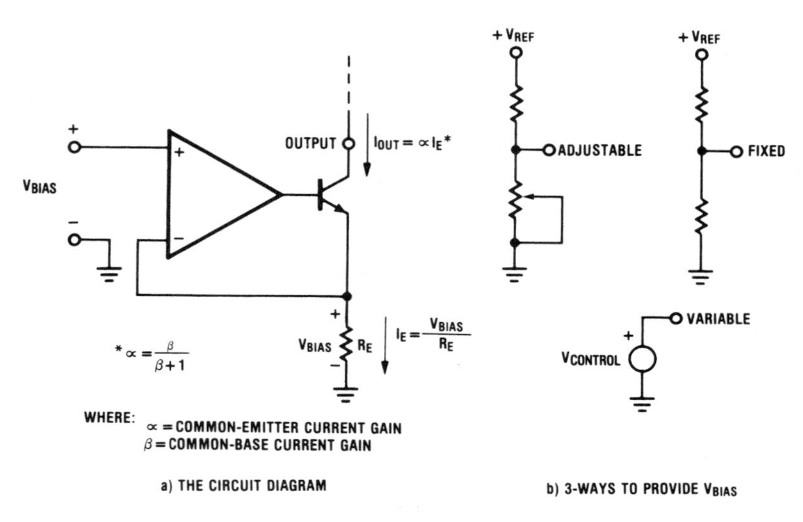

Fig. 5-28. A Single-Supply Current Sink

Notice that the emitter current is the controlled parameter and the collector current relates to the emitter current by the alpha, α, (the common-base current gain) of the transistor. This is an additional error source, but can be minimized by selecting an NPN transistor with a high current gain (either α or β) or by using two transistors in a Darlington connection.

Using JFETs. The current gain error of the previous circuit can be eliminated with the JFET circuits shown in Figure 5-29. The advantage of an FET is that the ratio of the drain current to the source current is very close to unity. At high drain-source voltages this current ratio may degrade because of leakage currents and at low currents and high output voltages a Darlington may therefore be a better choice.

For currents that are less than the I_{DSS} of the JFET, the circuit shown in Figure 5-29a can be used. The output voltage of the op amp must be able to swing negative far enough to control the FET. This will not be a problem with split supplies and a grounded resistor, R_D, but will require consideration when operating with a single supply. One solution is to limit the minimum value of V_{BIAS} to always be larger than the worst case pinch-off voltage of the JFET.

For higher currents, an NPN transistor can be used to boost the FET current (or a Darlington connection of NPNs can also be used) and maintain an alpha of 1, as shown in Figure 5-29b.

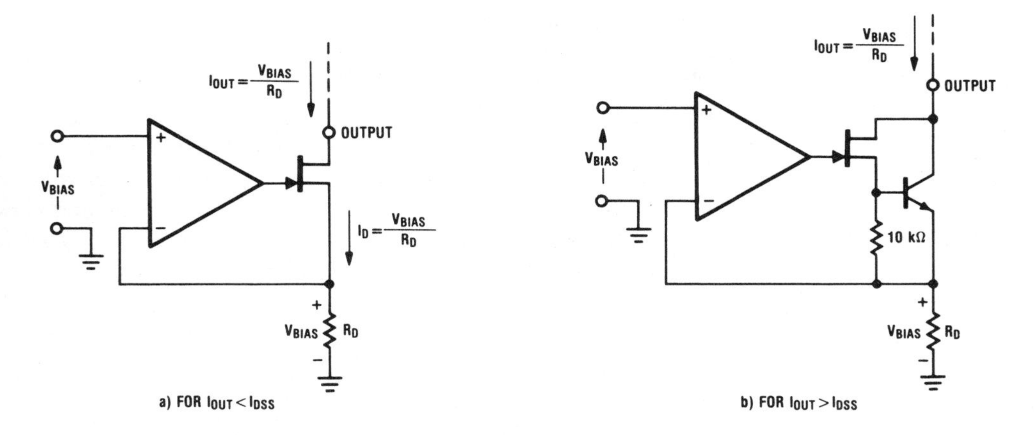

Fig. 5-29. Precision JFET Current Sinks

Multiple Current Sinks. A way to provide for multiple current sinks is shown in Figure 5-30. Notice that the op amp is only insuring that the first transistor, Q1, will be biased to conduct the proper collector current. The precision of the rest of the output currents depends upon matching the resistors, the alphas of the transistors, and the V_{BE}s of the transistors. Larger values of voltage drop across the resistors in the emitter leads will reduce the errors that result from V_{BE} mismatch.

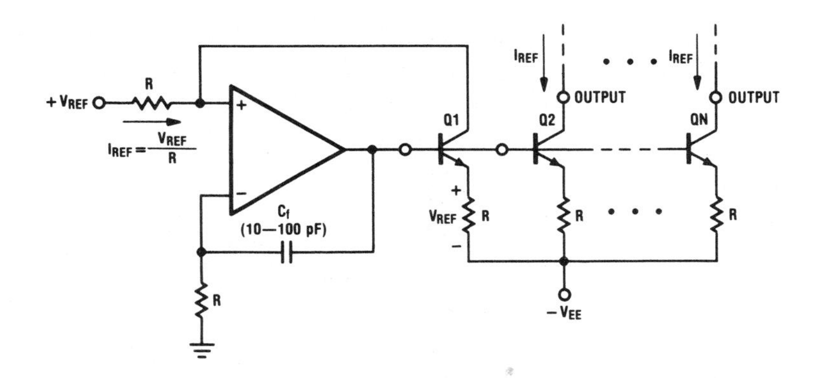

Fig. 5-30. Biasing Matched Transistors for Use as Multiple Current Sinks

Introducing a balancing resistor in the $V_{IN}(-)$ lead of the op amp will reduce errors and also provide a convenient way to stabilize the circuit with the added capacitor C_f.

This basic multiple current sink idea is often used to provide binary weighted currents as shown in Figure 5-31. The op amp drives the common *baseline* and the emitter resistors are reduced by one-half as each current becomes more significant. The emitter area of the transistors is also doubled as the current is doubled to match the V_{BE}s and to also keep the temperature drift of the V_{BE}s matched. (This transistor-consuming circuit technique is usually only used in IC designs.)

The reference technique shown in Figure 5-31 provides the best match for the smallest, or least significant current. Higher accuracy results from referencing either the most significant current (or at least a more significant fraction of the most significant current) by adding another identical emitter-area-scaled transistor with an identical emitter resistor and connecting the collector of this added transistor to the noninverting input of the op amp. (The resistor from the reference voltage source is also made correspondingly smaller.)

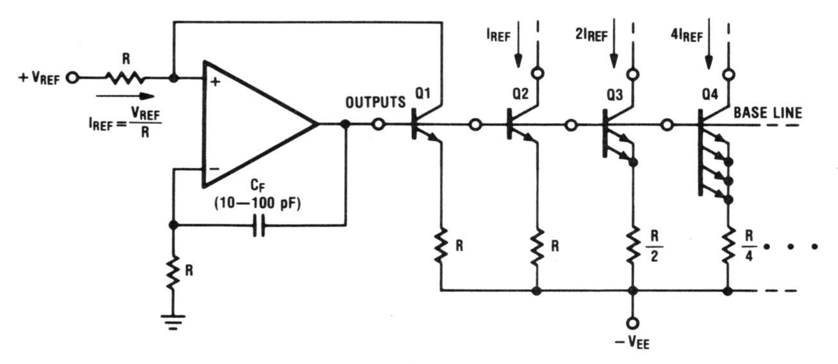

Fig. 5-31. Providing Multiple Binary-Weighted Current Sinks

Current Sources

A basic current source is shown in Figure 5-32. This is the PNP equivalent of the current sink that was discussed in the last section. To increase the precision, p-channel JFETs or a PNP Darlington can be used as was indicated with the previous circuits.

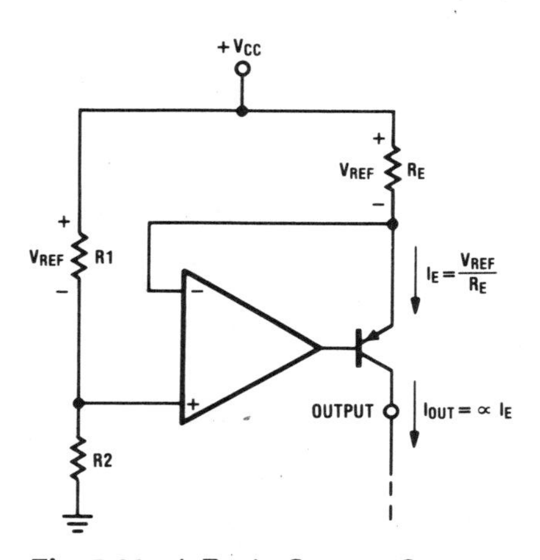

Fig. 5-32. A Basic Current Source

Current Pumps

In our classification, current pumps are circuits that can provide either direction of current flow. These circuits are usually voltage or current controlled so that they can be used to not only vary the magnitude of an output current, but also the sign or direction of the output current flow.

The Howland Current Pump. A single op amp current pump was devised many years ago by B. Howland of the Lincoln Laboratory of MIT. The unusual thing about this circuit, Figure 5-33, is that it is simultaneously making use of both positive and negative feedback. The output current depends upon the difference voltage, V2 - V1. The higher input impedance at the V1 input makes it desirable to ground V2 if this differential control feature is not needed.

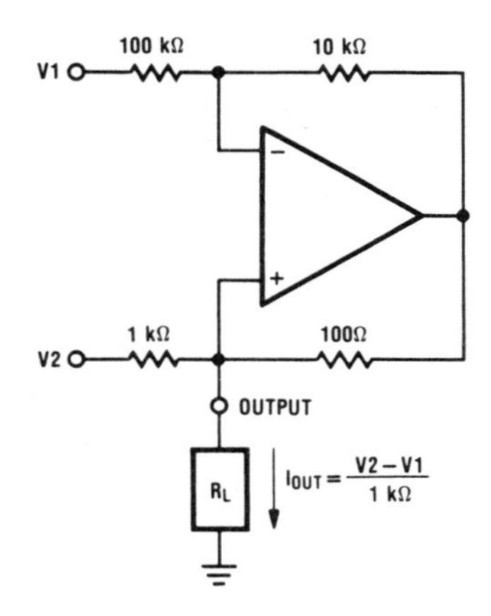

Fig. 5-33. The Howland Current Pump

To intuitively see how this unusual circuit works, conceptually start with V1 equal to zero and V2 equal to + 1 V_{DC} as shown in Figure 5-34a. Now, if R_L were equal to zero, $V_{R_L} = 0$ and the voltage at $V_{IN}(+)$ of the op amp would therefore be zero. The inverting input voltage, $V_{IN}(-)$, of the op amp would then also be zero, so the output voltage of the op amp, V_{OUT}, is 0V. This causes 1 mA to flow from V2 through the 1 kΩ input resistor which provides a 1 mA short-circuit current flow and $I_L = 1$ mA.

Notice that as the value of R_L is increased, the voltage across R_L also increases. This voltage increase at the noninverting input of the op amp forces $V_{IN}(-)$ to also increase. The feedback network that ties to $V_{IN}(-)$ will therefore cause V_{OUT} to raise an amount larger than this increase (in this example, $V_{OUT} = 1.1\ V_{R_L}$) so that *the output of the op amp will supply the added current to the load that no longer is supplied from V2,* because of the now, smaller, voltage drop across the V2 input resistor.

The situation for $R_L = 500\Omega$ is shown in Figure 5-34b. For further increases in R_L; when $V_{R_L} = V2$ as in Figure 5-34c, *the op amp will be supplying all of the load current.* Even further increases in R_L *will now cause current to flow into V2* (we will see a way to improve this in the next section), as shown in Figure 5-34d, where V_{R_L} = 10V. Over this complete range of changes in the voltage across R_L, the current through R_L, I_L, will remain constant, equal to the initial value that existed with R_L shorted out.

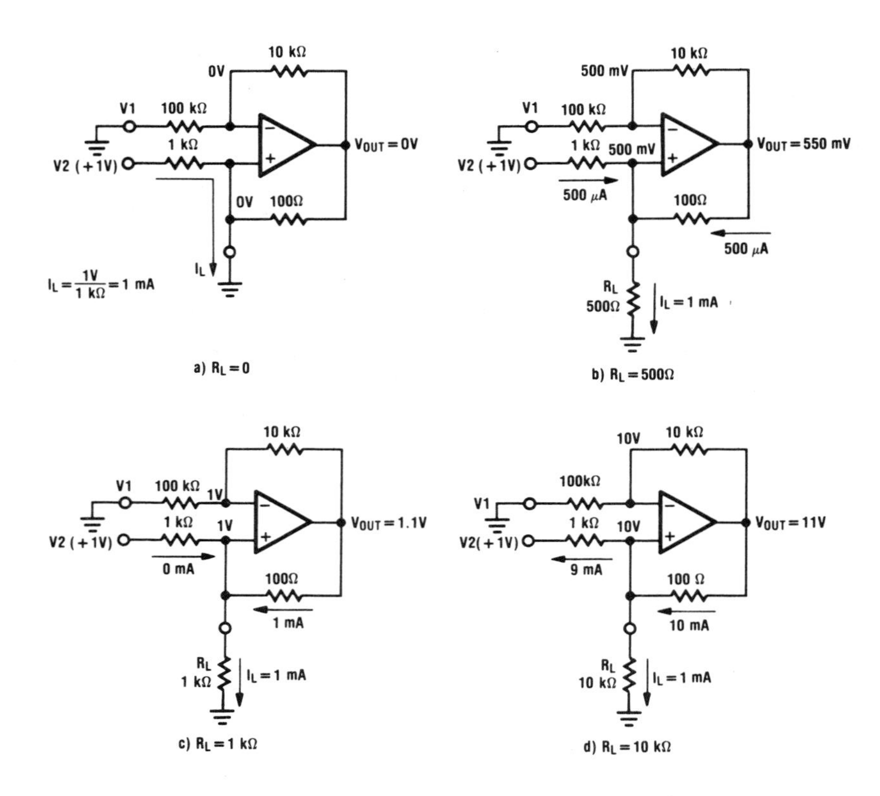

Fig. 5-34. The Circuit Action of the Howland Current Pump

If we now consider also applying a voltage at the V1 input, the output voltage of the op amp will be negative in the earlier example of Figure 5-34a, and therefore will reduce I_L, the short-circuit current. Superposition shows that this same current will always be subtracted from R_L.

Equations that describe this circuit operation can be developed as shown in Figure 5-35, where we have used the resistor ratio of N at the inverting input and a resistor ratio of K at the noninverting input so we can determine the effects on R_{OUT} when N is not exactly equal to K. The analysis of this op amp circuit starts with the determination of the voltage at the noninverting input. The trick here is to then proceed to calculate the output voltage of the op amp, V_{OUT}, that would result because of the R2 network. This expression for V_{OUT} will then be forced to equal a separate calculation for V_{OUT} that is made by assuming that the inverting input voltage is equal to the noninverting input voltage and then working from $V_{IN}(-)$ to V_{OUT}.

By using the circuit schematic diagram to keep track of the progress of the calculation you can prevent ending up in *algebraic oscillation*. (That's why each node voltage is identified with an equation in this figure.)

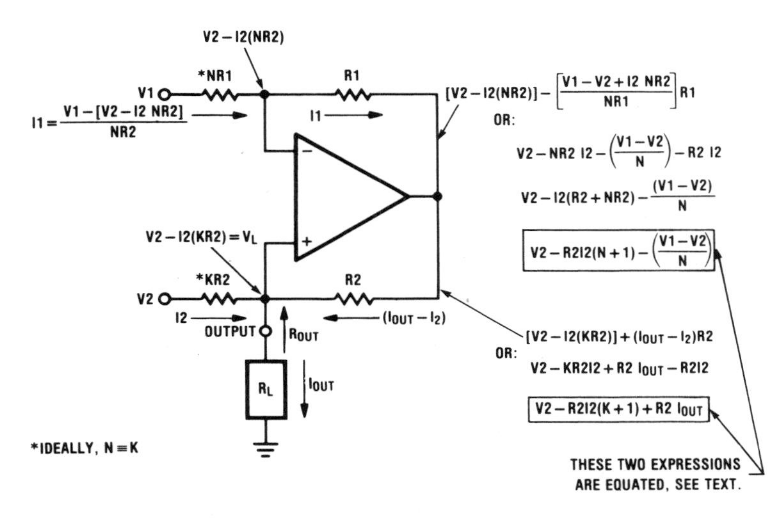

Fig. 5-35. Analyzing the Howland Current Pump

The two expressions for the output voltage of the op amp, from Figure 5-35, are equated to provide an equation for I_{OUT}; thus,

$$V2 - R2\ I2\ (N + 1) - \left(\frac{V1 - V2}{N}\right) = V2 - R2\ I2\ (K + 1) + R2\ I_{OUT}$$

or

$$I_{OUT} = \frac{V2 - V1}{N\ R2} + I2\ (N - K)$$

Note that the second term represents an error current that exists when the bridge is unbalanced ($N \neq K$). We can now determine the effects of this imbalance on the output resistance, R_{OUT}, of this circuit by looking at this second term as an error current, I_ϵ, or

$$I_\epsilon = I2\ (N - K)$$

From this figure we can find that

$$I2 = \frac{V2 - V_L}{K\ R2}$$

where V_L is the voltage across the load, R_L. Substituting this into the previous equation provides

$$I_\epsilon = (V2 - V_L)\left(\frac{N - K}{K\,R2}\right)$$

or

$$I_\epsilon = (V2 - V_L)\left(\frac{\frac{N}{K} - 1}{R2}\right)$$

and

$$\frac{1}{R_{OUT}} = \frac{\partial I_\epsilon}{\partial V_L} = \frac{-\left(\frac{N}{K} - 1\right)}{R2} = \frac{1 - \frac{N}{K}}{R2}$$

or

$$R_{OUT} = \frac{R2}{\left(1 - \frac{N}{K}\right)}$$

If 1% tolerance resistors were used, we could have a worst case with N high by 2% and K low by 2%, but even if we assume that typically we would find each resistor ratio to be within 1% we would then have

$$R_{OUT} = \frac{R2}{\left(1 - \frac{0.99N}{1.01N}\right)} = \frac{R2}{1 - 0.98}$$

or

$$R_{OUT} = 50\,R2$$

(Note that a negative output resistance can also result. The current can actually decrease as the output voltage is increased.)

In the previous circuit (Figure 5-35) with R2 = 100Ω, R_{OUT} would be only 5 kΩ. Therefore there is a benefit to raising the value of R2 and trimming or adjusting one of the four resistors to insure the N = K. (This is most easily done by using a floating current meter that is biased from a small lab power supply and then trimming one of the resistor ratios for a minimum current change as the output voltage, V_L, is changed.)

This circuit also depends on a good CMRR spec for the op amp to obtain a high output impedance, the source for the V2 input must supply the short-circuit current (if R_L is ever shorted out), and the op amp must not be allowed to saturate as a result of the current flow through R2. (This is aided by using a small value for R2.)

The positive feedback that is used causes the output voltage of the op amp to rise as the load voltage rises. An expression for the output voltage of the op amp, V_{OUT}, therefore becomes

$$V_{OUT} = V_L\left(1 + \frac{R1}{N\,R1}\right) - V1\left(\frac{R1}{N\,R1}\right)$$

or

$$V_{OUT} = V_L \left(1 + \frac{1}{N}\right) - V1 \left(\frac{1}{N}\right)$$

For N > 1, this approaches

$$V_{OUT} \cong V_{LOAD}$$

The Improved Howland Current Pump. In the previous example of the Howland current pump circuit action, Figure 5-34, it was seen that a large value of load resistance (10 kΩ) could cause the op amp to output a factor of 10 more current than was required in the load resistance. Most of this current also, unfortunately, was seen to be forced back into the source that drives the V2 control input. An improved Howland current pump, Figure 5-36, was therefore devised to remove this shortcoming of the original circuit. As can be seen, only one extra resistor is required, but the ratio of resistors must be maintained as indicated on this figure. (Typical values for the resistors are also shown and the equation for I_{OUT} also appears on the figure.)

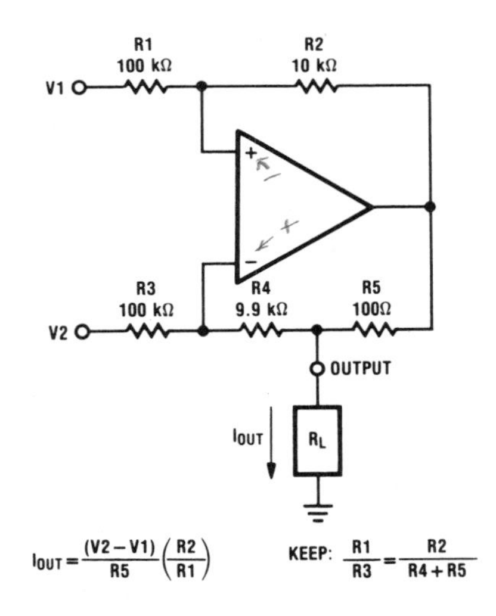

Fig. 5-36. The Improved Howland Current Pump

We can see the improvement if we return to the previous example of 1 mA supplied to a 10 kΩ load resistor (Figure 5-34d) and calculate the currents and node voltages that result with this new circuit, Figure 5-37. The improvement is caused by the relatively large valued resistor, R4, which prevents the high voltage across the 10 kΩ load resistor +10V as compared with the +1V control voltage at the V2 input from pushing a large current out of the V2 control input. This improved circuit is recommended when the voltage across R_L is larger than V2.

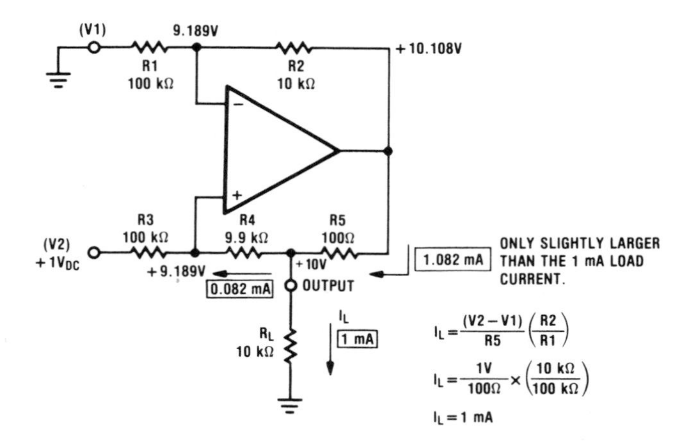

Fig. 5-37. The Improved Howland Current Pump With a 10 kΩ Load

Either of these Howland current pumps can be used to make a Howland Integrator, a noninverting integrator (with V1 = 0) or a difference integrator (V2 - V1), by simply using a grounded capacitor in place of R_L. (We will make use of this in the section on sine wave oscillators later in this chapter.) The output voltage of the op amp will be larger than the voltage directly across the capacitor because of the noninverting voltage gain that exists from the noninverting input of the op amp to the output.

A Voltage-Controlled Current Pump. A voltage-controlled current pump is shown in Figure 5-38. This presents a high input impedance to the control voltage, V_C. A positive V_C will produce a negative voltage across the load. This can be seen by inspection. Starting with V_C, there is a net inversion through the two op amp circuits, (+) then (-), to get to the output.

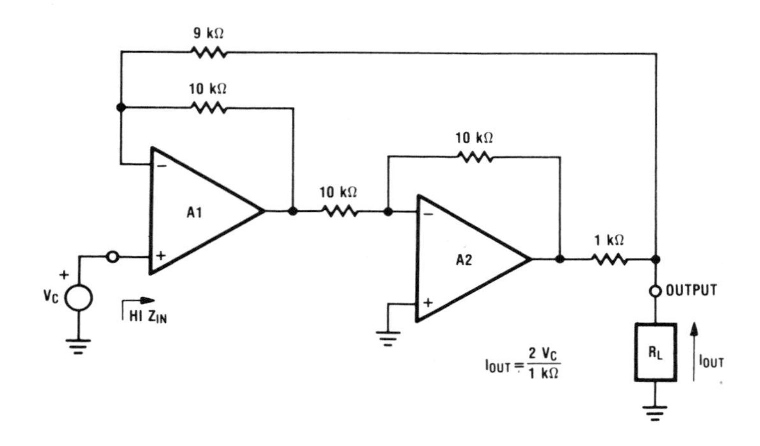

Fig. 5-38. A Voltage-Controlled Current Pump

An analysis of this circuit is shown in Figure 5-39. The two expressions (shown in the figure) for the voltage across the load, R_L, are equated to provide an equation for I_{OUT}, therefore

$$V_C - \left(\frac{V1 - V_C}{R}\right)(R - R_I) = -V1 + \left(\frac{V1 - V_C}{R} + I_{OUT}\right) R_I$$

expanding and canceling terms

$$V_C - (V1 - V_C) + \frac{R_I}{R}(V1 - V_C) = -V1 + \frac{R_I}{R}(V1 - V_C) + I_{OUT} R_I$$

collecting terms

$$2 V_C = I_{OUT} R_I$$

or

$$I_{OUT} = \frac{2 V_C}{R_I}$$

[A load (R_L) should always be kept in place, or replaced with a short-circuit, to keep the amplifier from saturating.]

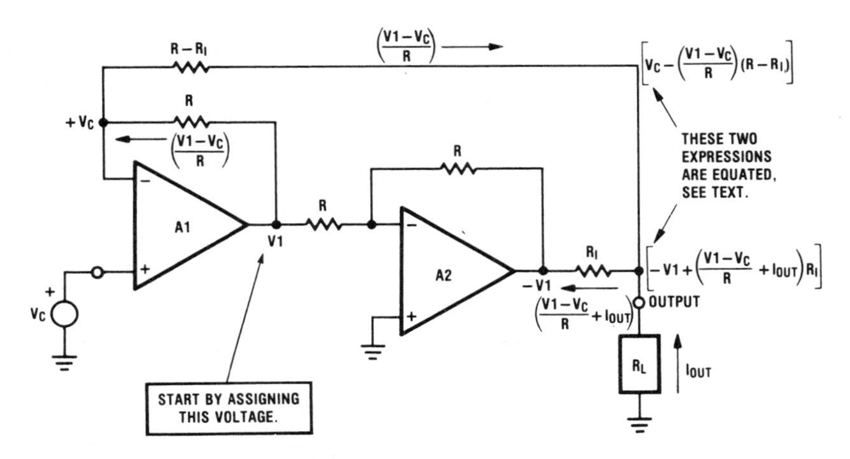

Fig. 5-39. Analyzing the Voltage-Controlled Current Pump

A Current-Controlled Current Pump. A current pump circuit that is basically current controlled, or voltage controlled with input resistors is shown in Figure 5-40. (This circuit was originally developed for chopper-stabilized amplifiers and therefore operates both summing junctions at 0 V_{DC}.)

When operated in the voltage-control mode, this circuit has a lower input resistance than that of the previous circuit. It has three useful features:

1. A positive control voltage, V_C, produces a current out of the circuit or a positive output voltage. It's noninverting because of the use of two inverting amplifiers.

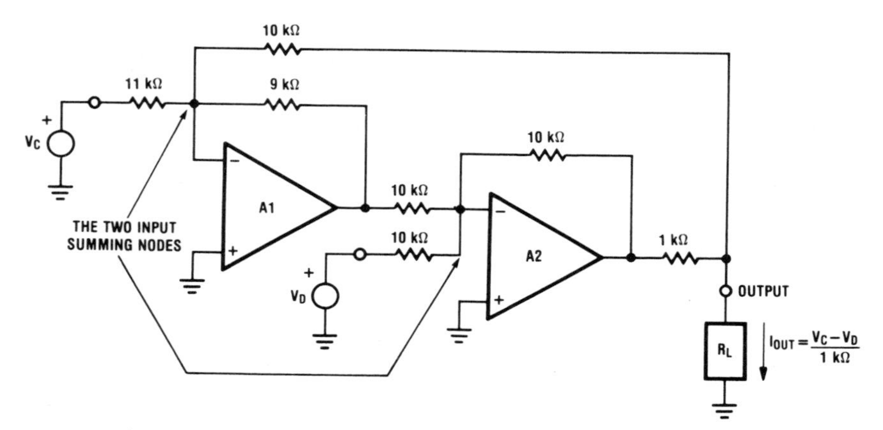

Fig. 5-40. A Noninverting Current-Controlled Current Pump

2. An additional control voltage, V_D, can be used at the input of A2 to provide a differencing feature.
3. The summing node at both control inputs easily allows multiple controls to be used, if desired.

The circuit analysis proceeds as shown in Figure 5-41. After chasing the node voltages and currents around this circuit, and working on equation (1) for I_{R1} to obtain equation (2) for I_{R1}, an expression for I_{OUT} can be obtained by equating the three currents at the inverting input to A1. Thus, from Figure 5-41,

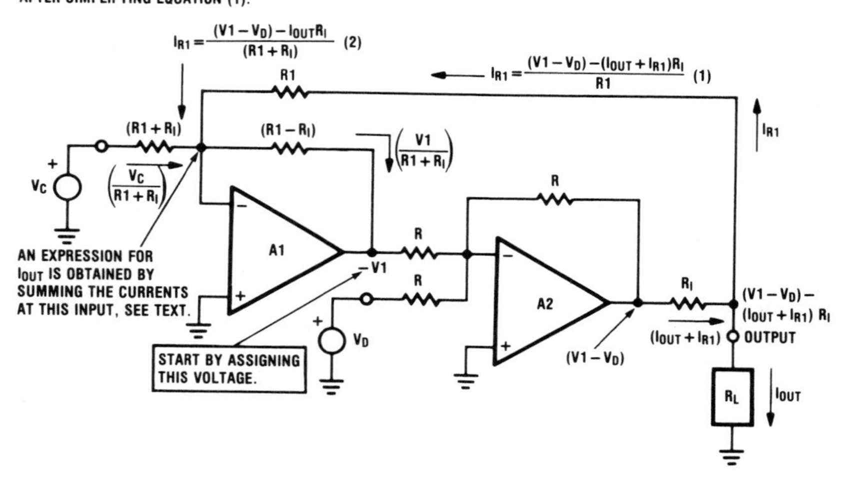

Fig. 5-41. Analyzing the Noninverting Current-Controlled Current Pump

$$\frac{(V1 - V_D) - I_{OUT} R_I}{(R1 + R_I)} + \frac{V_C}{(R1 + R_I)} = \frac{V1}{(R1 + R_I)}$$

$$(V1 - V_D) - I_{OUT} R_I = (V1 - V_C)$$

$$I_{OUT} = \frac{-[(V1 - V_C) - (V1 - V_D)]}{R_I}$$

or

$$I_{OUT} = \frac{V_C - V_D}{R_I}$$

A Precise Current Mirror

In some applications it is necessary to have the sign of an input current changed before it is entered into a summing junction. A circuit that accomplishes this (an input-current sign changer) is shown in Figure 5-42. The resistors labeled R_{B1} and R_{B2} at the noninverting inputs of each op amp equalize the dc resistance to reduce the dc errors.

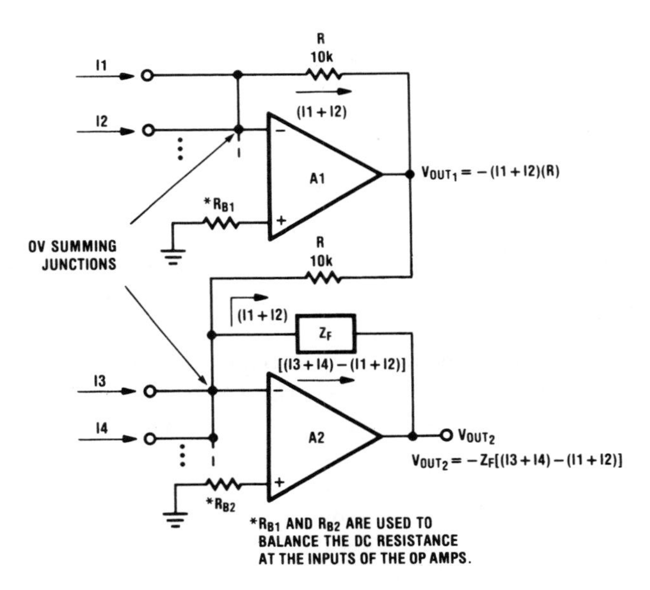

Fig. 5-42. A Precise Current-Mirror

5.5 BOUNDING CIRCUITS

Bounding circuits are voltage sensitive networks that are placed in parallel with the feedback resistor, Figure 5-43. *An ideal bounding circuit introduces no detrimental effects,* such as leakage current flow to the summing junction of the op amp. Bounding networks serve to bypass current and thereby limit the output voltage swing. They are used to prevent saturation of the op amp, to provide an accurate *bound* on the output voltage swing that is allowed for the op amp, to design curve-fitting circuits, or to provide protection when operating with excessive input voltage levels.

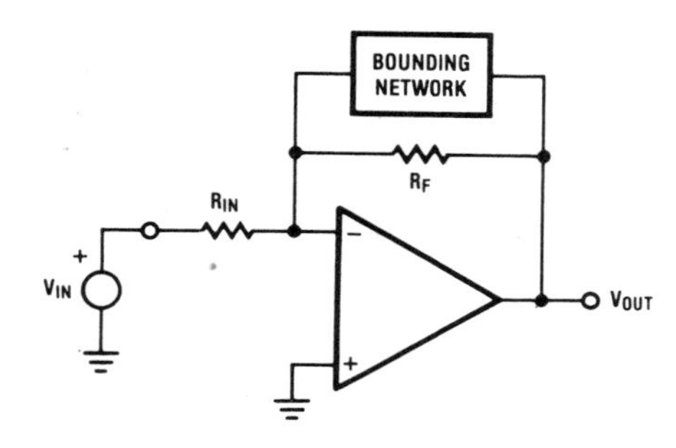

Fig. 5-43. Bounding Networks Shunt the Feedback Resistor

Op Amp Saturation Kills Speed

Among digital circuit designers, it is well-known that *high-speed circuits should not be allowed to saturate.* This has caused the development of the Schottky-TTL circuits and emitter-coupled logic, ECL.

This saturation problem is often overlooked in linear designs even though op amps take much more time to recover from saturation than the much shorter recovery time of a single, gold-doped, switching transistor. This creates recovery-time delays with op amp Schmitt-trigger circuits. (The added internal power dissipation during saturation can also create thermal settling problems in the output voltage response.) Saturation of the op amp will also introduce a phase lag in any type of oscillator circuit that can make the resulting frequency unpredictably low and also temperature dependent.

When an op amp is driven into saturation, the compensation capacitor, in time, assumes an unusually large, or an unusually small voltage. The recovery from a saturated state requires that a proper voltage once again be established on this internal frequency compensation capacitor. This causes unusually long recovery time delays for the op amps which have a slow slew rate spec and those op amp designs that allow a large voltage to exist across the comp cap. (Some op amp designs inherently limit the voltage swing across the comp cap to between zero and two-diode forward voltage drops.) A large input voltage error can therefore exist for a relatively long time interval (10's of μsec) for some op amps.

If the time delay that results from saturation can be tolerated, the next concern is the uncertainty in the magnitude of the output voltage swing that results when the op amp is saturated. If all this is still OK, then let the op amp saturate. If not, some form of antisaturation circuit must be used. This is the job of *bounding networks. They are the Schottky clamps for the op amp application circuits.*

Zener Bounding Circuits

A collection of zener bounding circuits is shown in Figure 5-44. Improvements in performance are achieved when the zener diode is initially biased, because this eliminates the rounding or soft-breakdown effects (Figure 5-44c and d).

In the double bounded circuits of Figure 5-44b and c, unsymmetrical voltage bounding can be easily achieved. For example, one zener diode and the associated biasing resistor in Figure 5-44c can be removed and the remaining diode directly connected to the output to provide a single polarity of output voltage swing.

The symmetry of double bounding is greatly improved and the cost is also reduced by the use of the circuit of Figure 5-44d. (Symmetrical clipping also reduces the even harmonic content when bounding is used to control the amplitude of sinusoidal oscillator circuits.)

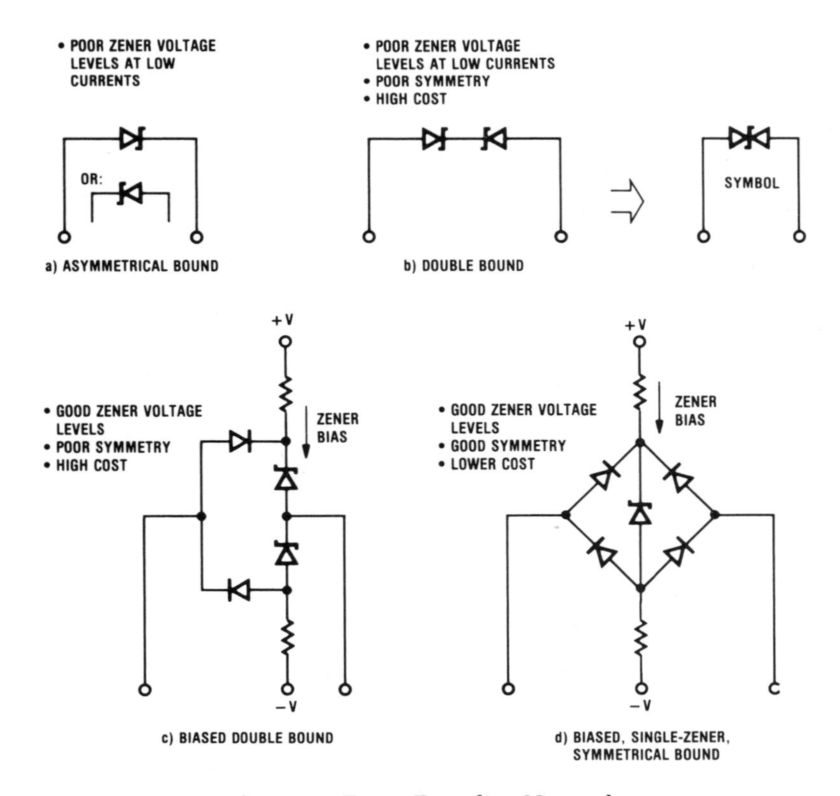

Fig. 5-44. Zener Bounding Networks

Diode Bounding Circuits

A collection of diode bounding circuits is shown in Figure 5-45. The simple single diode and double diode circuits are useful in voltage comparators. (They will be described later in this section).

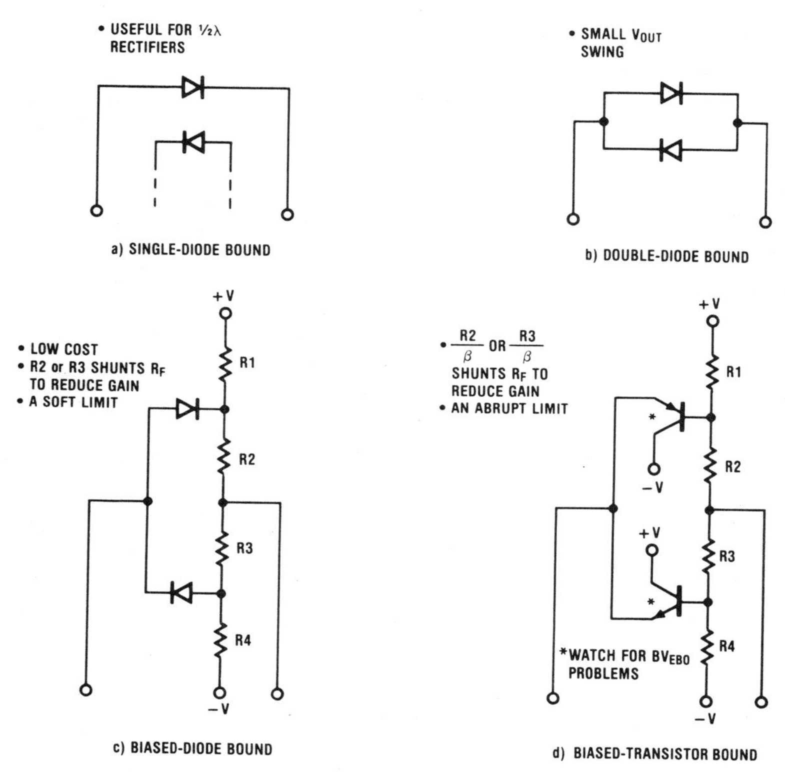

Fig. 5-45. Diode and Transistor Bounding Networks

The biased diode bounding circuit of Figure 5-45c is useful to create a gain change at a specified value of the output voltage swing for curve fitting applications. The resistor values for this circuit can be easily calculated by noting that the upper diode is biased OFF for an output voltage of 0V and will conduct when the output voltage swings to a predetermined negative voltage. By assuming the desired negative output voltage to exist at the bottom end of R2, the proper ratio of R1 to R2 can be calculated for a given reference voltage, + V (or positive power supply voltage), by requiring −0.6V to result at the tap point of R1 and R2. This small negative voltage will bring the diode into conduction at the correct value of negative output voltage. (Sometimes extra circuitry is added to temperature compensate this diode voltage.)

In a similar manner, the resistors R3 and R4 can be calculated. For symmetrical limiting, and if $|+V|$ equals $|-V|$, then let R4 equal R1 and R3 equal R2.

Once the ratio of these resistors has been calculated, the value for R2 (or R3) can be selected to provide the proper shunting of the feedback resistor, R_F, to produce the desired gain reduction. Using the previously calculated value for the ratio of R1 to R2, the value for R1 can then be determined.

As an example, the circuit shown in Figure 5-46 demonstrates a gain reduction from -5 to -1 when the output voltage is greater than ± 10V.

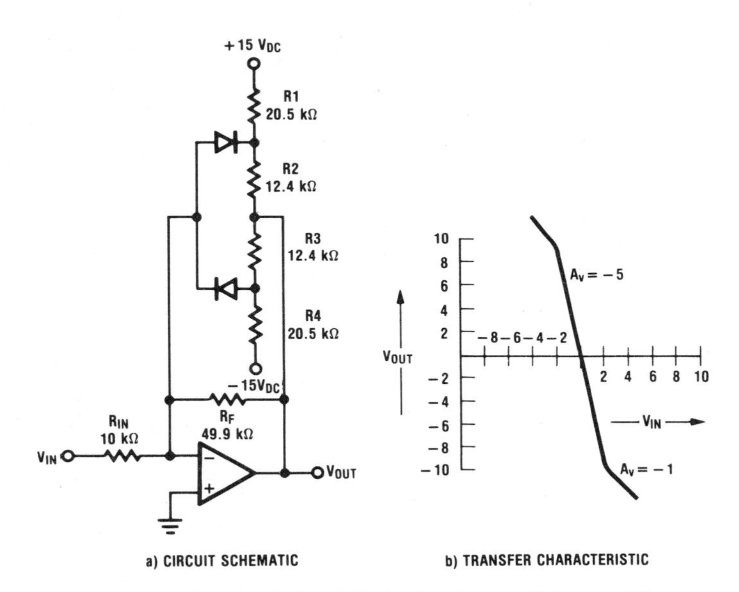

Fig. 5-46. Reducing Voltage-Gain for Large Values of V_{OUT}

The calculation of R1 and R2 proceeds as follows: Because of symmetry, let R4 = R1 and R3 = R2. A simplified circuit for this calculation is shown in Figure 5-47. From this figure we want the tap voltage between R1 and R2 to equal -0.6V (to cause D1 to be ON). The equation for this tap voltage is

$$-0.6V = -10V\left(\frac{R1}{R1 + R2}\right) + 15V\left(\frac{R2}{R1 + R2}\right)$$

Multiplying by (R1 + R2) and dividing by R2 provides

$$-0.6V\left(\frac{R1 + R2}{R2}\right) = \frac{-10\,R1 + 15\,R2}{R2}$$

simplifying

$$-0.6\left[1 + \frac{R1}{R2}\right] = 15 - 10\left(\frac{R1}{R2}\right)$$

collecting terms

$$\left(\frac{R1}{R2}\right)(10 - 0.6) = 15 + 0.6$$

or

$$\left(\frac{R1}{R2}\right) = \frac{15 + 06}{10 - 0.6} = \frac{15.6}{9.4} = 1.66$$

To limit A_v to -1 with $R_{IN} = 10\ k\Omega$
We want: $R2||R_F = 10\ k\Omega$

or

$$\frac{R2\ R_f}{R2 + R_f} = 10\ k\Omega$$

Normalizing both resistors to kΩ, this becomes

$$R2\ R_f = 10\ R2 + 10\ R_f$$

or

$$R2\ (R_f - 10) = 10\ R_f$$

and we have initially selected $R_f = 50\ k\Omega$ to provide the maximum gain of -5, so

$$R2 = \frac{10\ R_f}{R_f - 10} = \frac{10\ (50)}{50 - 10} = 12.5\ k\Omega$$

and

$$R1 = 1.66\ (12.5\ k\Omega) = 20.75\ k\Omega$$

These resistors (using closest 1% standard values) were shown in Figure 5-46.

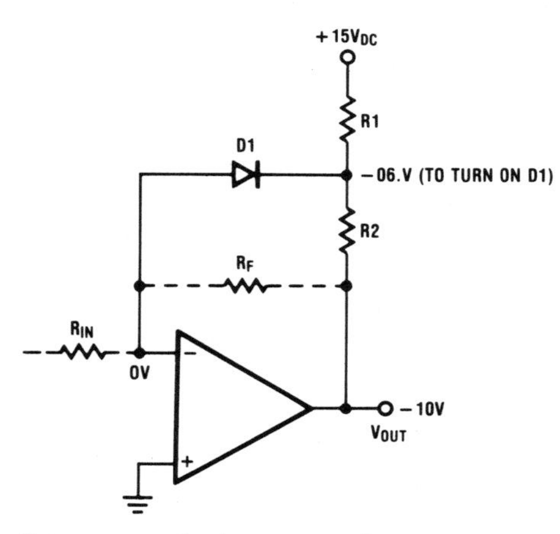

Fig. 5-47. Circuit Conditions to Calculate R1 and R2 of a Diode Bounding-Network

The transistor circuit that was shown in Figure 5-45d is used to abruptly reduce the gain. The effective values for R2 or R3 are reduced by the current gain, β, of the respective transistor to create an abrupt gain loss once the limit voltage is reached.

A major problem with this circuit is that the emitter-base, EB, junction of each transistor is placed in a reverse-biased, breakdown-inducing mode. Modern silicon transistors have EB breakdowns of 2 to 7 volts, with a few as high as 10V. This limits the peak-to-peak output voltage swing that can be obtained with this circuit. Transistors for this application should be selected to meet the required EB breakdown specification.

Reducing the Effects of Leakage Currents

A disadvantage of bounding circuits is that they are tied directly to the critical summing junction of the op amp. This, unfortunately, allows any small leakage currents that are produced by the components that are used in the bounding circuit (when it should be dormant), to create dc errors in the output voltage of the op amp. To prevent this, a leakage current isolation circuit can be added to any of the previously described bounding networks.

The idea of leakage isolation, Figure 5-48, is to provide a low resistance path (the R shown) to ground for the leakage current. Now, the added diodes will have essentially zero volts on both sides of them. The summing junction is at ground and the voltage across R is also essentially ground. As a result, the diodes will neither be forward biased nor create leakage currents of their own. This leakage isolation circuit is recommended for wide operating temperature range and high precision, where the leakage contamination of the bounding network may not be tolerated.

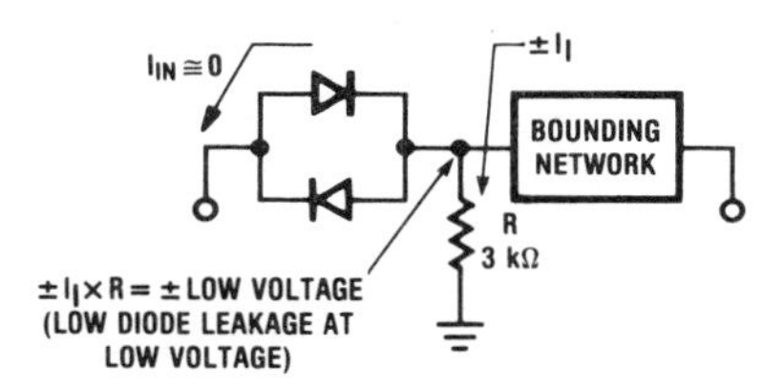

Fig. 5-48. A Leakage-Current Isolation Circuit

Diode leakage varies with the type of diode selected. For example, the fast diodes, such as the 1N914 or 1N4148 are relatively leaky (approximately 12 nA at 30°C). Planar diodes, like the 1N484 are neither quick nor have very low leakage. The collector-base junctions of silicon planar transistors will usually be found to leak approximately 20 pA at 10 V_{DC}. This is lower leakage current than that of most commercially available diodes. The diode connection of a silicon planar transistor (achieved with a collector-base short) also has low leakage but is limited to only 2 to 4V reverse bias. For dc precision, low leakage diodes are recommended. A few measurements of the leakage currents vs. reverse voltage of your favorite diodes and silicon transistor junctions can be very revealing. (We will discuss this in Chapter 6. Measuring pA currents is not easy and teflon or air are the only recommended insulators to be used in the test circuit).

An Unusual Circuit Application. An unusual use of a bounding network, that also uses a leakage current isolation circuit, is shown in Figure 5-49. The idea here is to sharpen up the output voltage transition of a current-comparator circuit. This results because the bounding diodes, D2 and D4, are biased by the current flow through R2. Forward-biasing these diodes eliminates the gradual turn ON that would otherwise result. The transfer function, Figure 5-49b, shows the improvement that is obtained when compared to the use of only the bounding diodes. This circuit significantly raises the *gain* in the vicinity of the output voltage transition.

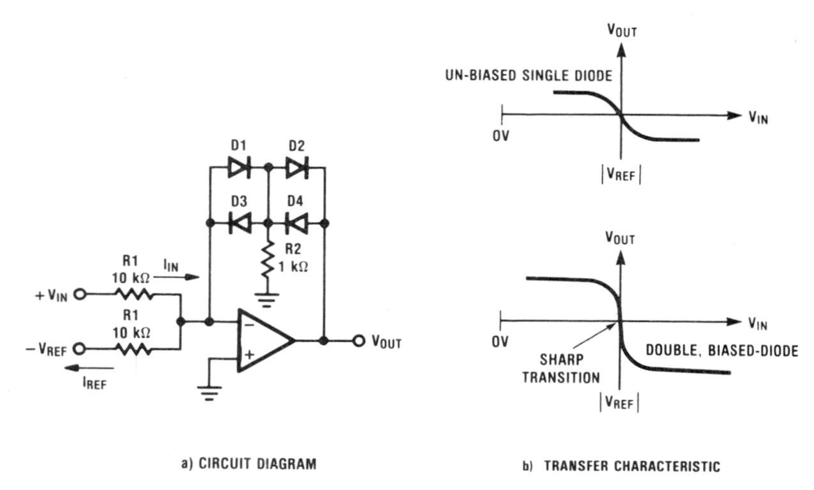

Fig. 5-49. Current-Biased Current-Crossing Detector

The Half-Wave Rectifier Is Only Half-Bounded

Unsymmetrical bounding will provide a half-wave rectifier, Figure 5-50, that can be used to respond to input voltages of only one polarity. The performance of this relatively simple circuit is actually very precise at low frequencies because *the diode is within the loop.* The transfer characteristic, Figure 5-50b, has a clean break at zero volts and there are no problems with soft turn ON of the diode.

The modifications shown in Figure 5-50c can be made to adapt this circuit for rectifier service. Note the use of the 100Ω isolating resistor to prevent the capacitor from causing amplifier instability and the feedback capacity, C_f, that is used to bypass the output circuitry at high frequencies.

Providing Gain. Gain can also be provided in a half-wave rectifier, Figure 5-51. One or both of the outputs can be used. If only one output is needed, the resistor for the unused output can be eliminated and the associated diode is then connected directly from the output to the summing junction.

For rectifier service, a capacitor can be added across the feedback resistor. This RC network provides an averaging circuit that will provide a rectified dc output voltage.

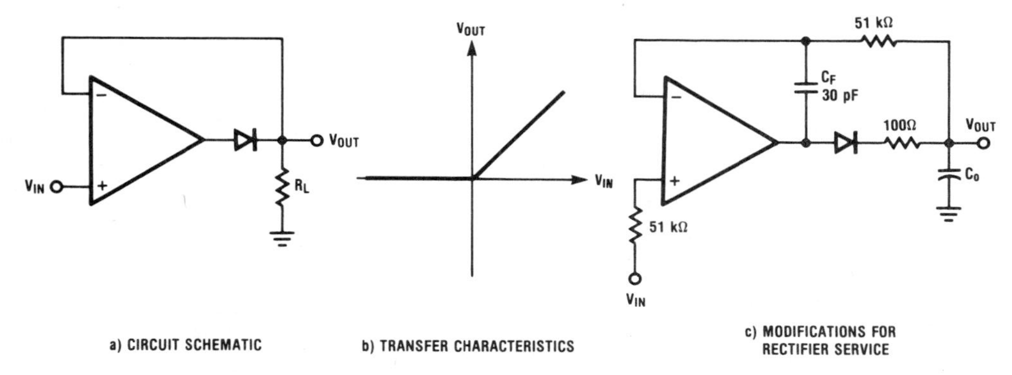

Fig. 5-50. A Precision Rectifier for Low Input Frequencies

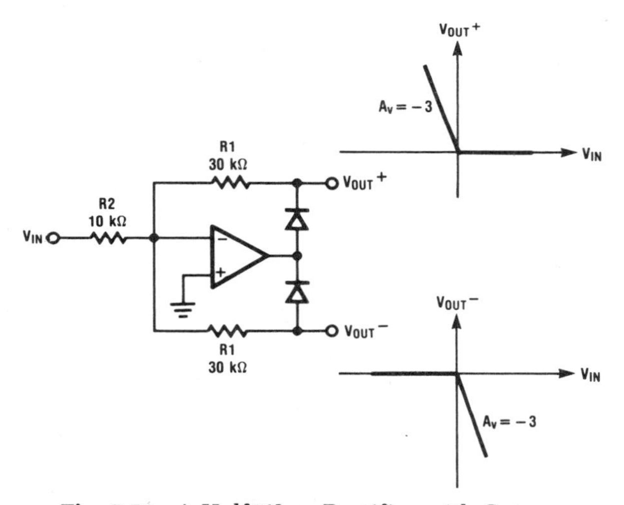

Fig. 5-51. A Half-Wave Rectifier with Gain

A Precision Analog Switch Using Forced Bounding

An analog switch, Figure 5-52, that can select either of two inputs will result by forcing the amplifier associated with the undesired input signal into a bounded state.

Precision in this circuit requires good resistor matching, small values of diode leakage currents, and small values of the dc noise sources of the op amps. The resulting output signals are provided without inversion and gain can easily be taken by increasing the value of the feedback resistor that is used with A4.

The Limiter, a Precise Bounding Circuit

Another slight modification of the half-wave bounding circuit provides a precise limiter, Figure 5-53. This circuit can be used to control the maximum value of the output voltage of an op amp.

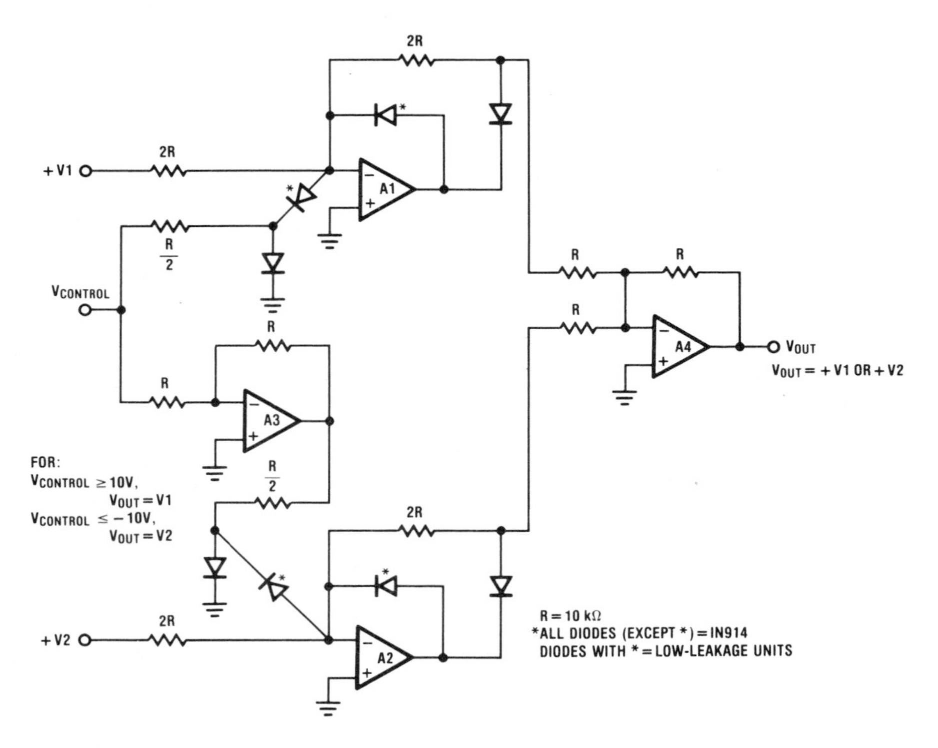

Fig. 5-52. A Precision Analog Switch Using Forced Bounding

Rather than forcing the bounding state, this circuit makes use of voltage references ($+V_{REF}$ and $-V_{REF}$) that supply a fixed amount of *bounding current* at the summing junctions of the bounded op amps, A1 and A3. This causes the outputs of these op amps to be bounded at 0V until the input signal, V_{IN}, becomes large enough to satisfy this bounding current. For a slightly larger V_{IN} signal, the bounded op amp will *wake up* and start to supply an output voltage.

As long as the input voltage creates an input current that is smaller than either of the bounding currents, the outputs of both of these bounded op amps will remain at 0V. Over this input voltage range, the bounding amplifiers can be removed from consideration because they are supplying a 0V input signal to the output summing amplifier, A2. This provides a unity-gain inverter from V_{IN} to V_{OUT}.

If either of the bounding op amps becomes active, it generates an inverted input signal at its output. This now cancels the effects of any further increases in the input signal. Therefore the output voltage of the final summing amplifier, A2, will remain locked at a value of voltage that is the negative of the input signal that existed just as the bounding op amp came to life.

The performance of this circuit is independent of the characteristics of the diodes that are used. The voltages where limiting occurs will only be as stable as the reference voltages that are used. (Note that the power supply voltages were used for the circuit example.) If desired, load resistors (smaller in value than R) can be added to each of the bounding ampli-

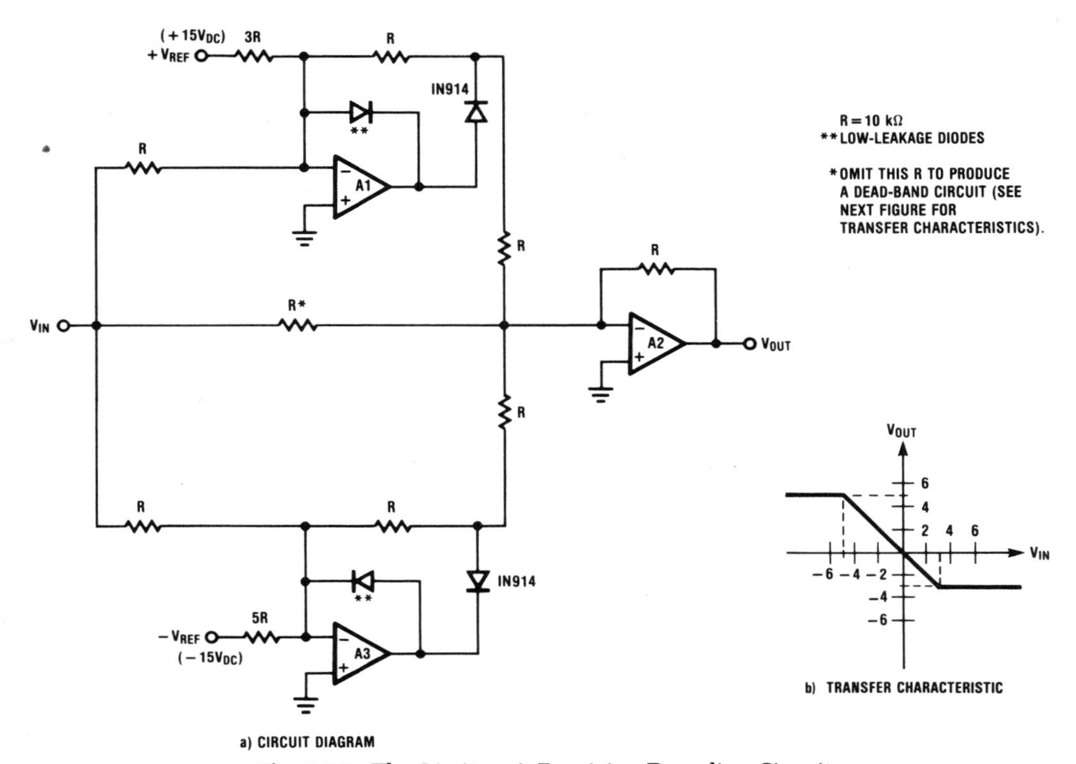

Fig. 5-53. The Limiter, A Precision Bounding Circuit

fiers (from the output side of each diode, where the feedback resistor and the input resistor to A2 are tied to ground) to attenuate the V_{OS} of each of the bounding amplifiers that would otherwise exist at these nodes and be presented to the main amplifier A2.

Converting to a Dead Band Circuit. The previous circuit can be easily modified to produce a dead band circuit if the resistor (shown as R* on the previous figure), that feeds from V_{IN} directly to the summing junction of A2, is removed. Now, for V_{IN} voltages within the range that previously provided a gain of −1, the gain becomes 0 and V_{OUT} therefore stays equal to 0V until the established break-voltages are reached. The gain then becomes equal to + 1, as shown in Figure 5-54. This circuit can be used to prevent taking action in a control system until the error has become significant.

Fig. 5-54. The Transfer Characteristic of a Dead-Band Circuit

Another circuit that can be used to provide a dead band in an amplifier characteristic is shown in Figure 5-55. This is an adaptation of a previous circuit (Figure 5-20, A Tri-State Window Comparator) and shows how a basic circuit idea can be used in different ways to achieve different results.

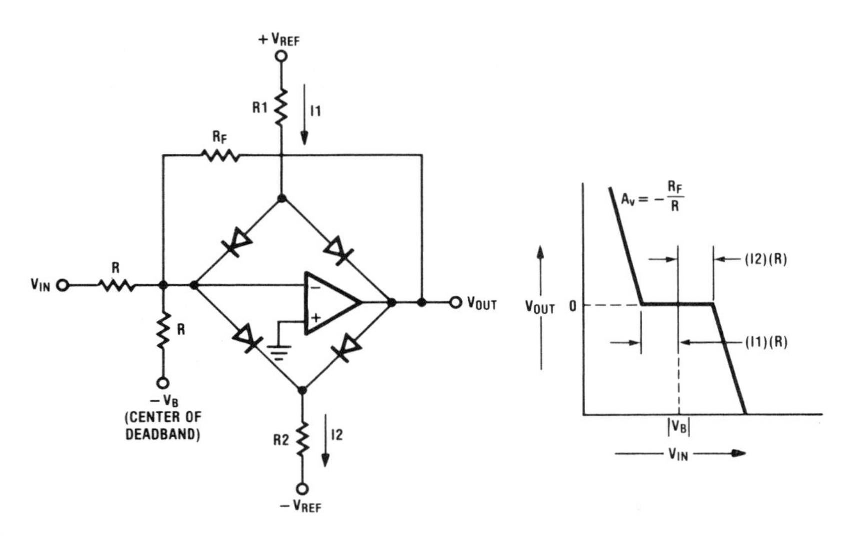

Fig. 5-55. An Amplifier Circuit with a Dead-Band

Full-Wave Rectifiers, the Absolute Value Circuits

Full-wave rectifiers are useful to provide a dc voltage in response to an ac input signal. They will also insure that an output signal has a constant polarity, regardless of the polarity of the input signal: the absolute value property. Additionally, they can be used in a clever way: they will provide a frequency doubling function when the input is a bipolar triangle waveform, Figure 5-56.

A Low Cost Circuit. A low cost full-wave rectifier, shown in Figure 5-57, uses biasing resistors (100 kΩ) to provide equal biasing currents in the diodes.

For a positive input voltage, D1 conducts and the circuit operates as a noninverting voltage follower. A negative input voltage causes D1 to turn OFF (D3 then turns ON) and the circuit now becomes a unity-gain inverting amplifier. In this mode of operation, D3 and R/2 supply the proper voltage for the noninverting input of the op amp.

There is a slight rounding of the transfer characteristic, shown in Figure 5-57b, near the origin as the circuit shifts over in response to a change in the polarity of the input voltage.

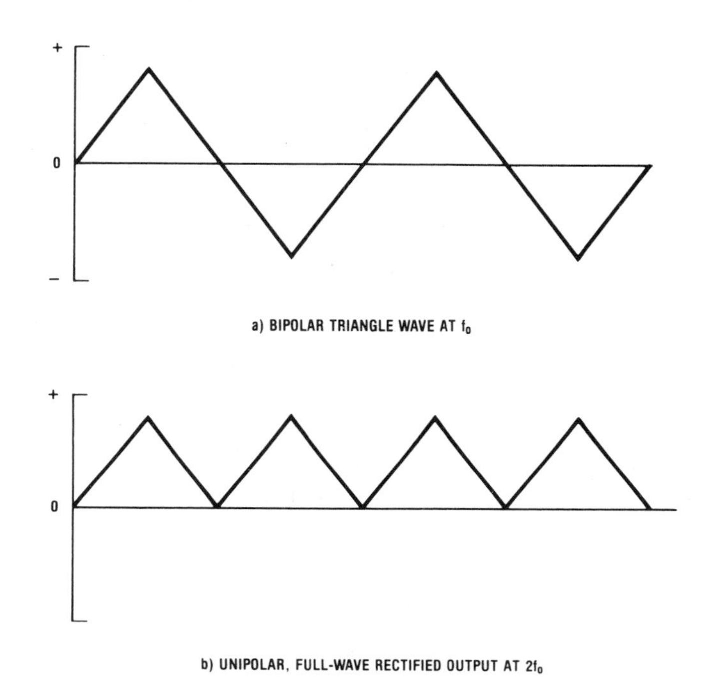

Fig. 5-56. Absolute-Value Circuits Can Be Used for Frequency Doubling

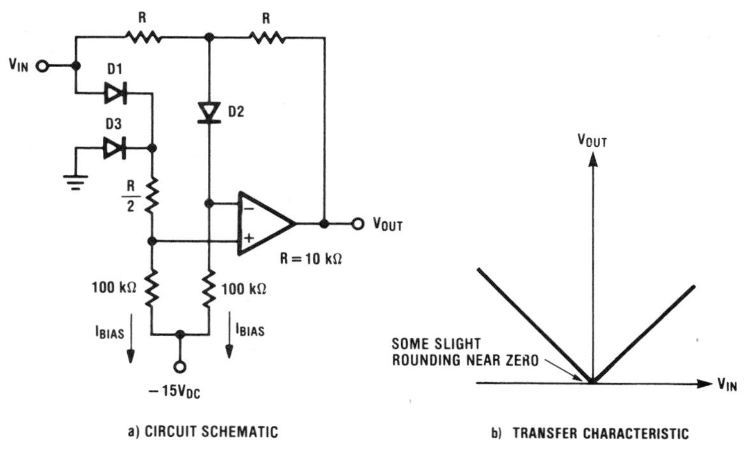

Fig. 5-57. A Matched-Diode Absolute-Value Circuit

Putting the Diodes in the Loop. Performance can be improved by making use of the more complex circuit of Figure 5-58. All of the diodes are now in the loop and there is no longer a problem owing to diode mismatch.

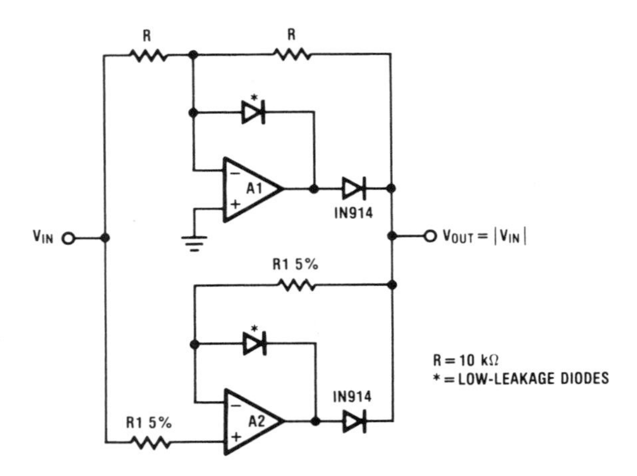

Fig. 5-58. An Absolute-Value Circuit With Only Two Matched Resistors

This circuit requires only two matched resistors (shown as R on the figure) because R1 is simply used to prevent loading of A1 during positive excursions of the input voltage waveform. A balancing resistor in the noninverting input of A2 is used to reduce the bias current error.

There may be a slight glitch due to dc noise source differences between the two op amps as the output control exchanges from A1 to A2 and vice versa. Some op amps may require R (or RC) networks to raise the noise-gain to improve stability. Additionally, this circuit does not have a high input resistance and there is no easy way to add a capacitor *within the loop* if a dc output voltage is desired. The voltage that would result across a single output capacitor to ground *will not be within the loop*, so external circuit loading of this capacitor can be a problem.

Allowing for Input-Signal Summing. A circuit that provides a convenient, single, summing-node input for multiple-signal summation and requires a few more matched resistors is shown in Figure 5-59.

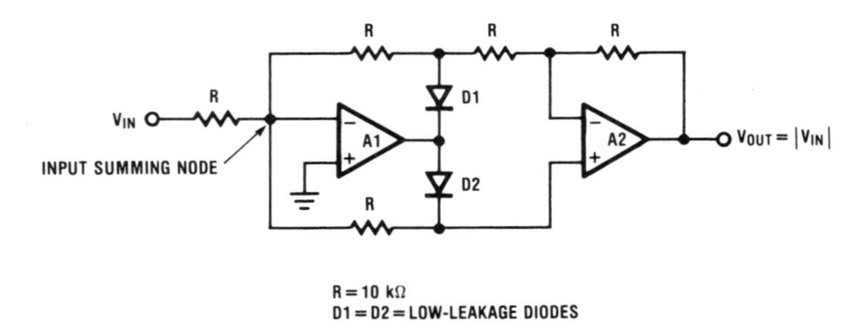

Fig. 5-59. An Absolute-Value Circuit With Input Summing Capability

This circuit operates by driving the output op amp, A2, as either a unity-gain inverting amplifier for V_{IN} positive or something that, at first, may appear incorrect for V_{IN} negative.

Let's take a closer look at what happens to the node voltages of this circuit when V_{IN} is negative, Figure 5-60.

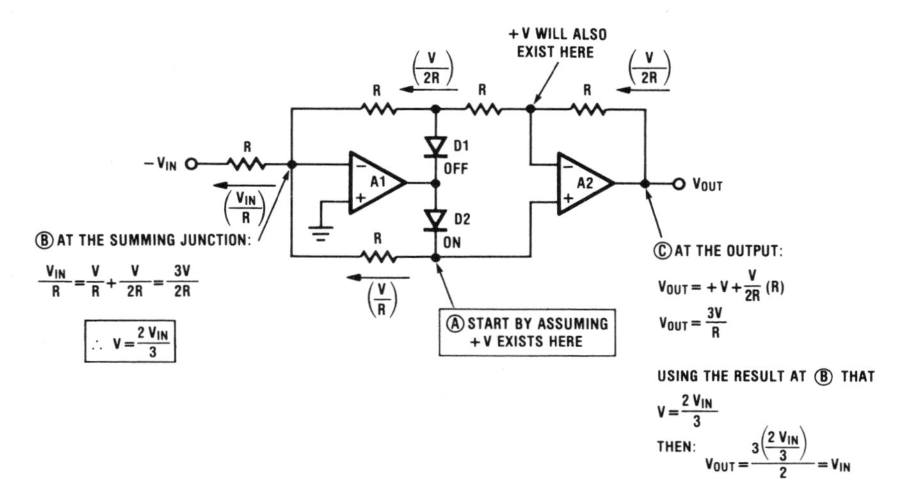

Fig. 5-60. Circuit Conditions for V_{IN} Negative

By starting with the assumption that the noninverting input of op amp A2 will have an initially unknown voltage (represented by V, step A on the figure) we can then determine the currents that would flow at the summing junction of op amp A1 (step B). This evaluates V.

Next (step C), we write an expression for the output voltage of op amp A2 as a function of the unknown voltage, V. By substituting the previously determined value for V, the output voltage can be determined. (Note how op amp application circuits are somewhat like the old *pea and shells game.* Unlikely looking circuits often provide the proper output voltage.)

A capacitor cannot simply be added across the feedback resistor of op amp A2 to obtain a dc output voltage because the summing junction of this op amp undergoes large voltage changes.

Handling Large Input Voltage. An absolute value circuit that can handle large input voltages or eliminate CMRR problems is shown in Figure 5-61. This circuit has an input impedance that is equal to the two input resistors in parallel. Multiple-inputs cannot easily be summed because there are now two input-summing nodes, which undesirably doubles the number of input resistors needed.

For a positive input voltage, the voltage of the R/2 input resistor (of A2) is the inverted input voltage, $-V_{IN}$. The larger current flow, $-2\ V_{IN}/R$, through this small valued resistor causes the output voltage of A2 to rise to V_{IN} to establish a balanced current condition at the inverting input of A2. (The current V_{IN}/R is also supplied by the input voltage source.)

For a negative input voltage, the voltage presented to R/2 is 0V, causing the circuit to operate as a unity-gain inverting amplifier.

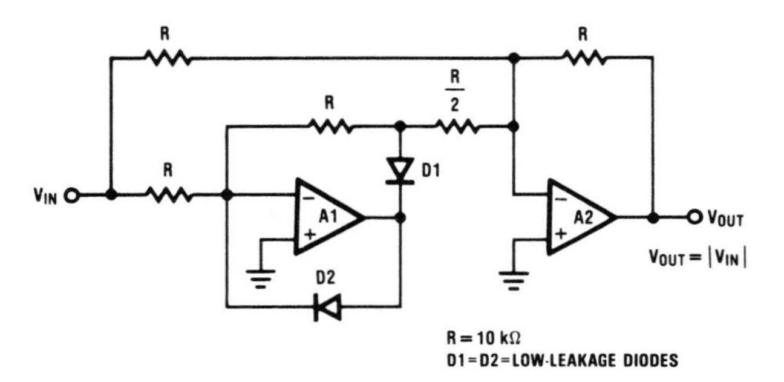

Fig. 5-61. Reducing the Common Mode Input Voltage Swing on an Absolute-Value Circuit

A dc output voltage can be obtained, if desired, by placing a capacitor across the feedback resistor of the second op amp, A2. (Only the circuits that operate with the noninverting input at a fixed voltage, such as ground, allow this ease of obtaining a dc output voltage).

Increasing the Input Resistance. For a higher input resistance, the circuit of Figure 5-62 can be used. This operates in a slightly different manner. When the input voltage is positive, all circuit nodes, except for the output voltage of A1, will bias at the same value as the input voltage.

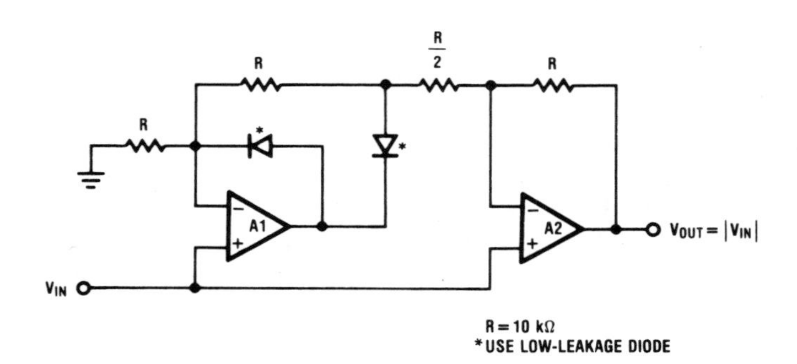

Fig. 5-62. A High Input Impedance Absolute-Value Circuit

For a negative input voltage, the voltage at the input side of R/2 becomes twice this negative input voltage value. The current through R/2 is then 2 V_{IN}/R. This causes a drop of 2 V_{IN} across the feedback resistor, R. When this positive drop of twice V_{IN} is added to the $-V_{IN}$ at the inverting input, V_{OUT} again becomes equal to V_{IN}.

Notice that a capacitor cannot simply be placed in parallel with the feedback resistor of op amp A2 to produce a dc output voltage because the voltage at the noninverting input is not constant.

Another Circuit Possibility. The final absolute value circuit in this collection, Figure 5-63, also presents a high impedance to the input signal source. This has been included for completeness and to show another way to provide an absolute value circuit. A gain of (N + 1) is provided and this, again, uses two amplifiers to alternately supply the output signal voltage.

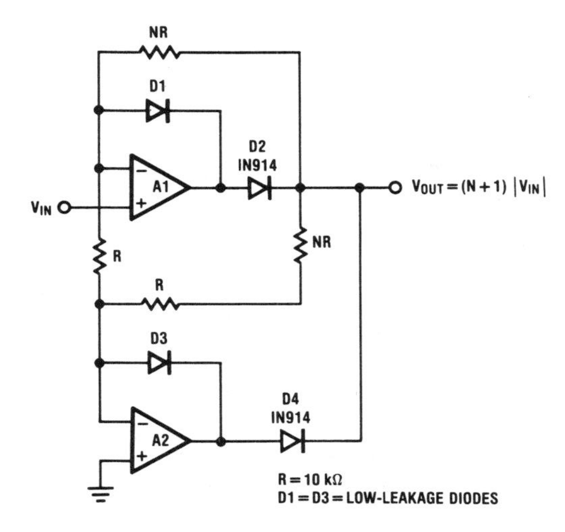

Fig. 5-63. A High Input Impedance Absolute-Value Circuit With Gain

To simplify this circuit, it has been redrawn in Figure 5-64 where a positive input signal is first considered. The main development can be followed by starting at the point labeled A and proceeding to E. All of the node voltages and branch currents are shown on the figure.

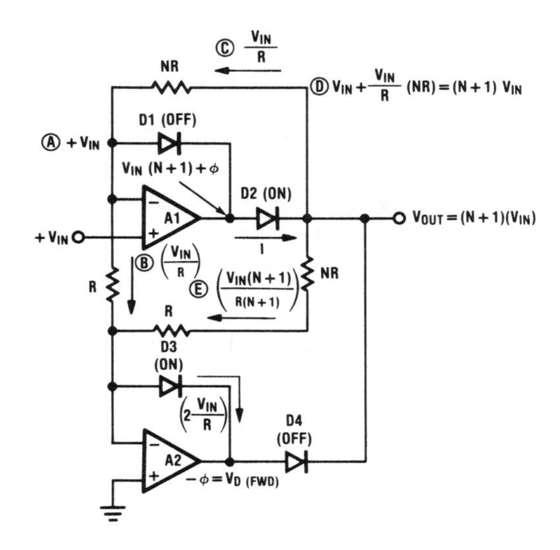

Fig. 5-64. For V_{IN} Positive

The circuit conditions for a negative input voltage are shown in Figure 5-65. Again, the labels indicate the major flow of the analysis.

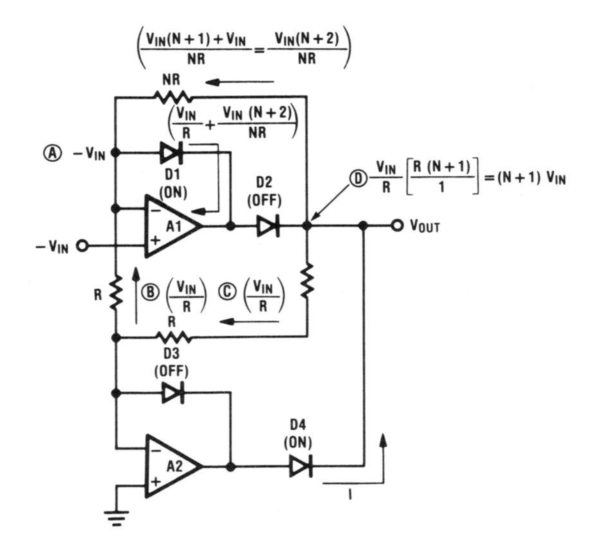

Fig. 5-65. For V_{IN} Negative

All of the previous absolute value or half-wave reactifier circuits used op amps to obtain good dc accuracy. Unfortunately, these circuits tend to be useful at only relatively low frequencies. For high-frequency operation, simple diode rectifier circuits are faster, but suffer from the changes in the magnitude of the forward-voltage drop of the diode as a function of both the varying operating current levels and temperature induced changes in the diode forward voltage drop. The type of circuit shown in Figure 5-66 is good to keep in mind because it can solve many of these problems and it also increases the high-frequency performance. If larger dc offset voltage errors can be tolerated, the two current sources can be replaced by resistors. (The reverse-breakdown limitations of the base-emitter junctions of the transistors should be observed when operating at large input-voltage levels.)

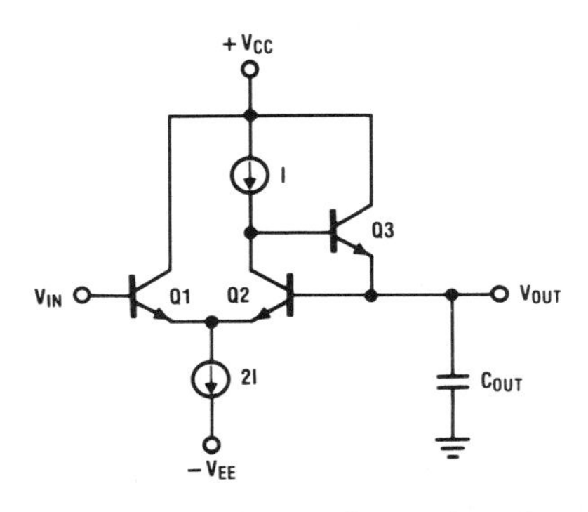

Fig. 5-66. A High-Speed Rectifier Circuit

Part of the justification for the large number of circuits in this section is to indicate the almost limitless variety of application circuit possibilities that exist to accomplish the same end result. There are subtle differences in all of these circuits and, for this reason, usually one of these circuits will be better suited for a particular set of circumstances. Unfortunately, this adds to the complexity of analog circuit design.

Waveform Generators

Many circuits have been published showing how op amps can be used to produce many different types of voltage waveforms. In this section we will include only a few of the more interesting circuits.

Squarewave Generators. Most of the simple circuits that produce squarewaves, Figure 5-67, allow the op amp to saturate. This circuit uses few components and can be easily adapted to operate from a single power supply voltage by adding an additional resistor (equal to 2R) from the $V_{IN}(+)$ of the op amp to the $+V_{CC}$ supply. A comparator can also be used in this same circuit. (A pull-up resistor may be needed from V_{OUT} to the $+V_{CC}$ supply.)

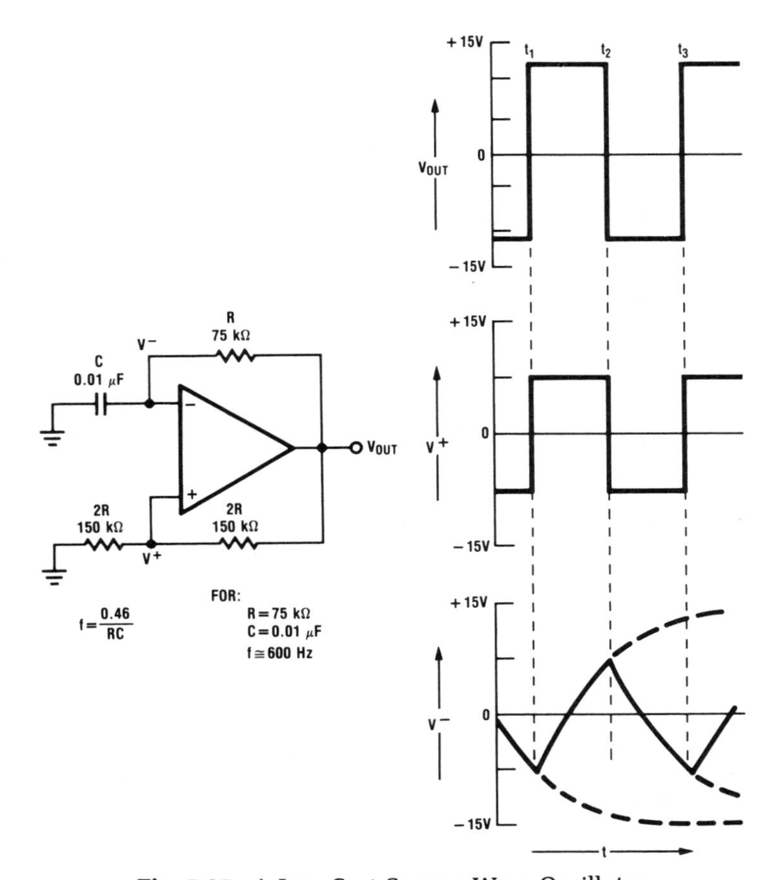

Fig. 5-67. A Low-Cost Square Wave Oscillator

This circuit makes use of rapid positive feedback and delayed negative feedback. Just as the delayed voltage waveform across the capacitor slightly exceeds the voltage at the positive input, the output voltage regeneratively changes state.

To improve both the speed and the stability of the frequency produced, the op amp should be bounded to prevent saturation. A circuit that achieves this (a practical application of a bounding network) is shown in Figure 5-68. The first op amp, A1, is bounded and the second op amp, A2, operates with unity gain (–1), so it will not saturate either.

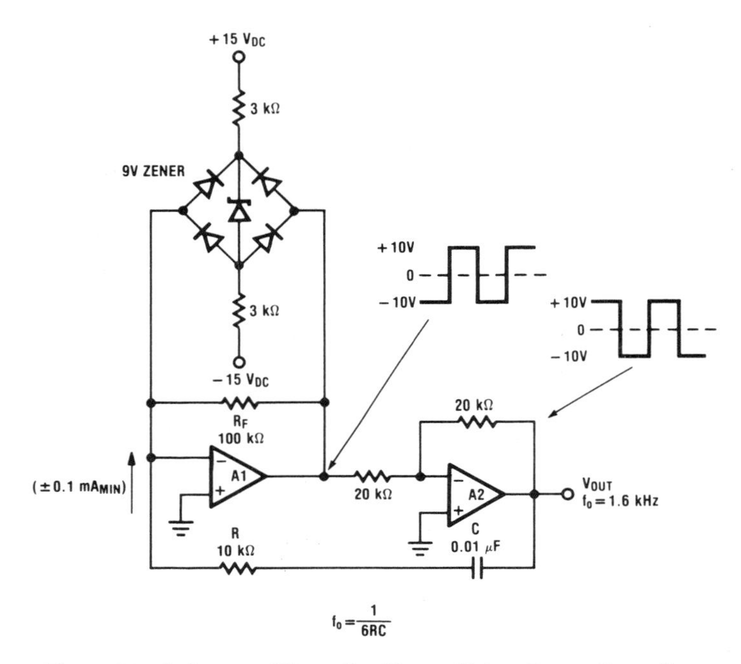

Fig. 5-68. A Square Wave Oscillator Using Zener Bounding

The RC network provides the timing by inputting an RC charging current into the summing junction of A1. Immediately following an output voltage change, this RC network will input a current that is equal, initially, to 20 V/10 kΩ, or 2 mA. This large value of input current would have driven A1 into saturation, but excessive input current is now shunted around R_f by the action of the bounding network. (Only 0.1 mA passes through the feedback resistor.)

As the capacitor charges to the new voltage that now exists at the output of A2, the charging current will exponentially decrease. (This changing current flow through the diodes of the bounding network will cause a change of a few hundred millivolts in the dc levels of the output squarewave). At the time this decaying charging current is just less than 0.1 mA, the output voltage of A1 can no longer be held at 10V. This drop in voltage at the output of A1 propagates around this overall positive feedback loop and regeneratively drives the circuit to the opposite voltage level. The capacitor now starts recharging to the new output voltage of A2 to provide the next half-period of the output squarewave.

The time taken for each half-period can be determined from the exponential decay of the charging current that flows into the timing capacitor. This current starts at 2 mA and holds A1 against the bound as long as it is greater than 0.1 mA, because 0.1 mA of current flow through the 100 kΩ feedback resistor of A1 is needed to keep the output voltage at 10V. This time can be determined by using the RC, exponential charging equation:

$$i_c(t) = \frac{V_{MAX} - 1V}{R} \exp(-t/\tau)$$

where V_{MAX} represents the 20V change at the output of A2 and the 1V is the 0.1 mA drop that exists across the 10 kΩ timing resistor, R, at the time of switching. The time, t, for $i_c(t)$ to equal 0.1 mA can be found from

$$0.1 \text{ mA} = \left(\frac{19V}{10 \text{ k}\Omega}\right) \exp(-t/RC)$$

$$\frac{0.1}{1.9} = \exp(-t/RC)$$

$$-t/RC = \ln(0.05)$$

or

$$t = -RC(\ln 0.05)$$

$$t = 3 \text{ RC}$$

For a 1.6 kHz squarewave, the half-period is 313 μsec, so

$$C = \frac{t}{3R}$$

$$C = \frac{313 \times 10^{-6}}{3(10^4)} \cong 0.01 \ \mu F$$

To temperature compensate this circuit, the zener diode used in the bounding network should have a breakdown of approximately 8 to 9V to provide a + TC of approximately + 4 mV/°C. When the approximately −2 mV/°C each of the forward voltage drops of the two diodes within the bounding network are added to this, a nearly zero change with variations in ambient temperature will exist in the squarewave voltage levels.

Amplitude-Bounded Sine Wave Oscillators. The classic analog computer network (discussed in Chapter 1) that was used to provide sine wave voltage-waveforms for a short time duration can be modified, as shown in Figure 5-69, by increasing the loop gain to be larger than one, because of the added positive feedback resistor, R_f, and adding a bounding network to produce a limited-amplitude, steady-state oscillator. This basic circuit became very popular as an RC active filter because it provides good frequency stability. Changes in the values of the passive components do not greatly shift the operating frequency or other performance specs of the filter (gain and Q). Both sine and cosine waveforms are simultaneously available from the low output impedance of fed back op amps. None of the amplifiers saturate and the bounding network introduces resistive damping which prevents uncontrolled build-up in the amplitude of oscillations.

ing network introduces resistive damping which prevents uncontrolled build-up in the amplitude of oscillations.

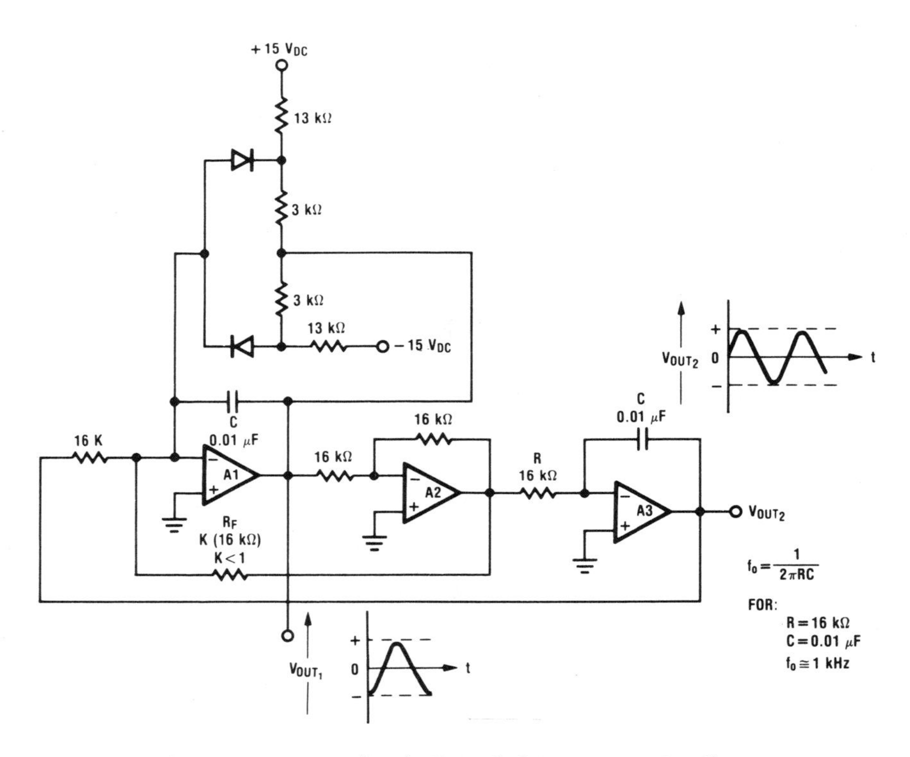

Fig. 5-69. An Amplitude-Bounded Sine Wave Oscillator

The positive feedback resistor, R_f, should be selected to provide reliable start-up but not excessive overall loop-gain.

This same type of oscillator circuit can be realized by making use of the noninverting Howland Integrator, with a "gain" of 2, to replace both a "unity-gain" inverting integrator and the unity-gain sign-changing amplifier of the previous circuit. The overall loop now has a gain of 2, so oscillations will exist without adding the extra resistor, R_f, that was included in the previous circuit. This simpler circuit is shown in Figure 5-70. A further circuit simplification is to use a single bound to prevent uncontrolled amplitude build-up. This circuit has been biased to operate with a single power supply voltage.

An Amplitude-Regulated Sine Wave Oscillator. A high performance sine wave oscillator can be realized by adding additional circuitry to measure and then control the amplitude of the oscillation. This will reduce the distortion and can provide temperature independence of the output voltage amplitude. A block diagram of this type of oscillator is shown in Figure 5-71. The idea is to purposely provide a positive feedback loop and then to electronically regulate

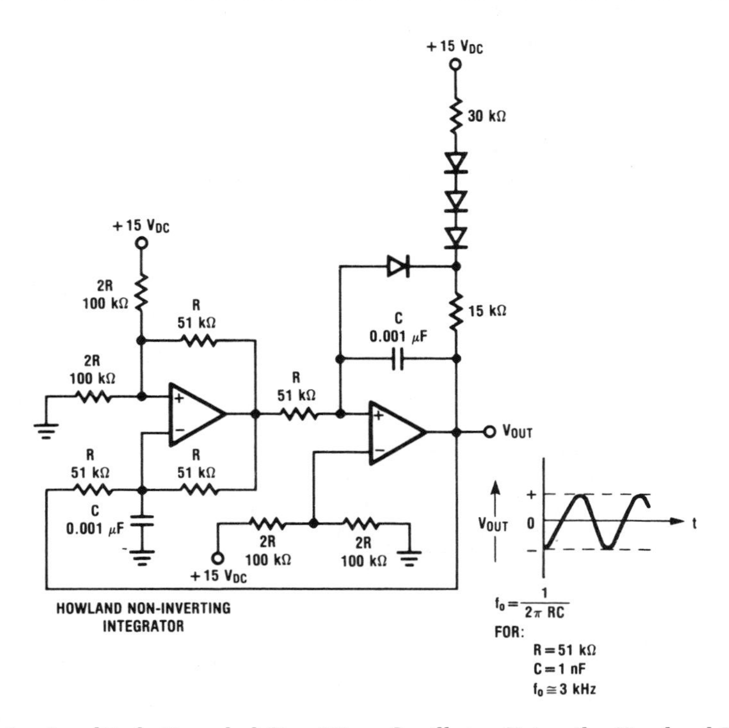

Fig. 5-70. An Amplitude-Bounded Sine Wave Oscillator Using the Howland Integrator

the overall gain of this loop. This gain control is based on measuring the amplitude of the resulting output sine wave signal, comparing this measurement to a reference voltage, and then automatically adjusting the gain of the positive feedback loop to keep the output sine wave voltage proportional to the reference voltage.

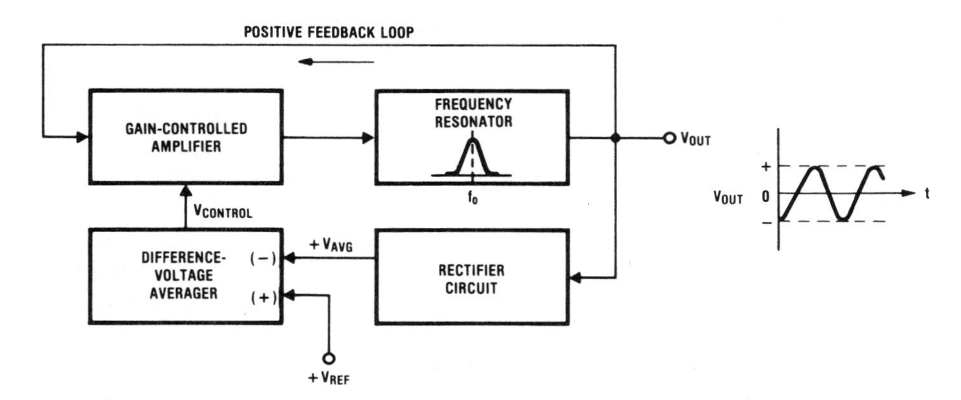

Fig. 5-71. The Diagram of an Amplitude-Regulated Sine Wave Oscillator

One of the many circuit implementations to achieve this type of an oscillator is shown in Figure 5-72. This interesting, although rather complex, system uses a single, 14-pin package of Norton, or current-differencing amplifiers (the LM3900). The three main sections of this oscillator are the RC Active Bandpass Filter, the Difference Averager, and the Gain-Controlled Amplifier. We will look at this oscillator design technique rather closely to provide an indication of how this type of design proceeds.

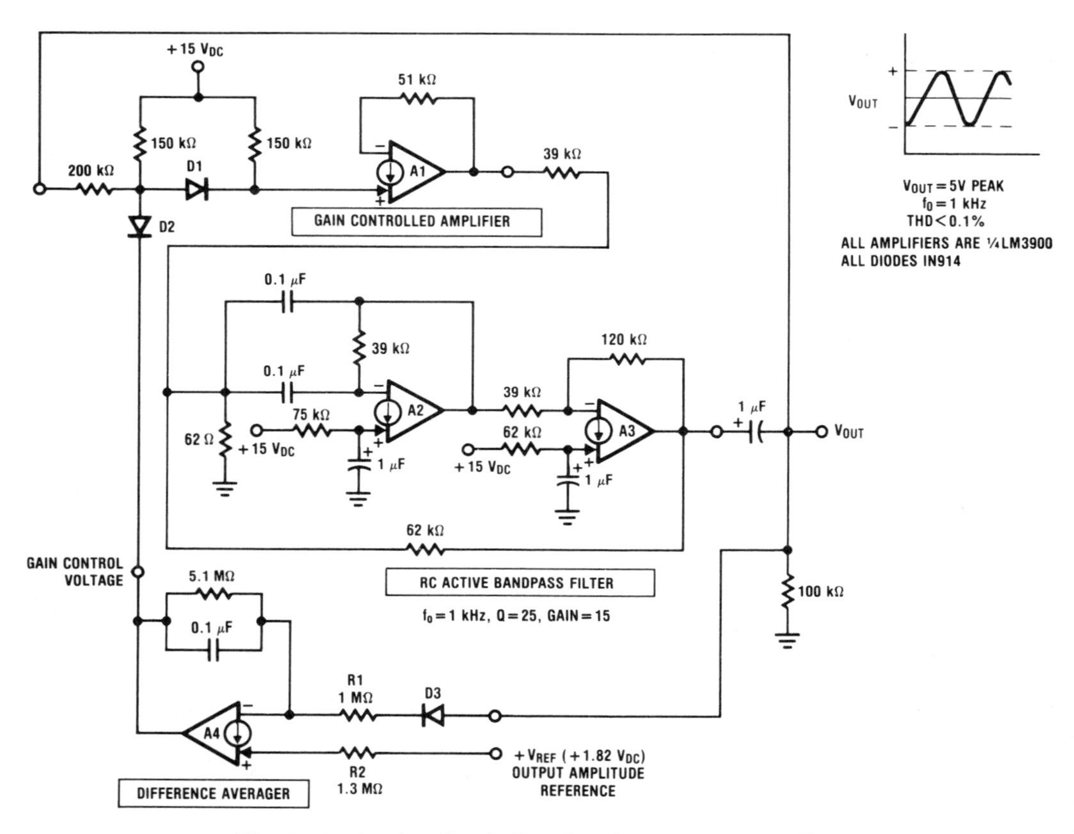

Fig. 5-72. An Amplitude-Regulated Sine Wave Oscillator

The average value of the half-wave rectified output sine wave is compared to V_{REF} at the noninverting input of A4. The diode that exists at each of the inputs of A4 is used to provide temperature compensation for the rectifier circuit. (This will be considered shortly.)

The RC Active Bandpass Filter is a standard two-op amp circuit, which we will discuss in the next section, that uses positive feedback to increase the attainable Q. This was selected because it provides an overall noninverting signal phase.

The Gain-Controlled Amplifier is also noninverting, so these two circuits (the Gain-Controlled Amplifier and the RC Active Bandpass Filter) can be connected in a loop to provide positive feedback for the oscillator configuration. The selectivity of the active filter will only allow a frequency that is equal to the center frequency of this bandpass filter to propagate around this overall positive feedback loop.

The Gain-Controlled Amplifier is realized by adding diodes D1 and D2 at the noninverting input of A1. These diodes work in conjunction with the diode that exists at this input, within the IC, to provide a series-shunt, diode attenuator network that is controlled by the dc voltage at the cathode of D2. This gain-controlled voltage is supplied by the output voltage of A4.

The maximum gain of this controlled amplifier is one-quarter and, with an active filter gain of 15. The maximum overall loop-gain is therefore 3.8. This insures that the oscillator will initially start. The two-150 kΩ biasing resistors on either side of D1 are used to reduce the dc output voltage changes of A1 that would otherwise result as the gain is changed.

The Difference Averaging Circuit places two diodes, instead of the expected single, one-half-wave rectifying diode in series with the ac signal input. One of these is D3 and the other exists internal to the IC at the inverting input of A4. The output amplitude reference voltage, V_{REF}, is applied to the input resistor, R2. A single diode (the one at the input of the IC) is in series with R2. The reason for this selection of diodes can be determined by considering the circuitry at the input of A4, as shown in Figure 5-73.

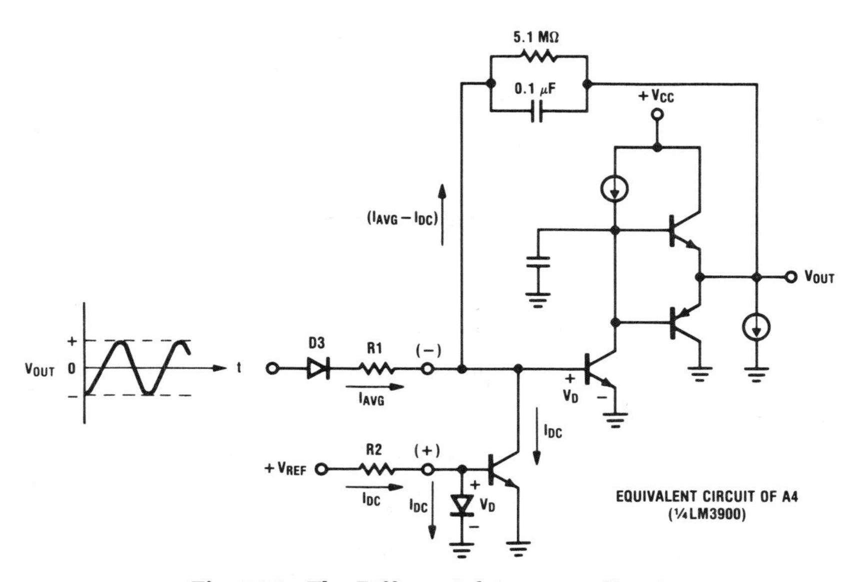

Fig. 5-73. The Differential-Averager Circuit

This equivalent circuit of one of the amplifiers of the LM3900 shows that a diode voltage drop, V_D, exists from each input to ground. This will therefore subtract V_D from any voltages that are applied to the input resistors R1 and R2.

The ac output voltage of the oscillator is applied to the series combination of the diode, D3, and the resistor, R1, at the inverting input of the difference averager. The resulting half-wave rectified voltage has a voltage loss of 2 V_D because of the forward voltage drop of D3 and also the V_D bias level that exists at the input of the amplifier. This 2-V_D loss must be considered in the determination of the average current flow, I_{AVG}, at this point.

To account for these diode voltage losses, we will start with an ideal, half-wave rectified ac waveform, Figure 5-74, and indicate the 2-V_D loss by the line that intersects this curve at X1 and X2. The average value of this reduced half-loop of the sine wave is indicated by the shaded area and can be evaluated as

$$V_{AVG} = \frac{1}{2\pi}\left[\int_{X1}^{X2} V_p \sin x \, dx - (X2 - X1)(2\,V_D)\right] \tag{5-1}$$

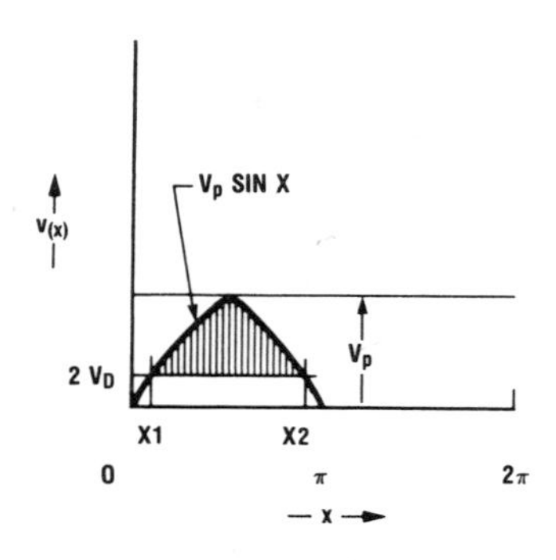

Fig. 5-74. The Half-Wave-Rectified Sine Wave With Diode Losses

The value of X1 can be found from the equation for a sine wave:

$$V_{(X)} = V_p \sin x$$

where:

$$V_{(X)} = 2\,V_D \quad \text{at } X = X1$$

so

$$2\,V_D = V_p \sin X1$$

or

$$X1 = \arcsin\left(\frac{2\,V_D}{V_p}\right) \cong \frac{2\,V_D}{V_p} \text{ radians}$$

and, therefore, because of symmetry,

$$X2 = \left(\pi - \frac{2\,V_D}{V_p}\right) \text{ radians}$$

Returning now to equation (5-1) for V_{AVG}, we find (after performing the indicated integration),

$$V_{AVG} = \frac{1}{2\pi}\left\{V_p\,[-\cos x]_{\frac{2V_D}{V_p}}^{\left(\pi - \frac{2V_D}{V_p}\right)} - \left(\pi - \frac{4V_D}{V_p}\right)(2V_D)\right\}$$

and evaluating [-cos x] at the limits indicated:

$$[-\cos x]_{\left(\frac{2V_D}{V_p}\right)}^{\left(\pi - \frac{2V_D}{V_p}\right)} = \left\{\left[-\cos\left(\pi - \frac{2V_D}{V_p}\right)\right] - \left[-\cos\left(\frac{2V_D}{V_p}\right)\right]\right\}$$

For $V_D = 0.6V$ and assuming $V_P = 5V$, this evaluates to 1.94, so V_{AVG} becomes

$$V_{AVG} = \frac{1}{2\pi}\left\{1.94\,V_p - \left[\pi - \frac{4\,(0.6)}{5}\right]2\,V_D\right\}$$

or

$$V_{AVG} = \left[\frac{1.94\,V_p}{2\pi} - \left(\frac{\pi - 0.48}{\pi}\right)V_D\right]$$

further provides,

$$V_{AVG} = \left[\frac{1.94\,V_p}{2\pi} - 0.85\,V_D\right] \qquad (5\text{-}2)$$

The amplitude control loop will keep the average input current at the inverting input terminal of amplifier A4 equal to the dc input current at the noninverting input terminal, or

$$I_{AVG} = \frac{V_{AVG}}{R1} = \frac{V_{REF} - V_D}{R2} \qquad (5\text{-}3)$$

Substituting in the previous value for V_{AVG} from (5-2),

$$\frac{1.94\,V_p}{2\pi R1} - \frac{0.85\,V_D}{R1} = \frac{V_{REF} - V_D}{R2} \qquad (5\text{-}3)$$

and for a temperature independent V_p we can force the terms involving V_D on both sides of the equation to cancel if

$$\frac{0.85\,V_D}{R1} = \frac{V_D}{R2}$$

or if

$$R1 = 0.85\,R2$$

(The value of V_D did appear in the evaluation of X1 and X2, but this will not create a significant temperature drift error.)

Using this resistor ratio in the expression for V_p from equation (5-3), provides

$$\frac{1.94\ V_p}{2\pi\ (0.85)\ R2} = \frac{V_{REF}}{R2}$$

or

$$V_p = V_{REF} \left[\frac{2\pi\ (0.85)}{1.94}\right] = 2.75\ V_{REF}$$

For the assumed $V_p = 5V$ the V_{REF} to use becomes

$$V_{REF} = \frac{V_p}{2.75} = \frac{5V}{2.75} = 1.82\ V_{DC}$$

An alternative to going through the above analysis is to use one of the previously described rectifier circuits that *place the diode within the loop.* This exercise indicates the computational benefits that result from working with idealized rectifier circuits.

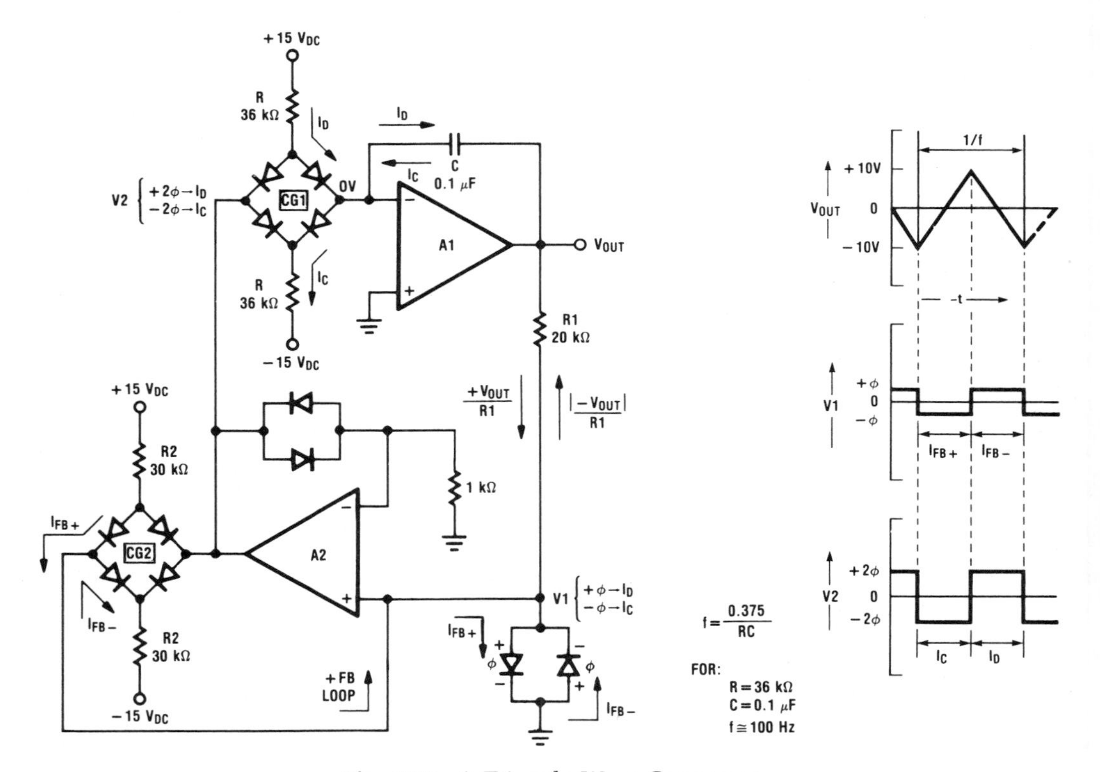

Fig. 5-75. A Triangle Wave Generator

A Trianglewave Generator. An interesting application of diode current gates provides the trianglewave generator circuit of Figure 5-75. The voltage ramping of the output waveform is provided by the integrator circuit that makes use of op amp A1. The input current for this integrator is supplied by the first current gate, CG1. The second op amp, A2, is used with a second current gate, CG2, to regeneratively control the input current to the integrator and also to monitor the output voltage, via R1. When the current through R1 becomes large enough, it "unlocks" the present state of A2 and causes the circuitry to switch to the other quasi-stable state.

Robert A. Pease is credited with originating this circuit and many additional derivatives have also been developed: frequency-, period-, and amplitude-modulated triangularwave generators; a two-phase triangularwave generator; a phase-modulated, dual triangularwave generator; and a universal, voltage-controlled triangularwave generator. (See reference #6 for further details of these circuits.)

A Voltage-to-Frequency Converter

Some more of RAP's (Robert A. Peases') handiwork has provided the charge-dispensing voltage-to-frequency converter shown in Figure 5-76. The interesting part of this circuit is the circuitry involving R1, C1, and C2. (The R3, R4, C3 network is used to provide positive feedback to aid the circuit transitions and to establish a time interval in which to dispense the charge. The circuitry associated with Q1 prevents circuit latch-up.)

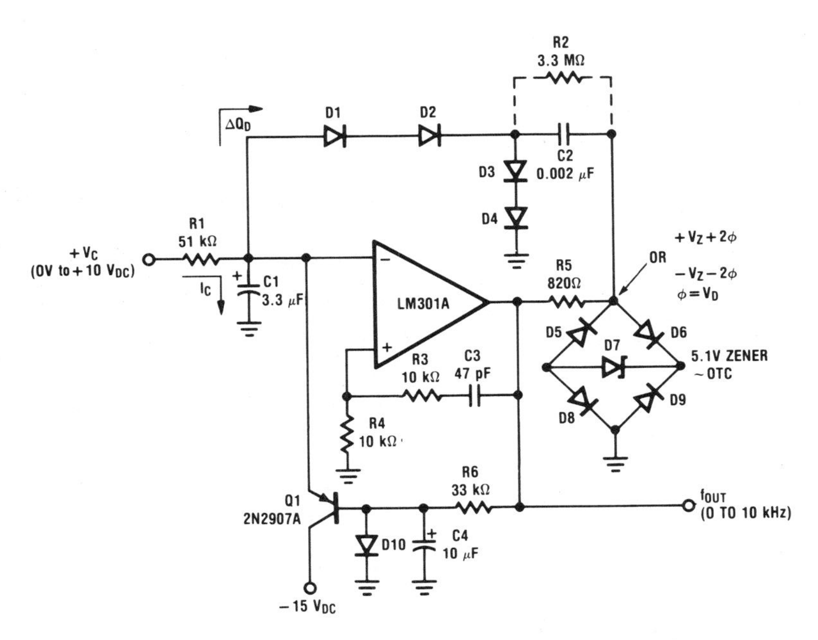

Fig. 5-76. A Charge-Dispensing Voltage-to-Frequency Converter

The key idea behind this circuit is that a dc control voltage, V_C, is used to input a current, via R1, that charges C1. The smaller valued capacitor, C2, is used to remove packets of charge, ΔQ_D, from C1. (This is the *charge-dispensing* action.) After each ΔQ_D removal, the voltage across C3 is left slightly negative. The op amp responds to this resulting negative voltage across C1 by swinging positive and recharging C2. When V_{C1} recovers from the slightly negative voltage (a time duration that depends on the magnitude of V_C) and again rises to the vicinity of 0V, the output of the op amp is regeneratively (because of C3) driven to negative saturation. (For the LM301A uncompensated op amp this can take less than 1 μsec.) This negative output voltage swing then removes another ΔQ_D of charge from C1.

The quantity of charge, ΔQ_D, is given by

$$\Delta Q_D = 2\ V_Z\ C2$$

and in this example

$$\Delta Q_D = (10V)\ (0.002\ \mu F) = 2 \times 10^{-8} \text{ coulombs}$$

This charge loss from C1 causes a voltage loss of ΔV_{C1} that is given by

$$\Delta V_{C1} = \frac{\Delta Q_D}{C1} = \frac{2 \times 10^{-8} \text{ coulombs}}{3.3 \times 10^{-6}\ F} = 6.06 \text{ mV}$$

The time rate of this charge loss constitutes a discharge current, I_D, flow out of C1 as

$$I_D = \frac{\Delta Q_D}{\Delta t}$$

or

$$I_D = \Delta Q_D\ f_{OUT}$$

The op amp does the accounting to automatically insure that, on the average,

$$I_C = I_D$$

or

$$\frac{V_C}{R1} = \Delta Q_D\ f_{OUT}$$

from which

$$f_{OUT} = \frac{V_C}{R1\ \Delta Q_D}$$

$$f_{OUT} = \left(\frac{1}{2\ V_Z\ R1\ C2}\right) V_C$$

where the terms in the parenthesis represent the "gain" of this V/F circuit.

Using the values from the circuit example,

$$f_{OUT} = \left[\frac{1}{2\,(5.1)\,(51 \times 10^{3})\,(0.002 \times 10^{-6})}\right] V_C$$

or

$$f_{OUT} \cong 1000\ V_C$$

so if $V_C = 10\ V_{DC}$, $f_{OUT} \cong 10$ kHz.

The diodes D1 through D4 compensate for the diodes D5, D6, D8, and D9 and allow C2 to precharge to V_Z (when the output voltage of the Zener bridge is $+V_Z + 2\phi$) and to discharge to $-V_Z$ as it removes ΔQ_D from C1 (when the output of the Zener bridge is $-V_Z - 2\phi$). (The resistor R5 is used to limit the Zener current.)

The 3.3 MΩ resistor (R2, shown in dotted lines shunting C2) can be added to increase the linearity of the transfer function.

Positive feedback and the control on the pulse width of the charge-dispensing cycle is established by the R3, R4, C3 network. The extra resistor (R3) is added to attenuate the approximately 26V (+ 13V to −13V or vice-versa) differentiated waveform that may otherwise exist directly at the noninverting input. This circuit was chosen to provide a time duration of approximately 5 μsec, so the noninverting input voltage is fully recovered, to 0V, for the negative pulse width at the output of the op amp.

This circuit was included, at the suggestion of Jim Williams, to stimulate your thinking because it shows a slightly unusual use of the summing junction of an op amp. A voltage-to-frequency converter function is available as an IC (the LM331) and there are a wide range of system applications for this product: from analog-to-digital conversion to frequency modulation or demodulation.

5.6 ACTIVE FILTERS

The idea behind active filters is to make use of a feedback loop to move poles created with RC networks off of the negative real axis of the complex-frequency plane and to cause them to become complex. The motivation is to eliminate the larger and more costly inductors and to reduce the capacitor values that are needed to realize complex pole locations for a desired, passive, LC filter function. (These benefits become especially important at low frequencies.) Further benefits of active filters are low cost, and the high input impedance and low output impedance of each circuit that allows active filter stages to be cascaded without interaction problems.

The electronic simulation of filters has been known for a long time. In 1955, before the first IC appeared, R. P. Salen and E. L. Key published their now classic paper on designing RC active filters in the Institute of Radio Engineers (now the IEEE) Transactions on Circuit Theory. They developed practical design methods for filters where the active elements were vacuum tubes.

An electric filter can be of two general types: the common *frequency-selective circuits* or circuits where the *phase response* is the important factor such as the *all-pass* filters or the *time-delay filters*. Our main concern in this chapter will be to introduce the frequency-selective filters.

The Filter Approximation Problem

All filters are based on approximations of some desired or ideal response characteristic. Many frequency-selective filters use the "brickwall" response as the ideal and the resulting three filter types: low-pass, high-pass, and bandpass are shown in Figure 5-77.

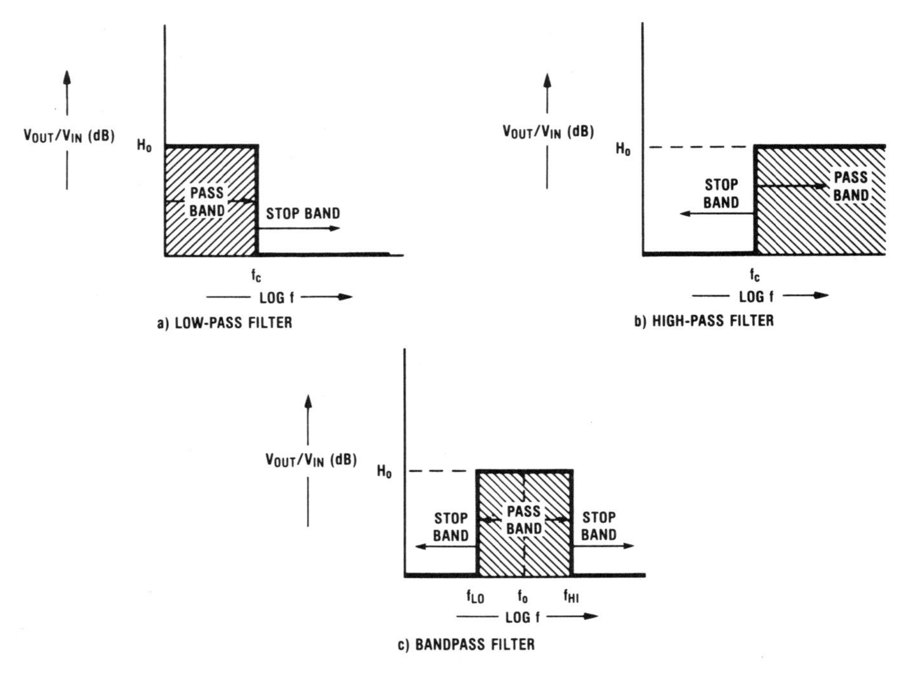

Fig. 5-77. Basic Ideal Brickwall Filter Responses

Notice that as long as the input signal frequency is within the *passband*, the voltage gain, H_0, of the filter is constant. Further, the transition to the *stopband* is abrupt and there is no output signal voltage provided for input frequencies that are within these stopbands.

The approximation problem for frequency-selective filters is to find ways of building actual filters that will "nearly" have these ideal brickwall response characteristics. The two standard brickwall filter approximations, Butterworth and Chebyshev, achieve this goal in different ways. The low-pass Butterworth filter, for example, approximates the ideal brickwall response by maintaining the frequency response flat with zero error, for the widest possible range of frequencies, and is therefore also called the *maximally-flat-magnitude filter*. All of the approximation error is lumped at the upper band edge of this filter and the amplitude at the corner frequency is always −3 dB down, independent of the *order* (the *number* of poles) of the filter. This low-pass Butterworth characteristic is shown in Figure 5-78. A Butterworth filter also produces a *monotonic* response, the maximally-flat characteristic, in the stopband. Notice that this approximation has a *transition band* (from f_c to f_1) where the response smoothly shifts from the passband to the stopband. The passband ends, for this Butterworth filter, when A1 is −3 dB down from the low-frequency response, A0, and the stopband is entered when the response drops to some predetermined value, A2. Higher order filters are used to provide a better approximation to the ideal brickwall response in the frequency domain. These higher order filters have an increasingly poorer step response as the order of the filter increases. In general, the transition band decreases as the order of the filter increases.

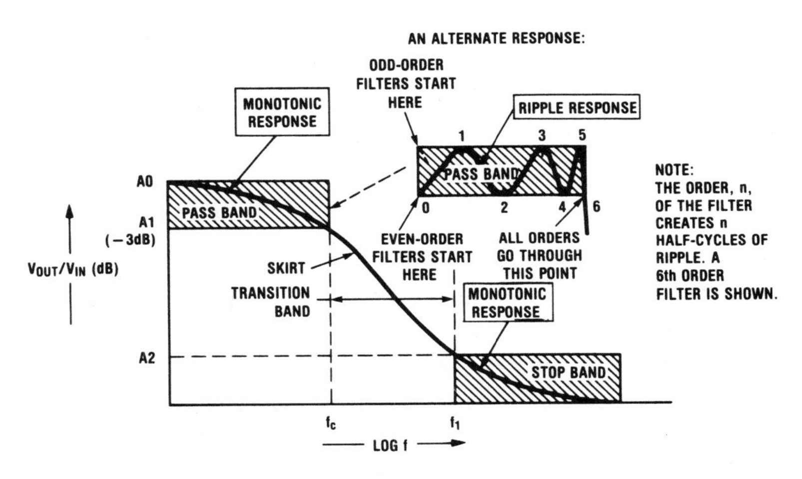

Fig. 5-78. The Butterworth Low-Pass Filter Response Characteristics

A different approach is taken with the Chebyshev (Tchebyscheff) brickwall approximation: *the amplitude error* of this filter *is equalized within the passband.* The upper band edge is defined as the highest frequency that still falls within this allowed error band. This is properly called the "tolerance bandwidth" of this filter. This *equi-ripple* (ripple response) is shown as an alternate response in Figure 5-75 where the 6 half-cycles of ripple that are shown in the passband result from a sixth-order Chebyshev filter. In general, for the same order of filter, the Chebyshev filter has steeper *skirts* (a smaller transition band) than a Butterworth.

Both of the previous filters are called all-pole responses, because there are no zeros available to control the response in the *stopband* of the filter. The *elliptic-function filter* is popular because it also introduces zeros within the stopband and therefore can simultaneously control both the passband and stopband performance. (Both will have a ripple-response characteristic.)

Filters are available that provide either monotonic or ripple response in either the passband or the stopband. These four permutations are listed by name in Table 5-1.

TABLE 5-1. Characteristics of the Classic Filter Types

Filter Type	Passband Response	Stopband Response
Butterworth	Monotonic	Monotonic
Chebyshev	Ripple	Monotonic
Inverse Chebyshev	Monotonic	Ripple
Elliptic	Ripple	Ripple

Determining the Number of Poles Needed for a Butterworth Filter

To meet the design goals of a particular Butterworth filter, the complexity or order, n, (the number of poles needed) can be determined from

$$n \geq \frac{\log\left(10^{\frac{\alpha 2}{10}} - 1\right)}{2 \log (f_1/f_c)}$$

where: n is chosen as the next larger integer
$\alpha 2$ is the stopband attenuation (in dB)
and f_c and f_1 are defined on the previous figure

For example, if we wanted a Butterworth low-pass filter with a cutoff frequency, f_c, of 1 kHz and also wanted at least -60 dB of attenuation (a relatively large attenuation) at 2 kHz (only a factor of two above the cutoff frequency) the order, n, can be found as

$$n \geq \frac{\log\left(10^{\frac{60}{10}} - 1\right)}{2 \log (2\text{ kHz}/1\text{ kHz})} \geq 9.97$$

Therefore, using the next larger integer, n = 10: a tenth-order filter is needed. (Design equations are available, see Reference #10, to determine the order needed for the other classic filter types.)

A final common filter is the Bessel (or Thompson) filter. Here the phase is controlled to approximate an ideal linear phase versus frequency characteristic, but the resulting magnitude response is a much poorer approximation to the ideal brickwall than either the Butterworth or the Chebyshev designs. The Bessel filter is most useful for pulse applications. The comparison of the step responses of these filters is shown in Figure 5-79.

Cascading to Provide a High-Order Filter

A basic RC active filter stage generally provides a second-order (a two-pole) response characteristic. (We will discuss these circuits later in this chapter.) Higher performance, even-order filters are realized by cascading these second-order building blocks and a passive, RC, single-pole filter is added to realize an old order, overall filter response as shown in Figure 5-80. (Notice, in Figure 5-80a, that you should never admit to using something as common as an "RC low-pass," it is worth much more if is called a "first-order Butterworth low-pass filter.")

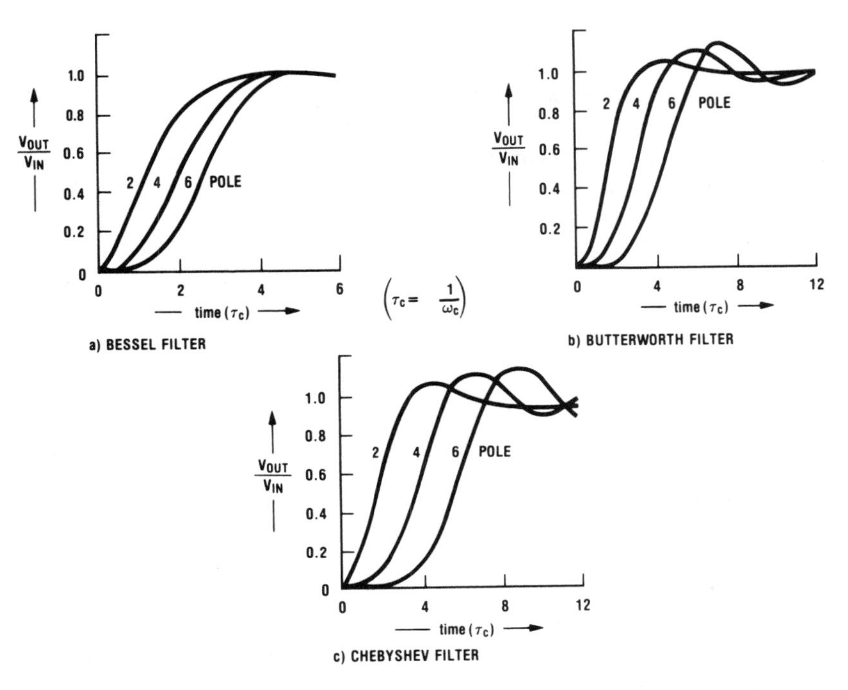

Fig. 5-79. The Step Response of Various Filters

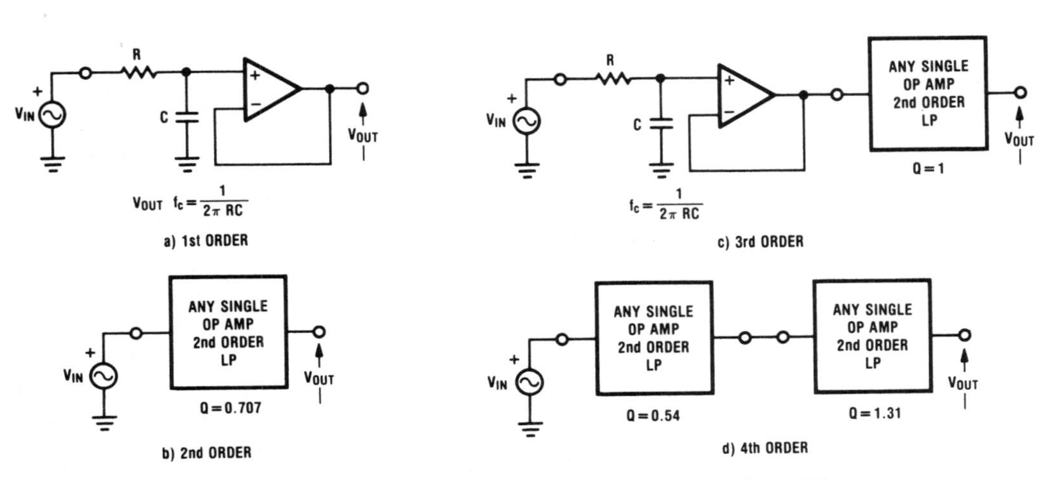

Fig. 5-80. Successively Higher-Order Butterworth LP Filters

The nice thing about the Butterworth filter is that all of the building blocks have the same cutoff frequency and gain. Only the Q of the second-order sections is changed, as indicated on the figure. These values of Q are determined from the pole locations (in the s-plane) that correspond to the Butterworth approximation function. It has been shown that this approximation locates all of the poles on a unit circle. (The radius of this circle is ω_C and the cutoff frequency of the resulting filter. This is why all of the basic sections of a Butterworth filter are chosen to have the same cutoff frequency.) These pole locations also have what is called "quadrantal symmetry," that is, if we also consider the half of the Butterworth unit circle that extends into the right half plane, the pole locations will be equally spaced around the unit circle. These symmetrical pole locations are shown for a first- through fourth-order Butterworth low-pass filter in Figure 5-81. The Qs of each of the second-order poles (we only consider the pole locations in the left half plane) indicate how close these poles approach the jω axis. (A closer approach indicates a higher Q.)

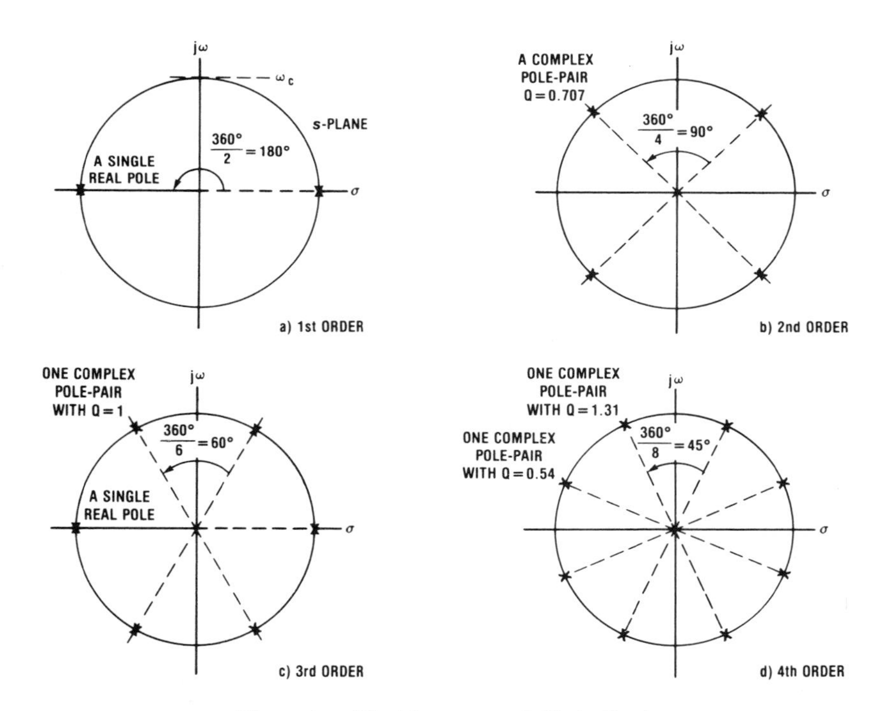

Fig. 5-81. The Butterworth Unit Circle

The pole locations for a Chebyshev filter can be derived from the pole locations of a Butterworth filter. To realize an n-th order Chebyshev response, we start with the pole locations of an n-th order Butterworth response and then shift all the poles over horizontally to the right, as shown in Figure 5-82. These shifted poles are no longer on a circle, they are now located on an ellipse. Because of this, each pole pair has a unique resonant frequency and a unique Q.

(Filter tables list the f and Q values that are needed for each section of an n-th order Chebyshev filter.) This shifting of the poles closer to the $j\omega$ axis increases the required Q for each pole pair over what was needed for the Butterworth response, providing steeper skirts for Chebyshev filters.

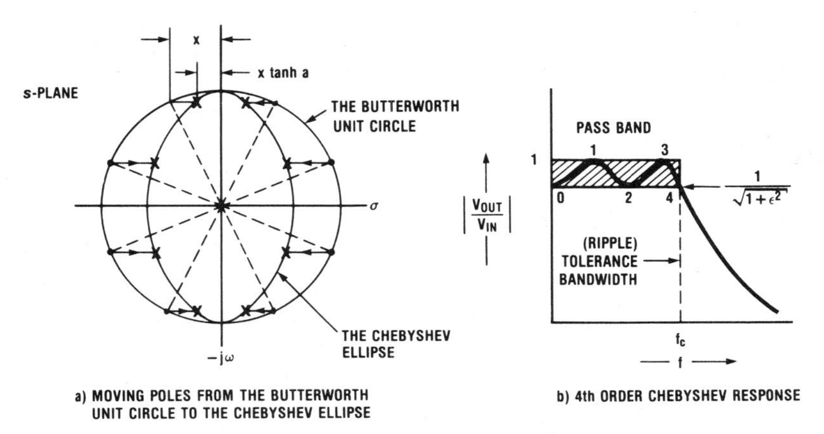

Fig. 5-82. The Chebyshev Ellipse

It is interesting that the amount of this pole shifting depends on both the order, n, and the allowed ripple in the passband, where the ripple width, RW, of a Chebyshev filter is given by

$$RW = 1 - \frac{1}{\sqrt{1 + \epsilon^2}} \cong \frac{\epsilon^2}{2} \quad \text{for } \epsilon << 1$$

This ripple width parameter, ϵ, can be determined from the allowed passband ripple, r (expressed in dB), as

$$\epsilon = \sqrt{10^{r/10} - 1}$$

For example, a 0.5 dB ripple width corresponds to an ϵ of

$$\epsilon = \sqrt{10^{0.5/10} - 1} = 0.3493$$

[The widest allowable ripple width of a Chebyshev filter is 3 dB ($\epsilon = 1$).]

Each pole is shifted by the factor

$$\tanh a$$

where

$$a = \frac{1}{n} \sinh^{-1} \left(\frac{1}{\epsilon}\right)$$

and

n = the order of the filter
ϵ = the previously determined ripple factor

(Now you can see why tables are generally used to design a Chebyshev filter.)

Filter design tables are available, with separate listings for each filter type, (see Reference #10) to show the individual Q, f_0, and gain that should be used in each second-order filter section of a cascade to achieve high-order filters of any of the standard filter types. It is not always an easy matter to determine which filter type is best for a given application, because sometimes the phase characteristics or the step response of the filter is also important.

When working with filters, a sinusoidal steady-state input signal is assumed. If the input sine wave is only present for a limited number of cycles, new problems are created. We will now consider an input signal that is only a short burst of sine waves.

Responding to a Few Cycles of a Sine Wave

In some applications of filters it becomes important to respond to a tone burst if only a few cycles of an input sine wave are supplied. If the frequency spectrum of this short burst is evaluated, it will be seen that a band of frequencies is produced that is centered at the frequency of the sine wave, ω_i. The width of this spectrum, ω_s, is given by

$$\omega_s = \frac{2}{N}\,\omega_i$$

where: N is the number of cycles of sine waves in the burst
and ω_i is the frequency of the input sine wave.

If this gated sine wave is fed to an RLC bandpass filter that is resonant at ω_i, the requirements on the Q of this filter that will allow passing the main portion of this input frequency spectrum is given by

$$Q = \frac{N}{2}$$

and not much frequency selectivity can be provided with the low values of Q that result if only a small number of cycles of the input sine wave are present. For example, if $N = 3$, the Q required would be only 1.5; if $N = 20$, the Q could be raised to 10.

There is also a build-up time for an input voltage to appear from this filter. If we use an output voltage response of at least two-thirds of the final value as a reference, this build-up time becomes approximately Q/3 cycles. With Q equal to N/2, this means that one-sixth of the cycles (or one-sixth of the input burst time) are required. More complex filters, with many more storage elements, will take a correspondingly longer time to respond.

Selecting the Passive Components

The quality of the passive components is very important in the performance of an active filter. For this reason, carbon composition resistors should be used only in noncritical filters (up to fourth-order) and are useful mainly for room-temperature breadboarding of initial concepts or for final trimming (or *tuning)* of the more stable metal film (use 1% tolerance for higher than fourth-order), or wirewound (used at low frequencies) resistors.

Capacitors present more problems and the ceramic types are generally not suitable (except for the NPO ceramics, which are excellent for values smaller than 2000 pF) for active filter applications because of their relatively poor performance characteristics. Mylar capacitors are commonly used, but polypropylene (or polystyrene) or even teflon capacitors are often needed in high-performance filters.

Capacitor values for active filter designs should be selected first and standard values should be used. A wider range of standard resistor values is generally available, and this higher resolution of values can then be used to obtain the desired filter performance.

Scaling the Impedance Levels

The impedance level of the passive components can be scaled without affecting the filter characteristics (multiply all resistor values and divide all capacitor values by the same factor). In an actual circuit, if the resistor values become too small ($< 2\ k\Omega$), excessive loading may be placed on the output of an amplifier. This will reduce accuracy and can exceed either the output current or the power dissipation capabilities of the amplifier. Excessive op amp loading can easily be checked by calculating *the impedance* that is presented to the output terminal of the amplifier at the highest operating frequency. (Capacitors can also cause overloading.)

A second limit sets the upper range of impedance levels, this is because of the dc noise sources of the IC op amp or stray, signal pick-up problems.

Sensitivity Functions

The performance of the op amp also limits the overall performance of the filter. Historically, the researchers that initially worked with active filters did not recognize this problem. Later contributors included both the nonideal gain and phase shift effects of a real op amp and also the effects of the tolerances of the passive components in what are called, *sensitivity functions*. These relate the percentage change in a particular parameter of the filter, such as center frequency, Q, or gain, to a percentage change in a particular passive component or a percentage change in a characteristic of the op amp, such as the unity-gain frequency, f_u.

As a point of reference, we will first consider the sensitivity function of a passive filter. The sensitivity functions of passive LC filters are very low because, for an LC parallel-resonant network, the resonant frequency, f_o, varies inversely as the square root of the LC product:

$$f_o = \frac{1}{2\pi \sqrt{LC}}$$

If the value of L, for example, were to be incorrect (either as a result of initial tolerance or as a result of changes with temperature or time) and a larger value, L′ were to exist, where

$$L' = (1 + \Delta) L$$

and Δ is the error in the component value, a new resonant frequency, f_o', would result where

$$f_o' = \frac{1}{2\pi \sqrt{L'C}}$$

This can be compared to the desired f_o by calculating the ratio of f_o' to f_o as given by

$$\frac{f_o'}{f_o} = \frac{2\pi \sqrt{LC}}{2\pi \sqrt{L'C}} = \sqrt{\frac{L}{L'}} = \sqrt{\frac{L}{L\,(1 + \Delta)}}$$

or

$$\frac{f_o'}{f_o} \cong \sqrt{1 - \Delta} \cong 1 - \frac{\Delta}{2}$$

This shows that f_o *decreases* by *one-half the increase* in the value of L, or the sensitivity of f_o to L is written as

$$S_L^{f_o} = -\,^1\!/_2$$

Many of the early RC active filter circuits had sensitivity functions that varied as Q^2. This caused major problems in high-volume production of RC active filter circuits. Many times an RC active oscillator resulted. The newer, multi-op amp, RC active filter circuits have solved this sensitivity problem and therefore these have been very successful circuits for high-volume production.

A subtle problem is also created if the operating frequency of the filter is high enough, or the output voltage swing is large enough, that slew distortion appears. This causes an effective phase lag in the signal that is fed back, which can result in Q enhancement, a decrease in f_o, or an amplitude-dependent oscillation in the active filter circuit.

The Effects of Q on the Filter Response

The Q of a second-order filter represents the Q of the complex pole-pair that is generated by this filter. There is sometimes confusion with this parameter in the design of simple, second-order, RC active filter networks. For the simple, second-order bandpass case, Q relates the –3 dB passband (the bandwidth, BW) to the center frequency, f_o, as

$$Q = \frac{f_o}{BW}$$

(More complex bandpass filters are possible that have a maximally-flat or an equal-ripple response in the passband. Elliptic-function bandpass filters also control the stopband response).

For the low-pass and high-pass filters, this Q parameter sometimes introduces conceptual problems. Whereas high Q was desirable for a narrow-band bandpass filter, high-Q in a simple low-pass or high-pass filter will create peaking in the frequency response near the band edge. This is usually not desired. For these filters, Q must be restricted to less than 0.707 to prevent peaking. A Q value of 0.5, *the critically-damped value,* will not have ringing in the step response or peaking in the frequency response. As Q increases beyond 0.707, the following equation can be used to predict the magnitude of the maximum peaking in the frequency response,

$$\text{Peaking (dB)}\Big|_{\text{for } Q > 0.707} = 20 \log \frac{Q}{\sqrt{1 - \frac{1}{4Q^2}}}$$

We will now briefly consider some single-, dual- and triple-op amp RC active filter circuits that produce a second-order filter response that is characterized by the pass band voltage gain, H_0, the corner frequency, f_C, (or center frequency, f_0, in the case of a bandpass filter), and the Q.

Single-Op Amp Filters

Active filters that make use of only a single op amp should be limited to applications that require only low Q ($\leq$ 10), low frequency ($\leq$ 5 kHz), and low gain ($\leq$ 10), or a small value for the product of gain times Q ($\leq$ 100). The op amp selected for these filters should have an open-loop voltage gain at the highest frequency of interest at least 50 times larger than the gain of the filter at this frequency. Further, the slew rate, SR, required for the op amp should be determined from

$$SR \geq \tfrac{1}{2} (\omega_H V_{OUTp\text{-}p}) \times 10^{-6} \text{ V}/\mu\text{sec}$$

where ω_H is, again, the highest frequency of interest. For example, for

$$\omega_H = 2\pi \text{ (5 kHz)}$$

and

$$V_{OUTp\text{-}p} = 20\text{V}$$

$$SR \geq \tfrac{1}{2} (2\pi) (5 \times 10^3) (20) (\times 10^{-6}) \geq 0.3 \text{ V}/\mu\text{sec}$$

The Bi-FET op amps are recommended for very low frequency circuits, because larger valued resistors can be used to keep the capacitor values reasonable, or for very high frequency circuits because of the generally wide bandwidths and high slew rates of the Bi-FET op amps.

We will consider both the inverting, infinite-gain multiple-feedback and the noninverting, voltage-controlled, voltage-source filters. Many applications use only one of these relatively simple single-op amp filters. The following circuit collection will therefore be useful as is and it is also good background material for those readers who may be sufficiently motivated to continue their study of RC active filters.

High-Pass Filters. An infinite-gain, multiple-feedback, single-op amp, high-pass filter is shown in Figure 5-83. (The input dc blocking capacitor, C1, suggests that dc is not amplified so this circuit can't be a low-pass filter.) This filter has an inverting gain characteristic and the design proceeds as:

Given: H_0, Q and $\omega_C = 2\pi f_C$
To Find: R1, R2, C1, C2, and C3

Let C1 = C3, and choose a standard value close to $(10/f_C)$ μF. (This initial value can later be changed, if necessary, to allow reasonable values for the resistors.) Then:

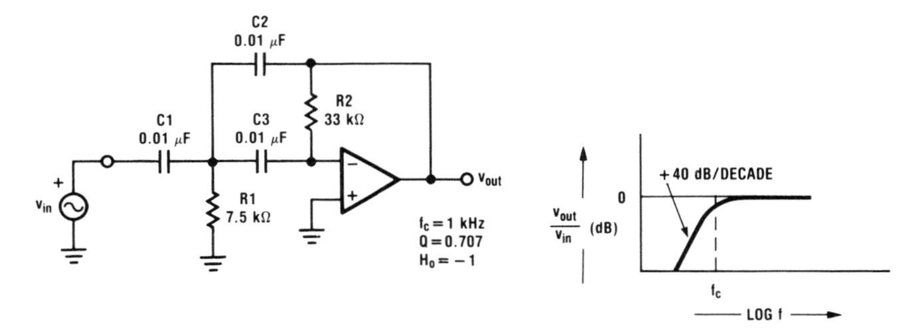

Fig. 5-83. An Infinite Gain, Multiple Feedback High-Pass Filter

$$R1 = \frac{|H_o|}{Q\ \omega_C\ C1\ (2\ |H_o| + 1)} \tag{5-4}$$

$$R2 = \frac{Q}{\omega_C\ C1}\ (2\ |H_o| + 1) \tag{5-5}$$

and

$$C2 = \frac{C1}{|H_o|} \tag{5-6}$$

[Remember that the resistors can also be scaled by a factor M (as MR) if the capacitors are also all scaled by this same factor (as C/M) without affecting the performance of the filter.]

As a design example,

Require: $H_o = -1$
$Q = 0.707$

and

$f_c = 1$ kHz ($\omega_C = 6.28 \times 10^3$ rps)

Start by selecting C1 = (10/1000) μF = 0.01 μF and then from equation (5-4)

$$R1 = \frac{1}{(0.707)\ (6.28 \times 10^3)\ (10^{-8})\ (2 + 1)}$$

$$R1 = 7.5\ k\Omega$$

and from the equation (5-5),

$$R2 = \left[\frac{0.707}{(6.28 \times 10^3)\ (10^{-8})}\right] \left[3\right]$$

$$R2 = 33.8\ k\Omega$$

and from equation (5-6)

$$C2 = \frac{0.01\ \mu F}{1} = 0.01\ \mu F$$

This filter presents a capacitive load, because C1 is essentially tied to the summing junction, as the input impedance. The signal source therefore must supply relatively large currents to this capacitor to create a large voltage swing at high frequencies. For example, a 10V peak signal at 16 kHz has a dV/dt given by

$$\frac{dV}{dt} = \omega V_p = 2\pi (16 \times 10^3)(10) = 1 \text{ V}/\mu\text{sec}$$

to create this voltage-rate-of-change across a 0.01 μF capacitor requires a peak current that can be found from

$$\frac{dV}{dt} = \frac{I}{C}$$

or

$$I = C \frac{dV}{dt} = (10^{-8}F) \left(\frac{1V}{10^{-6} \text{ sec}} \right) = 10^{-2}A = 10 \text{ mA}$$

(Supplying this large value of current can easily become a problem for the driving op amp).

An alternative single-op amp high-pass filter can be used to reduce this loading problem at the input. This filter, shown in Figure 5-84, is a voltage-controlled voltage-source realization

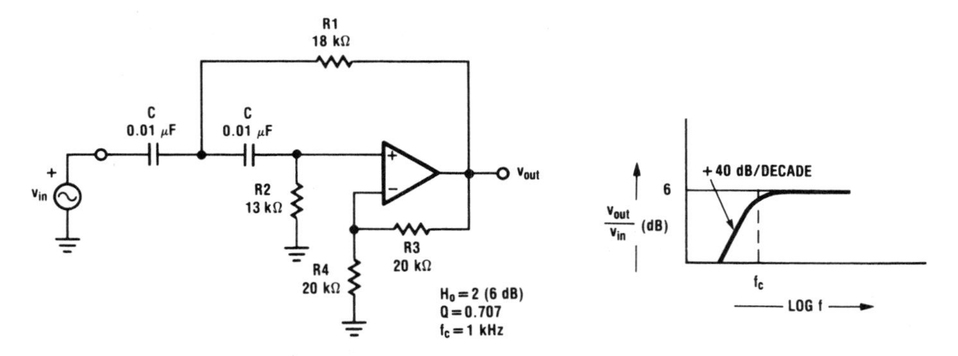

Fig. 5-84. A Voltage Controlled, Voltage Source High-Pass Filter

that provides a noninverting response. (If a gain of 1 is adequate, an emitter-follower can be used instead of an op amp.) The components are selected as follows:

Choose: $C = (10/f_c)\ \mu F$ (choose the closest standard value)

$$H_o = \left(1 + \frac{R3}{R4}\right) \geq 1$$

$$R1 = \frac{\frac{1}{Q} + \sqrt{\left(\frac{1}{Q}\right)^2 + 8(H_o - 1)}}{4\ \omega_c\ C}$$

$$R2 = \frac{4}{\omega_C C} \frac{1}{\left[\frac{1}{Q} + \sqrt{\left(\frac{1}{Q}\right)^2 + 8(H_o - 1)}\right]}$$

As a design example

Require: $H_o = 2$
$Q = 0.707$
$f_c = 1$ kHz

Start by determining values for R3 and R4 as,

$$H_o = 1 + R3/R4 = 2$$

Therefore, let R3 = R4 = 20 kΩ

Then select C as

$$C \cong \frac{10}{f_c} = \frac{10}{1000} = 0.01\ \mu F$$

Now R1 can be found as

$$R1 = \frac{\left(\frac{1}{0.707}\right) + \sqrt{\left(\frac{1}{0.707}\right)^2 + 8(2-1)}}{4(2\pi \times 10^3) \times 10^{-8}}$$

$$R1 = 18.2\ k\Omega$$

and

$$R2 = \frac{4}{(2\pi \times 10^3)(10^{-8})} \frac{1}{\left[\left(\frac{1}{0.707}\right) + \sqrt{\left(\frac{1}{0.707}\right)^2 + 8(2-1)}\right]}$$

$$R2 = 13.9\ k\Omega$$

Low-Pass Filters. An infinite-gain, multiple-feedback, single-op amp, low-pass filter is shown in Figure 5-85. The design of this filter proceeds as:

Given: H_o, Q, and ω_C
To Find: R1, R2, R3, R4, C1, and C2

Let C1 be a standard value close to $(10/f_c)$ μF, then select a value for K, such that

$$K > 4Q^2(|H_o| + 1) \tag{5-7}$$

then

$$C2 = KC1 \tag{5-8}$$

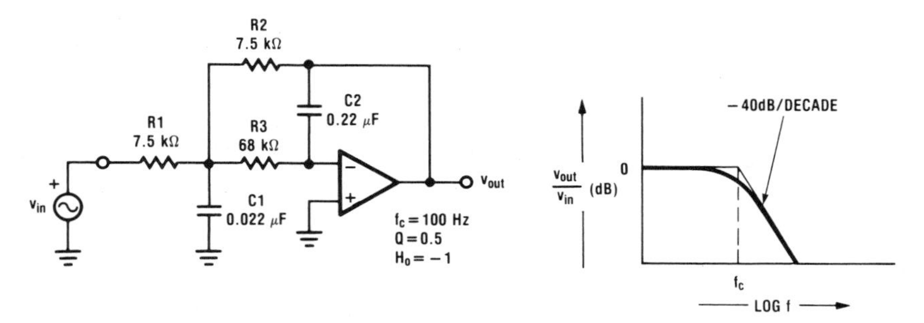

Fig. 5-85. An Infinite Gain, Multiple Feedback Low-Pass Filter

where K is a constant of the design and can be used to adjust component values, providing it satisfies the above inequality. [Notice that K should relate to the standard capacitor values that are available so equation (5-8) will be easy to satisfy.]

Then

$$R2 = \frac{1}{2Q\,\omega_C\,C1}\left[1 \pm \sqrt{1 - \frac{4Q^2\,(|H_o| + 1)}{K}}\right] \tag{5-9}$$

$$R1 = \frac{R2}{|H_o|} \tag{5-10}$$

and

$$R3 = \frac{1}{\omega_C^{\,2}\,(C1)^2\,R2\,K} \tag{5-11}$$

As a design example:

Require: $H_o = -1$

$Q = 0.5$

and

$f_C = 100$ Hz

Start by selecting $C1 = 10/100 = 0.1\ \mu F$ and see if $K = 10$ is allowed; as, from equation (5-7)

$$K > 4\,(0.5)^2\,(2)$$
$$10 > 2$$

so this value of K will provide a value for C2 as given by equation (5-8)

$$C2 = (10)\,(0.1\ \mu F) = 1\ \mu F$$

Now, from equation (5-9)

$$R2 = \frac{1}{2\,(0.5)\,(2\pi \times 10^2)\,(10^{-7})}\left[1 \pm \sqrt{1 - \frac{4\,(0.5)^2\,(2)}{10}}\right]$$

$$R2 = 159 \times 10^2\,[1 \pm 0.894]$$

$$R2 = 1.68\ k\Omega$$

and, from equation (5-10)

$$R1 = 1.68\ k\Omega$$

From equation (5-11)

$$R3 = \frac{1}{(2\pi \times 10^2)^2\,(10^{-7})^2\,(1.68 \times 10^3)\,(10)}$$

$$R3 = 15.1\ k\Omega$$

Notice that the values of the resistors R1 and R2 are rather low and therefore we will scale them to a larger impedance level by using 4.5 as a scaling factor. (This is determined from C1 and C2, so we will still obtain standard values.) The new component values are

$$C1 = 0.1\ \mu F/4.5 = 0.022\ \mu F$$

$$C2 = 1\ \mu F/4.5 = 0.22\ \mu F$$

$$R1 = R2 = (4.5)\,(1.68\ k\Omega) = 7.56\ k\Omega$$

$$R3 = (4.5)\,(15.1\ k\Omega) = 67.95\ k\Omega$$

A noninverting voltage-controlled, voltage-source, low-pass filter is shown in Figure 5-86. This provides a noninverting transfer function. Components for this filter are selected as

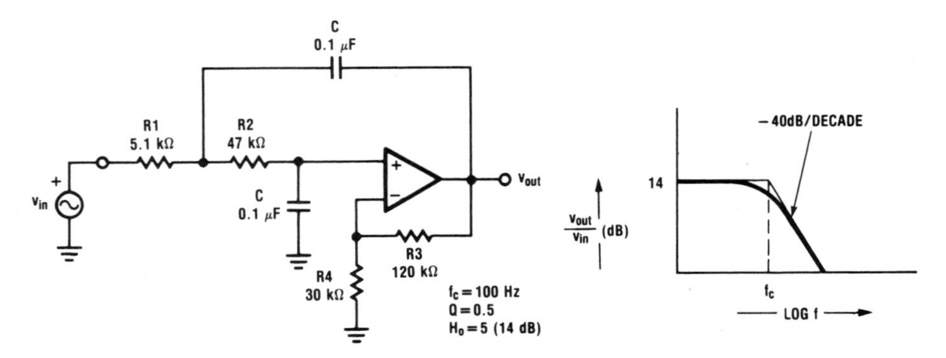

Fig. 5-86. A Voltage Controlled, Voltage Source Low-Pass Filter

$$H_o = \left(1 + \frac{R3}{R4}\right) > 2$$

$C = (10/f_C)\ \mu F$ (choose the closest standard value)

$$R2 = \frac{1}{2Q\ \omega_C\ C}\left\{1 + \sqrt{1 + Q^2\,[4\,(H_o - 2)]}\right\}$$

$$R1 = \frac{1}{\omega_C^{\ 2}\ C^2\ R2}$$

As a design example;

Require: $H_o = 5$
$Q = 0.5$
$f_c = 100$ Hz

Start by calculating R3/R4

$$H_o = 5 = 1 + \frac{R3}{R4}$$

$$\frac{R3}{R4} = 4$$

$$R3 = 4R4$$

Let

$$R4 = 30\ k\Omega$$

Then

$$R3 = 120\ k\Omega$$

select

$$C \cong \frac{10}{100}\ \mu F = 0.1\ \mu F$$

Then

$$R2 = \frac{1}{2\,(0.5)\,(2\pi \times 10^2)\,(10^{-7})}\left\{1 + \sqrt{1 + (0.5)^2\,[4\,(5 - 2)]}\right\}$$

$$R2 = 47.7\ k\Omega$$

and

$$R1 = \frac{1}{(2\pi \times 10^2)^2\,(10^{-7})^2\,(47.7 \times 10^3)}$$

$$R1 = 5.31\ k\Omega$$

A Bandpass Filter. An infinite-gain, multiple-feedback, single-op amp, bandpass filter is shown in Figure 5-87. The design of this filter proceeds as:

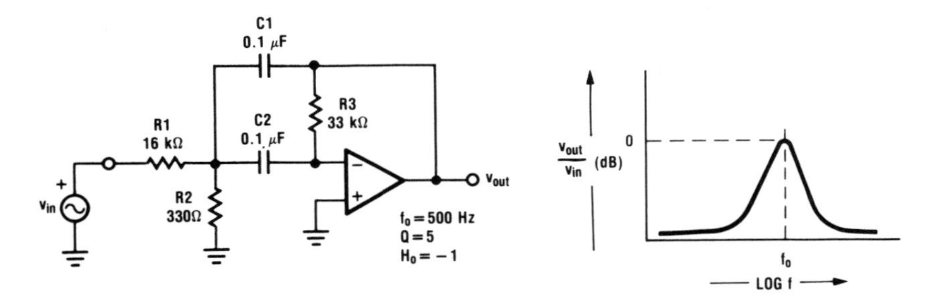

Fig. 5-87. An Infinite Gain, Multiple Feedback Bandpass Filter

Given: H_0, Q and ω_C
To Find: R1, R2, R3, C1, and C2

Let C1 = C2, and select a standard value near $(10/f_0)$ μF.

Then

$$R1 = \frac{Q}{|H_o| \omega_o C1} \tag{5-12}$$

$$R2 = \frac{Q}{(2Q^2 - |H_o|)\omega_o C1} \tag{5-13}$$

and

$$R3 = \frac{2Q}{\omega_o C1} \tag{5-14}$$

As a design example;

Require: $f_o = 500$ Hz
$H_o = -1$
$Q = 5$

Start by selecting

$$C1 = C2 = (10/500)\ \mu F = 0.02\ \mu F$$

Then, using equation (5-12)

$$R1 = \frac{5}{(1)(2\pi \times 500)(2 \times 10^{-2})(10^{-6})}$$

$$R1 = 79.6\ k\Omega$$

and, using equation (5-13)

$$R2 = \frac{5}{[2\,(5)^2 - 1]\,(2\pi \times 500)\,(2 \times 10^{-8})}$$

$$R2 = 1.62\ k\Omega$$

from equation (5-14)

$$R3 = \frac{2\,(5)}{(2\pi \times 500)\,(2 \times 10^{-8})}$$

$$R3 = 159\ k\Omega$$

Again we will use impedance scaling, this time to reduce the resistor values. An impedance scaling factor of 1/5 will allow standard capacitor values and the new component values become

$$C1 = C2 = 5\,(0.02\ \mu F) = 0.1\ \mu F$$

$$R1 = (1/5)\,(79.6\ k\Omega) = 15.92\ k\Omega$$

$$R2 = (1/5)\,(1.62\ k\Omega) = 320\Omega$$

$$R3 = (1/5)\,(159\ k\Omega) = 31.8\ k\Omega$$

This filter has some significant noninteraction benefits: the gain can be varied by changing R1, Q can be adjusted via R2, and the center frequency can be shifted if both R2 and R3 are simultaneously changed by the same percentages.

A voltage-controlled, voltage-source, bandpass filter is shown in Figure 5-88. The components for this filter are determined by again selecting a value for C as

$$C \cong (10/f_o)\ \mu F$$

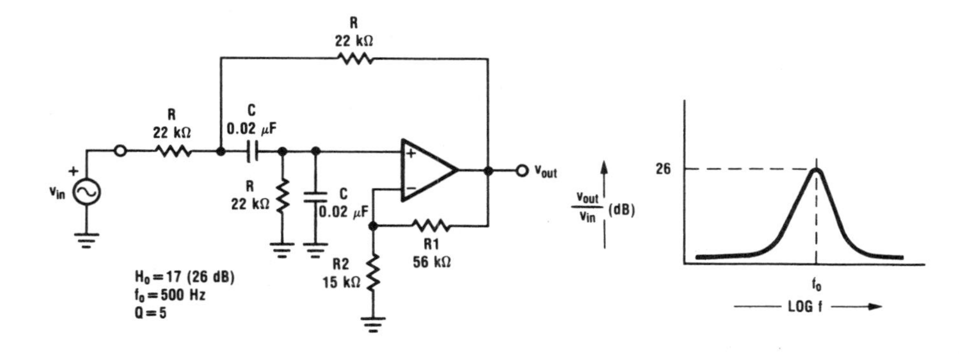

Fig. 5-88. A Voltage Controlled, Voltage Source Bandpass Filter

The values for R1 and R2 are determined from

$$\frac{R1}{R2} = 4 - \frac{\sqrt{2}}{Q}$$

and

$$R \mp \frac{\sqrt{2}}{\omega_o C}$$

Then the gain that results (an uncontrolled parameter) is found from,

$$H_o = \frac{5Q}{\sqrt{2}} - 1$$

As a design example;

Require: $f_o = 500$ Hz
and $Q = 5$

Start by selecting C as

$$C \cong \frac{10}{f_o}\ \mu F \cong \frac{10}{500} \cong 0.02\ \mu F$$

then

$$R = \frac{\sqrt{2}}{(2\pi)(5 \times 10^2)(2 \times 10^{-2})(10^{-6})} = 22.5\ k\Omega$$

$$H_o = \frac{5\ (5)}{\sqrt{2}} - 1 = 16.7$$

and

$$\frac{R1}{R2} = 4 - \frac{\sqrt{2}}{5} = 3.72$$

$$R1 = 3.72\ R2$$

Therefore, let

$$R2 = 15\ k\Omega.$$

Then

$$R1 = 3.72\ (15\ k\Omega) = 55.8\ k\Omega.$$

A Two-Op Amp Bandpass Filter

To allow higher Q (between 10 and 50) and higher gain, a two-op amp filter is required. This circuit, Figure 5-89, uses only two capacitors. A positive feedback loop is used to improve the performance characteristics.

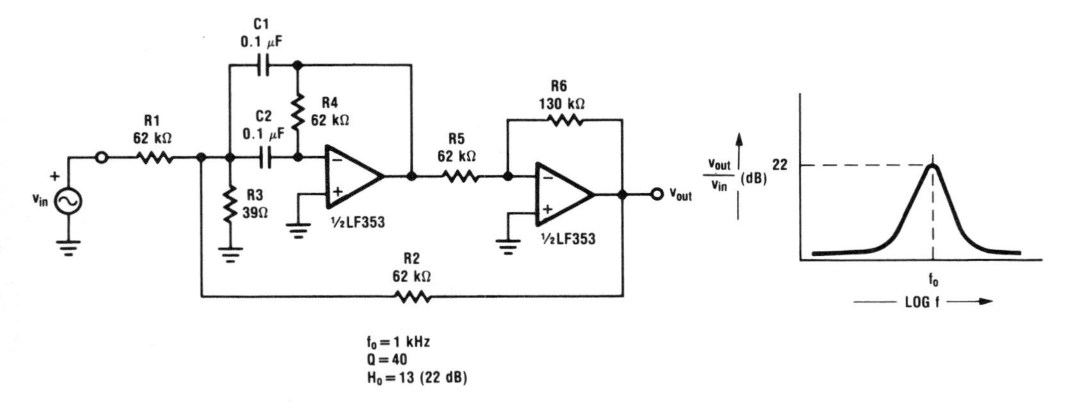

Fig. 5-89. A Two Op Amp Bandpass Filter

The design procedure is as follows:

Given: Q and f_o
To Find: R1 through R6, C1, and C2

Let C1 = C2 and choose a standard value near $(100/f_o)$ μF. The parameter K (1 to 10, typically) can be used to reduce the spread of the component values.

Then

$$R1 = R4 = R5 = \frac{Q}{\omega_o C1} \tag{5-15}$$

$$R2 = R1 \frac{KQ}{(2Q - 1)} \tag{5-16}$$

$$R3 = \frac{R1}{Q^2 - 1 - \frac{2}{K} + \frac{1}{KQ}} \tag{5-17}$$

$$R6 = K\, R5 \tag{5-18}$$

and

$$H_o = \sqrt{Q}\, K \tag{5-19}$$

As a design example:

Require: Q = 40 and f_o = 1 kHz
Select: C1 = C2 ≅ (100/1000) μF = 0.1 μF

and

$$K = 2$$

Then from equation (5-15)

$$R1 = R4 = R5 = \frac{40}{(2\pi \times 10^3)(10^{-7})} = 63.7\ k\Omega$$

and from equation (5-16)

$$R2 = (63.7\ k\Omega)\left[\frac{(2)(40)}{2(40) - 1}\right] = 64.5\ k\Omega$$

equation (5-17) then provides

$$R3 = \frac{63.7\ k\Omega}{(40)^2 - 1 - \left(\frac{2}{2}\right) + \frac{1}{2(40)}} = 39.9\Omega$$

and equation (5-18) evaluates the last resistor value,

$$R6 = 2(63.7\ k\Omega) = 127\ k\Omega$$

The gain is given by equation (5-19)

$$H_o = \sqrt{40}(2) = 12.7\ (22\ dB)$$

A Three-Op Amp Bandpass Filter

A three-op amp filter circuit that is similar to the oscillator feedback loop of the early analog computers (discussed in Chapter 1), is very popular for high Q (up to 100) RC active bandpass filters. (This more complex circuit is generally not required for low-pass or high-pass applications because these are usually low Q, so we will restrict our consideration to only the bandpass application.) This three-op amp circuit has low sensitivity to component values and also to the characteristics of the op amp, as shown in Figure 5-90.

$$S_{f_u}^{Q} \cong -\frac{3Q^2 f_o}{f_u} \qquad S_{f_u}^{Q} \cong -\frac{3.2Q f_o}{f_u}$$

$$S_{f_u}^{f_o} \cong \frac{3Q f_o}{f_u} \qquad S_{f_u}^{f_o} \cong \frac{1.5 f_o}{f_u}$$

a) FOR A ONE-OP AMP RC ACTIVE FILTER b) FOR A THREE-OP AMP RC ACTIVE FILTER

Fig. 5-90. Comparison of Sensitivity Functions of One and Three Op Amp Filters

A slight circuit variation is provided by what is called the *Bi-Quad* (for bi-quadratic: because a transfer function is produced that is quadratic in s in both the numerator and the denominator).

We will consider a three-op amp, second-degree filter circuit, Figure 5-91, that can also provide a notch output if an extra op amp is used. (If this notch function is not desired, this fourth amplifier can be omitted.)

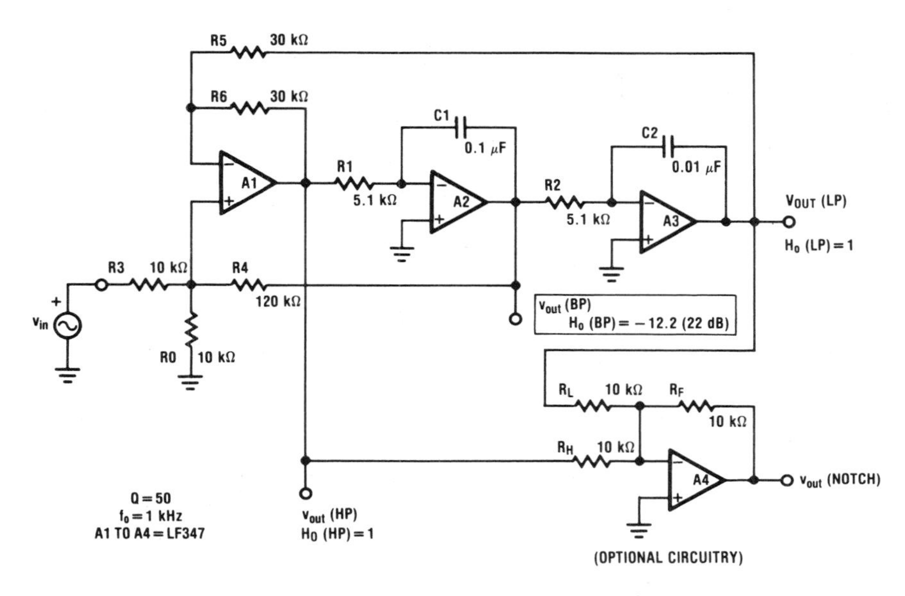

Fig. 5-91. A Three Op Amp, State Variable Filter

This circuit is known as a "state-variable" filter. This name results because the basic methods of realizing the analog computer circuits have more recently been rediscovered by system theorists who now consider them as part of *state-space theory* (or *state-variable theory*). (This theory, for example, uses independent inductance currents and the independent capacitance voltages to completely determine the *state* of an RLC network. These time-dependent variables are called *state variables* in the set of differential equations that are then used to describe the network.)

The general design equations for this bandpass circuit are:

$$f_o = \frac{1}{2\pi}\sqrt{\frac{R6}{R5}\left(\frac{1}{R1\ C1\ R2\ C2}\right)} \tag{5-20}$$

$$f_{NOTCH} = \frac{1}{2\pi}\sqrt{\frac{R_H}{R_L\ R1\ C1\ R2\ C2}} \tag{5-21}$$

$$Q = \frac{1 + \frac{R4}{R3} + \frac{R4}{R0}}{1 + \frac{R6}{R5}}\sqrt{\frac{R6}{R5}\left(\frac{R1}{R2}\ \frac{C1}{C2}\right)} \tag{5-22}$$

$$H_{o\ HP} = \frac{1 + \frac{R6}{R5}}{1 + \frac{R3}{R0} + \frac{R3}{R4}} \tag{5-23}$$

$$H_{o\ BP} = \frac{1 + \frac{R4}{R3} + \frac{R4}{R0}}{1 + \frac{R3}{R0} + \frac{R3}{R4}} \tag{5-24}$$

$$H_{o\ LP} = \frac{1 + \frac{R5}{R6}}{1 + \frac{R3}{R0} + \frac{R3}{R4}} \tag{5-25}$$

As a design example:

Require: A bandpass filter with $f_o = 1$ kHz and $Q = 50$

(Remember that such a large Q will cause excessive peaking in a LP or HP filter and therefore the high Q benefit of this circuit usually has no value for LP or HP filters.)

Let C2 = 0.01 μF
C1 = 0.1 μF (to help obtain large Q)
R5 = R6 = 30 kΩ
R1 = R2

from equation (5-20)

$$1 \times 10^3 = \frac{1}{2\pi\ R2} \sqrt{\frac{1}{(10^{-7})\,(10^{-8})}}$$

$$(2\pi \times 10^3)^2\,(R2)^2 = 10^{15}$$

or

$$R2 = \sqrt{\frac{10^{15}}{(2\pi \times 10^3)^2}}$$

$$R1 = R2 = 5.03\ \text{k}\Omega$$

from (5-22)

$$40 = \frac{\left(1 + \frac{R4}{R3} + \frac{R4}{R0}\right)}{2} \sqrt{10}$$

$$\frac{80}{\sqrt{10}} = 1 + \frac{R4}{R3} + \frac{R4}{R0}$$

or

$$1 + \frac{R4}{R3} + \frac{R4}{R0} = 25.3$$

$$\frac{R4}{R3} + \frac{R4}{R0} = 24.3$$

Let

$$\frac{R4}{R3} = \frac{R4}{R0} = 12.15$$

and select

$$R0 = R3 = 10\ k\Omega$$

Therefore

$$R4 = 121.5\ k\Omega$$

From (5-24), the bandpass gain becomes:

$$H_{o\ BP} = \frac{1 + 12.15 + 12.15}{1 + 1 + \left(\frac{1}{12.15}\right)}$$

$$H_{o\ BP} = 12.15$$

We will check to insure that the other outputs will not have higher gains and therefore cause unexpected saturation. (This is not usually a problem when the bandpass output is used, but it's a good idea to always check this because saturation of any one of the three amplifiers will seriously degrade the operation of this filter).

From equation (5-23), the gain at the high-pass output is

$$H_{o\ BP} = \frac{2}{1 + 1 + \left(\frac{10}{121.5}\right)}$$

$$H_{o\ HP} = 0.96$$

which is no problem, and equation (5-25) indicates the gain to the low-pass output will always be the same as that to the high-pass output, because we have selected R5 = R6.

The Effects of the Op Amp on the Filter Performance. For the three-op amp filter we have just considered, the actual center frequency, f_o', will be less than the design value because of the limited frequency response of the op amp. This actual center frequency is given by

$$f_o' = \frac{f_o}{1 + \frac{2\ f_o}{f_u}}$$

and if we were to allow a 10% error because of this, f_o' would equal 0.9 f_o. To determine the f_u needed for the op amp to only reduce f_o by 10%, we can solve the above equation as,

$$0.9\ f_o = \frac{f_o}{1 + \dfrac{2\ f_o}{f_u}}$$

or

$$0.9 \left[1 + \frac{2\ f_o}{f_u}\right] = 1$$

$$\frac{2\ f_o}{f_u} = \left[\frac{1}{0.9} - 1\right]$$

so

$$f_o = {}^1\!/_2 \left[\frac{1}{0.9} - 1\right] f_u$$

$$f_o \leq 0.06\ f_u$$

which is a limit of 60 kHz for a 1 MHz op amp. For less effect, f_o must be kept correspondingly smaller, as

$$f_o \leq 0.03\ f_u \quad \text{(for a -5\% error)}$$

and

$$f_o \leq 0.005\ f_u \quad \text{(for a -1\% error)}$$

The value of Q will also be *enhanced* because of the phase lag of the op amp. The actual value of Q, Q', to be expected is given by

$$Q' \cong \frac{Q}{1 - \dfrac{3.2\ (f_o \times Q)}{f_u}}$$

and calculations can be made to show the limit on the $f_o \times Q$ product, as:

$$f_o \times Q \leq 0.03\ f_u \quad \text{(for a 10\% Q enhancement)}$$
$$f_o \times Q \leq 0.02\ f_u \quad \text{(for a 5\% Q enhancement)}$$

and

$$f_o \times Q \leq 0.003\ f_u \quad \text{(for a 1\% Q enhancement)}$$

In our previous example, if we had used the LM348 (with f_u = 1 MHz) to build this filter (f_o = 1 kHz and Q = 50), we would find

$$f_o = \left(\frac{1\ \text{kHz}}{1\ \text{MHz}}\right) f_u = 0.001\ f_u$$

so we would expect a negligible error in f_o.

The $f_o \times Q$ product in this example filter is 1 kHz × 50 = 50 kHz

which becomes

$$f_o \times Q = 0.05\ f_u$$

so we can expect an error greater than 10% in the value of Q. (The actual Q will be larger than the design value.)

For less dependence on the op amp and more predictability, the LF347, with an f_u of 4 MHz, was therefore used. This op amp will reduce the above problem because

$$f_o \times Q = 0.01\ f_u$$

and the Q′ value will become

$$Q' = \frac{50}{1 - \left[\frac{3.2\ (50\ \text{kHz})}{4\ \text{MHz}}\right]}$$

$$Q' = 52$$

or a 4% enhancement. (In all of the above discussions, the passive components have been assumed to be perfect. We have considered only the effects of a nonideal op amp.)

Including Passive Filters

In harsh electrical environments, it is also advisable to include a passive filter ahead of an RC active filter to reduce the noise signals that are allowed to couple into the first op amp. This also prevents high frequency response beyond the bandwidth of the op amp. (A particular problem for a bandpass active filter.)

Attention also must be paid to the possibility of providing a balanced filter, as shown in Figure 5-92, if it becomes important to reject signals that exist on signal input lines in the common-mode as a result of noise pickup. The usual RC active filters are single-ended and therefore it is advisable to use a passive balanced filter ahead of the differential-to-single-ended conversion that is supplied by the instrumentation amplifier that is shown in this figure.

The New Switched-Capacitor Filters

Now that RC active filters are nicely along on their development cycle, it is almost too bad that a replacement for many of these applications is now possible. This new way to provide a filter is based upon a sampled data design instead of the continuous linear design that has been used for the previous RC active filters.

To appreciate this new filter technique, we show in Figure 5-93 how a switch and a capacitor are used to replace the input R of an RC integrator. Two of these basic integrators supply the negative-real (on the s-plane) poles that are relocated by the use of feedback to provide the desired filter function.

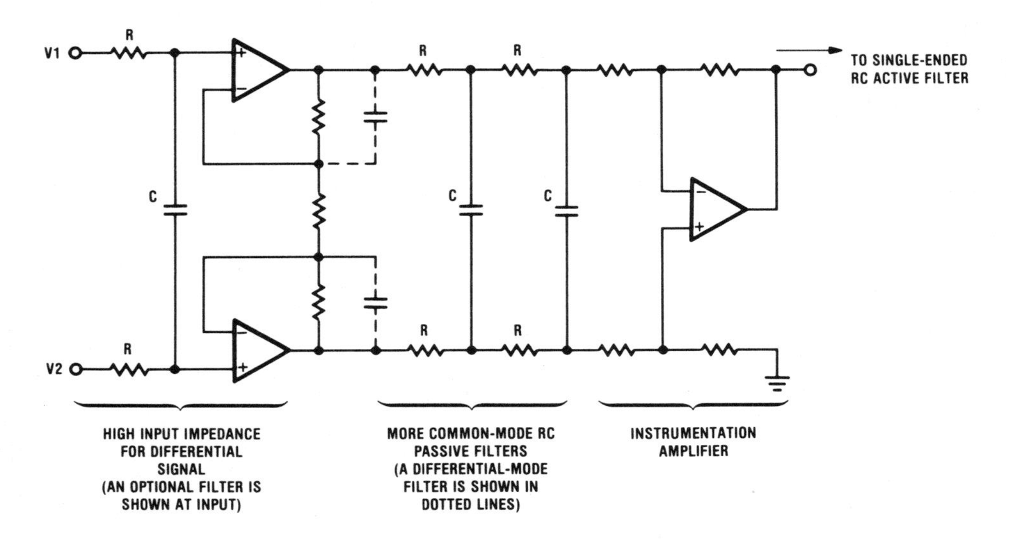

Fig. 5-92. Adding Common Mode Filtering Ahead of an RC Active Filter

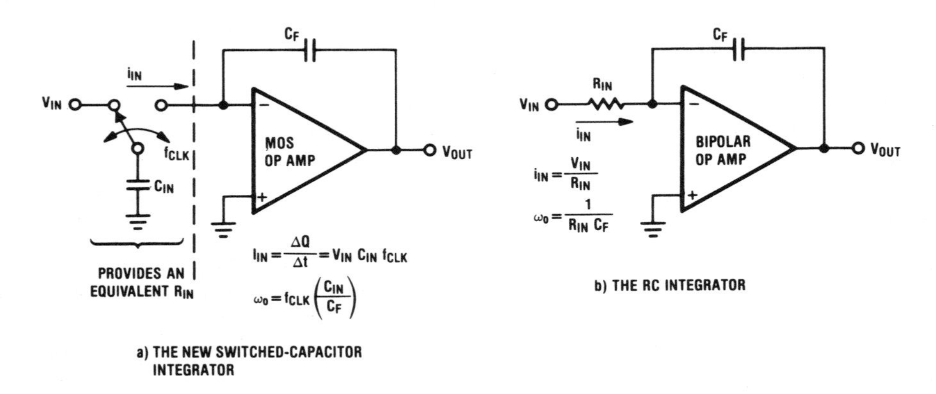

Fig. 5-93. Using a Small Capacitor and a Switch to Replace a Large Valued Resistor

The RC time constant of this switched-capacitor integrator is now dependent on the clock frequency, which can be easily supplied under computer control or it can be crystal controlled, and the ratio of two on-chip capacitors. *This is the basis for a new precision and predictability in filter performance and temperature stability with these switched-capacitor filters.*

The switches and the op amps for this filter are obtained with a CMOS process. This low cost filter building block is starting to replace a large number of the bipolar (or Bi-FET) op amps that have been previously used in active filter applications. The switched-capacitor filters also allow some new system designs that make use of this high accuracy, lack of tuning, and the repeatability in filter performance.

5.7 MACROMODELING THE IC OP AMP

Many users of op amps have attempted to do a computer simulation of an IC op amp. The idea is that this simulation can then be used to develop and evaluate op amp application circuits on the computer. With twenty to sixty transistors involved in a single IC op amp circuit, it rapidly becomes memory- and time-intensive to evaluate even the simplest multiple-op amp circuit. A way was needed to take a short cut. Don't model each active device of the op amp circuit, simply model the overall transfer function of the op amp. This trick greatly reduced the computer loading and newer macromodels are now available that can accurately simulate not only the dc and small-signal ac characteristics, but also the large-signal performance and even the anomalies that are unique to a particular IC op amp design.

As a basic idea of how a macromodel is developed, Figure 5-94 shows a simplified macromodel of the LM741 or LM348 op amps. The thing to notice is that models for each individual transistor do not appear because the intent is to model the overall transfer function. Only the input transistors are modeled, the rest of the circuit consists of components and dependent-generators that will provide the proper overall performance. Diodes are included to simulate the large-signal output voltage swing limits of the op amp.

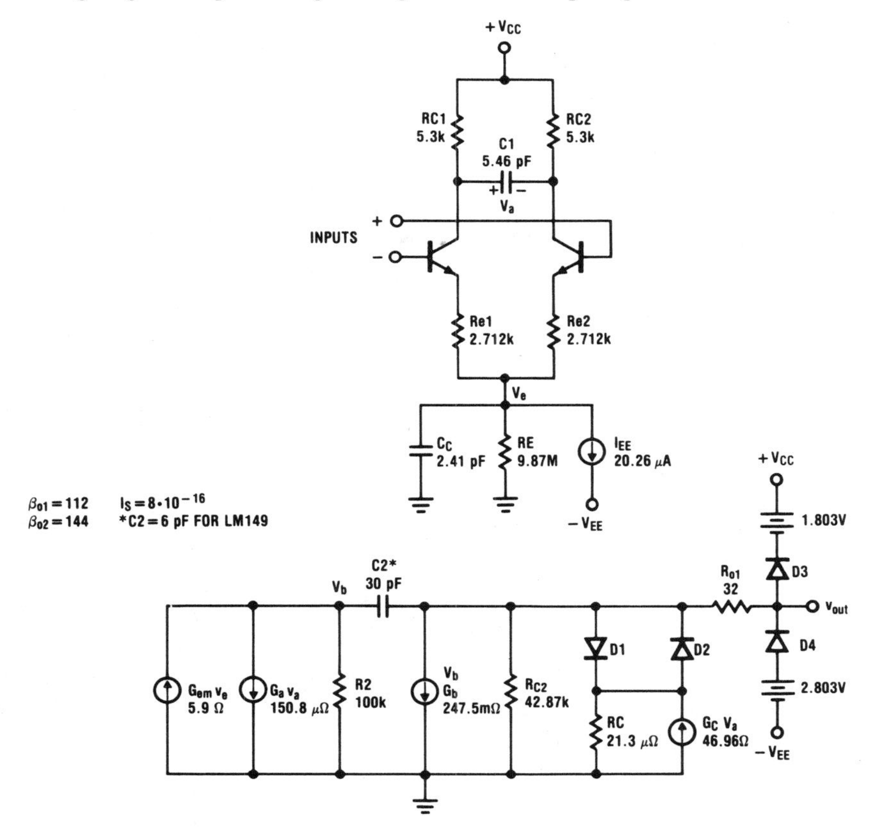

Fig. 5-94. A Simplified Macromodel for the LM741 and LM348 Op Amps

With the use of macromodeling, computer run times drop by a factor of up to 50 to 1. This high computer efficiency not only reduces the cost-per-run, it also allows the simulation of complex, multi-op amp application circuits.

The development of macromodels for all the popular op amps (and even most of the voltage comparator ICs) is largely done by specialist companies. The continual cost reductions of computer systems and the increasing capability of the new VLSI computer chips will allow this macromodeling of op amp application circuits to become a parlor game for modern teenagers.

Now that we have seen a number of op amp application circuits, we will turn our attention to some of the practical problems that have bothered many op amp users.

CHAPTER VI

Some of the Typical User Problems

A lot of the linear art is taken for granted by the practitioners, much is regarded as common sense, and, for these reasons, is usually never written down. When you are entering this magic linear world for the first time, it is worthwhile to pick up some tips from the people who have lived there. Much of what, at first, seems to be linear tricks are really techniques that are based on science and a good understanding of the physical world. The purpose of this chapter is to discuss some of the nonobvious traps that have caught many people and consumed valuable time while the victim cussed linear and vowed to never again leave the more forgiving land of digital electronics.

6.1 "WE HOLD THESE TRUTHS TO BE SELF-EVIDENT . . ."

Included in this section are many of the *holes* that either we have *fallen into* or have helped others *out of*. In retrospect, many seem obvious, and when this becomes the case it indicates progress has been made.

But There Were No Supplies Shown on The Application Circuit!

Power supplies are almost never shown on application circuits. This convention is also true for logic diagrams. This avoids unnecessary clutter and also helps indicate that the particular idea is not limited to only the specific power supply voltages that happen to be listed, but a power supply of some sort is always needed.

Compensate The Scope Probe

Problems exist when many people share equipment. Any time you start using an oscilloscope, get into the habit of checking the frequency compensation of the scope probe, especially whenever response times aren't looking correct. It can be very embarrassing when this turns out to be the cause for an extra day (or worse) of circuit debugging.

A further caution is to be careful of overdriving an oscilloscope. The vertical amplifier can be driven into saturation and a long recovery delay will result. This is most likely to happen when the vertical sensitivity is increased and you can't see the complete voltage waveform on the screen.

Finally, be aware that scope probes, and the probes of other measuring equipment, introduce a shunt capacitance to ground (as large as 60 pF) at the point that is being measured. This can reduce the bandwidth and thereby attenuate high frequency signals. This problem is most severe when measuring high impedance nodes and when using passive probes. Special FET-input probes can be used to reduce this stray capacitance to a low value (1.5 pF).

When You Can't Trust Ground

The emerging electronic systems combine digital and linear circuitry in the same enclosure and even sometimes on the same PC board. This creates extremely difficult signal coupling problems that involve common grounds and even common, restricted space because the relatively high cost of electronic enclosures tends to force the location of all circuitry to be within the same box.

Digital circuits tolerate this terrible environment; they even have such a thing as *noise margin* and it is a very large number, something like 400,000 μV. The problems of placing microvolt analog circuitry anywhere close to this mess are insurmountable; or at least very difficult: there are even problems for the less sensitive millivolt analog circuitry.

A carefully planned grounding scheme is mandatory. The use of three separate ground lines running to a single-point main ground, as shown in Figure 6-1, often becomes necessary. Early in the design phase, attention must be given to the grounding problem (use of twisted pairs and avoidance of ground loops, etc.) to prevent costly mechanical rework and to avoid slipping delivery schedules.

Power supply decoupling capacitors should be located to create the shortest possible path length from the IC power supply pin to the grounded side of the load that the op amp is driving. These local energy sources (charge reservoirs) will supply the high frequency and transient demands for load current and thereby reduce glitches on the system power supply and ground lines. Keep in mind the complete current path when locating bypass capacitors.

This key idea of considering the total current path also applies to the component layout examples shown in Figure 6-2. The innocent half-wave power supply shown in Figure 6-2a has very large recharging current pulses that are many times larger than the dc load current that is supplied. Low values of ripple voltage across the capacitor increase the magnitude of these

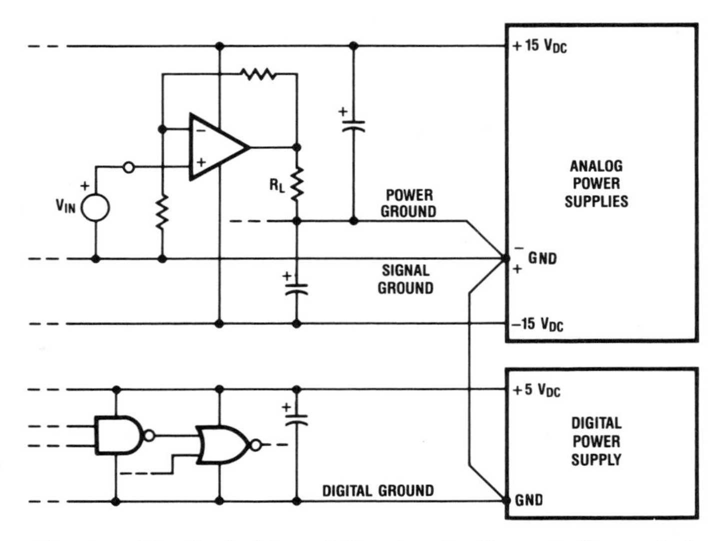

Fig. 6-1. The Basic Idea of Keeping the Grounds Separated

recharging current pulses. Rather than allowing these large current pulses to flow from point B to A through the system grounding, as shown in the figure, connect both the diode and capacitor as directly as possible to the secondary winding of the transformer and then run a separate wire to the system ground.

The LC resonant circuit shown in Figure 6-2b creates similar problems owing to high resonant circulating currents, where the circulating current is Q times the current flow through the tank network. As shown in this figure, the placement of the resonating capacitor C forces this circulating current to flow from B to A through the system grounding. The added radiation and RFI caused by this hookup can be eliminated by placing C directly across L.

Even logic circuits can malfunction if attention is not paid to the complete current path. In Figure 6-2c we show an open collector, or open drain logic circuit that is driving a remotely located load, R_L. If many of these open-collector circuits share a common ground line (as shown), the self inductance and the dc resistance of this ground line can cause a positive voltage pulse to result that will raise the zero level at R_L and prevent proper logic action. The solution here is to use as many individual ground return lines as possible to obtain a good zero level.

The equation for the inductance of a long conductor, provided in Figure 6-2c, shows a logarithmic dependence of the inductance on the diameter of the wire. For example, #22 gauge wire (0.025 inches in diameter) has 67 μH/100 ft, #10 gauge (0.101″ diameter) has 59 μH/100 ft, and #0000 cable (0.460″ diameter) still has 50 μH/100 ft.

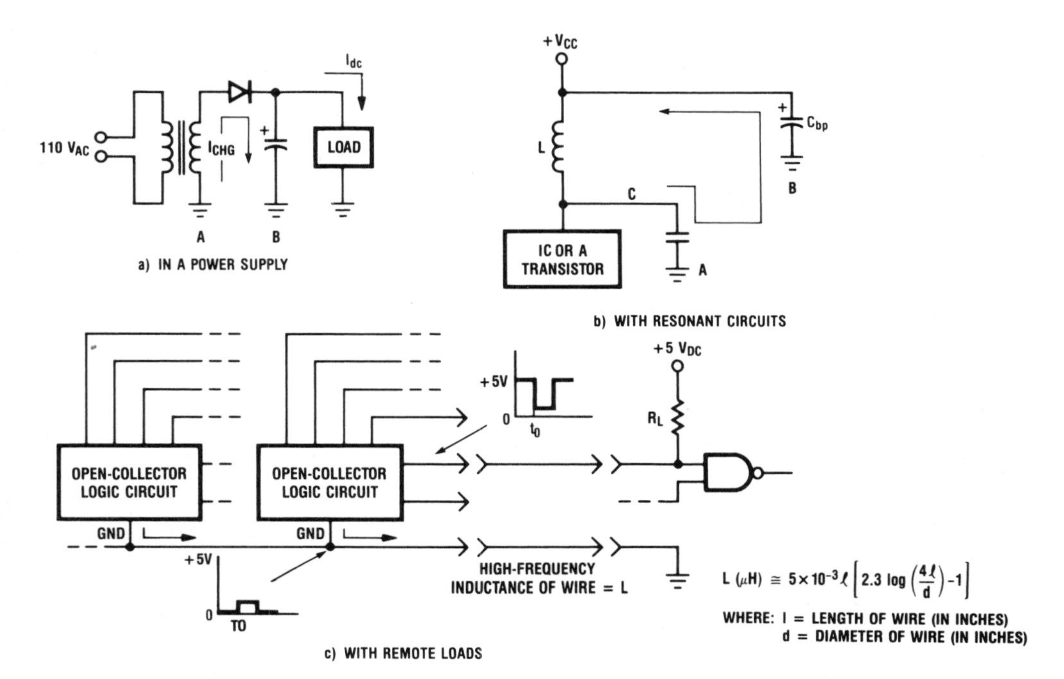

Fig. 6-2. Some Practical Component Layout Problems

Special shielded enclosures are usually required to isolate low-level analog circuitry. A grounded conductive shield can be used to prevent capacitive coupling of undesired noise frequencies that are greater than 100 Hz. To eliminate inductive coupling of noise signals (a problem for frequencies less than 100 Hz), high permeability magnetic shielding must be used, or sources of magnetic noise should be located far from the sensitive preamps (a foot of air is often better than 1/4 inch of steel or even $10.00 worth of mu metal).

Power line noise (60 Hz "hum" signals) can originate from the capacitive coupling between the windings of a power transformer. To reduce this, an electrostatic shield is added between the primary and secondary windings. Careful attention to grounding techniques is required to eliminate 60 Hz noise. In some cases a doubly-shielded power transformer, two-conductor shielded wire for the input signal, or a differential amplifier must be used. (For a more complete discussion of the grounding and shielding techniques for low-level analog signal processing see Reference 15 in the Bibliography.)

The present emphasis on analog circuitry that operates from a 5-volt power supply also places tougher requirements on shielding and grounding. When signals are a few-hundred volts in magnitude, noise can be allowed to be in the volt range and still a 40 dB signal-to-noise ratio is maintained. This was the secret of the old vacuum tube TV sets.

Use Short Lead Lengths to the Inputs

Any *input resistors* or *feedback resistors* of an op amp application circuit *should be located right at the op amp.* This prevents stray signals from directly coupling currents into the summing node of the op amp. The input pins of an op amp are very sensitive nodes and should not be exposed to the ever-present 60 Hz power line interference signals. Long leads should not be used on the op amp side of the components, especially if they are outside a shielded cable. Keep in mind how easy it is to create hum in your hi-fi audio system when only one-quarter inch of an input lead is *exposed. This 60 Hz pickup is a major problem in all low-level analog systems:* shielded two-conductor input lines are often necessary.

Determining the Minimum Supply Voltage

The minimum power supply voltage requirements of an op amp are established by the internal biasing circuitry and also by the ability of the op amp to provide a useful output voltage swing and a useful common-mode input voltage range. An idea of the magnitude of the minimum power supply voltage for a particular IC op amp can generally be obtained by checking the curves that are provided in the **Typical Characteristics** section of the data sheet. Many of these curves show power supply voltage as the independent variable and the degradation in performance that results with low power-supply voltage can also be seen.

6.2 BEING UNKIND TO AN IC OP AMP

Questions usually come up regarding the consequences of exceeding the manufacturers absolute maximum ratings of an IC. The amplifiers that are most likely to be subject to this harsh treatment are at the input or the output end of a system (the intermediate circuitry is generally protected from the people that cause accidental improper electrical connections).

There are many bad things that can happen when taking IC pins outside the supply voltage range. In general, the junction temperature can be exceeded and even the melting point of silicon can be exceeded, + 1420°C!, during a safe-operating-area or 2nd-breakdown failure in a power transistor. This is not a significant problem in standard IC op amps. Large currents can flow that will melt or fuse the bonding wires or the aluminum interconnect-metal on the surface of the IC chip. Transistors can also be forced into collector-emitter breakdown and thereby allow large current flow. The emitter-base junction of an input differential amplifier can be driven into a zener (avalanche) breakdown that permanently degrades V_{OS}, the current gain,and also the I_B and offset current specs. Even the power dissipation of an IC can be exceeded during a fault mode. Before arbitrarily exceeding these absolute maximum ratings it is interesting to look inside an IC and see why there are so many possible ways of getting into trouble.

Here is a list of the major fault-mode problems of IC applications:

Triggering an SCR
Causing metal fusing
Exceeding the E-B breakdown of the input stage
Exceeding C-E breakdown
Exceeding the power dissipation

Many of these fault mechanisms depend upon parasitic circuitry of the IC and are not necessarily the same with identical IC part numbers that are supplied by different manufacturers.

We will now show how some of these parasitic circuits are possible on an IC chip. This should aid the understanding of this complex problem of operating ICs in fault modes.

The Parasitic Circuitry is Not Shown

There are many PN, PNP, NPN, and even PNPN and NPNP possibilities in an IC layout. To appreciate some of these, we can start by taking a look at the way PN junctions are used to create the intended devices. We can then also see how easy it is to create additional parasitic devices. Many op amp designers make use of these parasitic devices within the circuit. The only difference is that these parasitics are then intentionally turned ON and special care is used in the layout to insure that nothing unusual will happen. Fault modes can turn ON unexpected parasites and create some interesting and harmful effects.

If we quickly review the way active devices are configured on an IC chip, we can start to understand the reasons for all of the possible parasitic devices that can be created. (For a more complete discussion see the author's previous book, Intuitive IC Electronics.) Just the simple act of isolating the many epi tubs on an IC, Figure 6-3, creates the possibility of parasitic lateral-NPN transistor action between not only adjacent epi-tubs, but also between any emitting tub and *every other tub on the IC!*, Figure 6-4. (The electron flow that is shown on this figure is opposite to the direction of *conventional current* flow.) Those tubs that are remotely. located from an emitting tub will collect smaller values of current. These epi tubs can be used for p-diffused resistor islands, P-channel JFET gate regions, the collectors of NPN transistors, or the bases of PNP transistors.

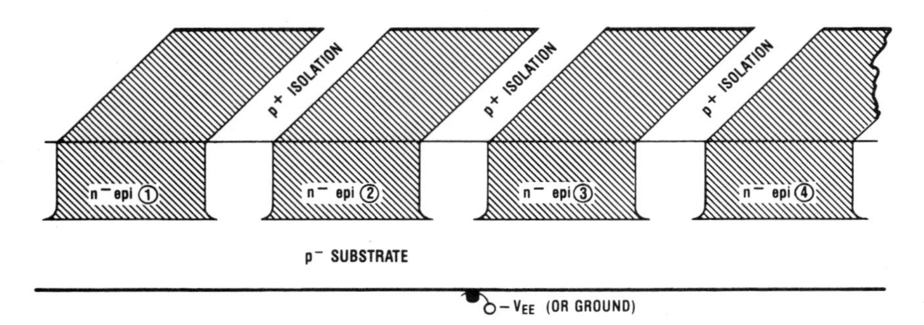

Fig. 6-3. The Isolated Epi Tubs on an IC Chip

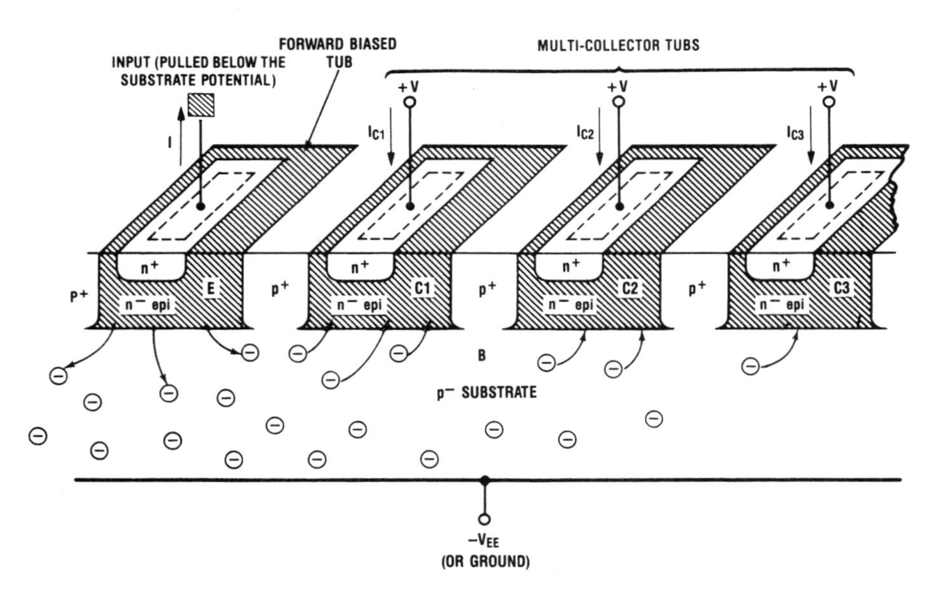

Fig. 6-4. Forward Biasing the Epi Substrate Diode of an IC

The parasitic lateral-NPN structure is shown in transistor symbol form in Figure 6-5. This is not a very good transistor and the common-base current gain, α, is low, especially to a remotely located collector. This is the parasitic device that is activated when an input lead, that is normally tied to the base of a lateral or substrate PNP input stage, is pulled below ground (or $-V_{EE}$). *The whole IC chip can be affected by this parasitic transistor action*, but an SCR is not necessarily involved, so the circuit will usually recover, once the injecting emitter lead is once again brought to a potential that is larger than the substrate potential. Input currents during this fault mode should be externally limited so they don't exceed the current density limits in the aluminum interconnect metal (1 mA per square micrometer, μm, at 25°C). Also, the power dissipation limits of the package should not be exceeded.

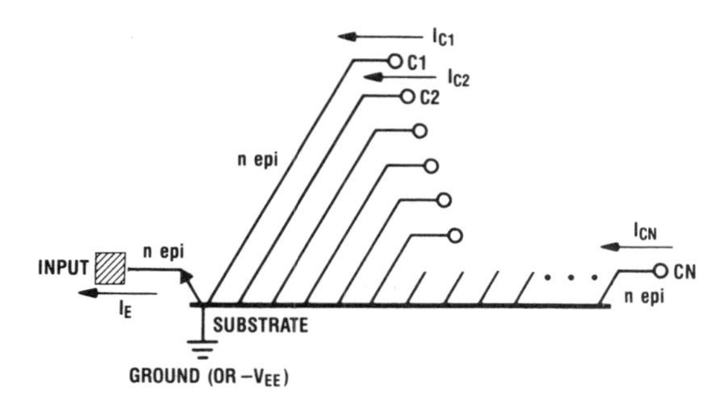

Fig. 6-5. The Parasitic Multi-Collector Lateral-NPN Transistor

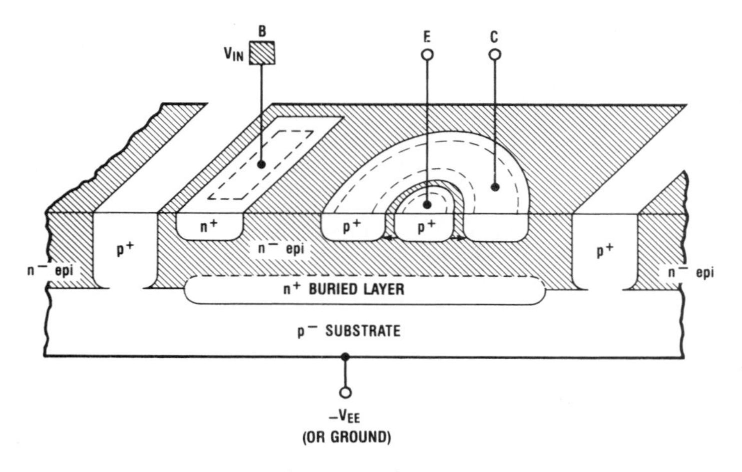

Fig. 6-6. The Lateral-PNP Transistor

Getting Access to an Epi Tub. To show how an input pin can directly tie to an epi tub, we have indicated the structure of a lateral PNP in Figure 6-6. This transistor makes use of a P-diffusion that is normally used to form the base regions of NPN transistors, as both the central P-type emitter and the surrounding P-type collector ring.

If a vertical PNP is used at the input of the op amp, this structure also allows access to an epi tub via the input pin, as shown in Figure 6-7. The normal transistor action for this device is now in the vertical direction, although some current is also collected by the sidewalls of the isolation diffusion.

A last way that an input pin can access an epi tub is provided when a P-channel JFET is used as an input transistor in a Bi-FET op amp, for example (Figure 6-8). The gate contact of this device ties to an epi tub. It is the leakage current of this epi tub to substrate, reverse biased PN diode that causes the input current, I_B, of the Bi-FET op amp. This PN junction will be turned ON if the input pin is taken to a voltage that is more negative than the negative power supply voltage, $-V_{EE}$.

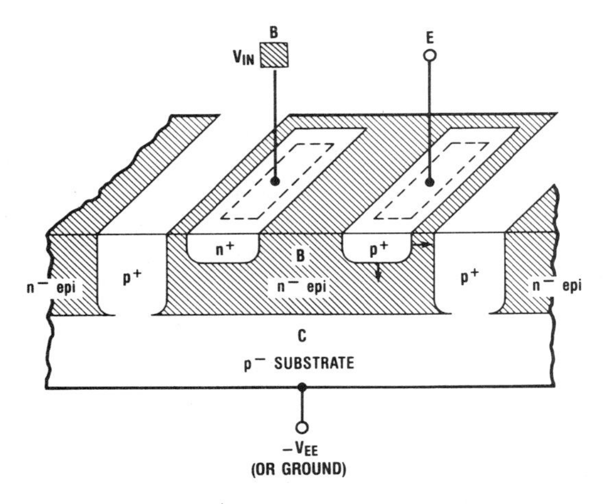

Fig. 6-7. The Vertical-PNP Transistor

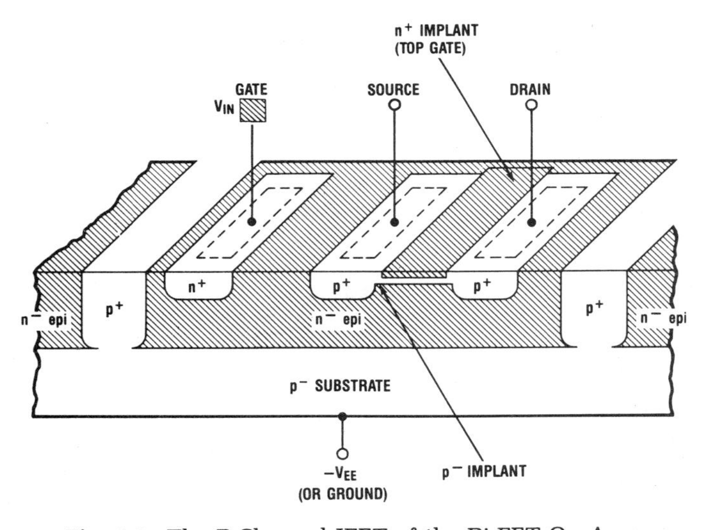

Fig. 6-8. The P-Channel JFET of the Bi-FET Op Amp

The standard IC NPN transistor, Figure 6-9, can be turned into a vertical PNP when the base-collector junction becomes forward biased any time this transistor is saturated.

The Parasitic SCR. The final parasitic structure we will consider, the silicon controlled rectifier, SCR, requires a four-layer sequence (either NPNP or PNPN) and can be considered as the interconnection of an NPN and a PNP transistor (called a hook connection) as shown in Figure 6-10.

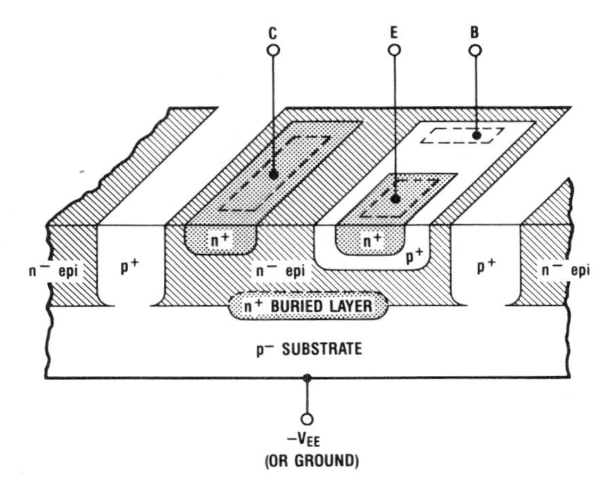

Fig. 6-9. The NPN Transistor

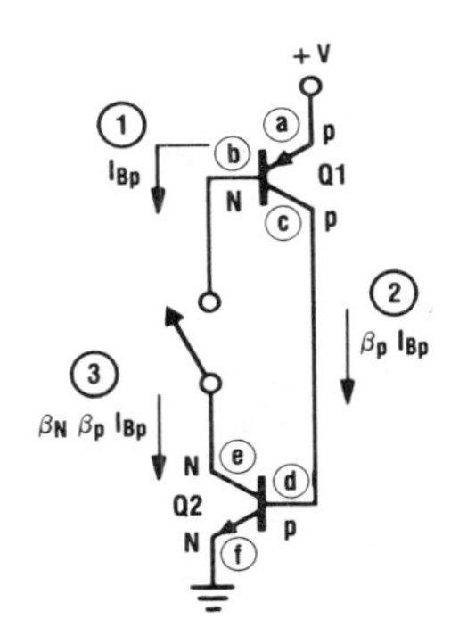

Fig. 6-10. The SCR or Hook Connection

The regeneration or latching of this two-transistor circuit can be understood by considering the events that happen with the switch open when the base-emitter junction of the PNP (Q1) is turned ON by extracting a base current. A collector current results that is larger than the base current by a factor equal to the current gain of the PNP, β_P. Note that this increased current enters the base of the NPN transistor. It is now increased again, this time by the current gain of the NPN transistor, β_N, when it appears in this collector. Because the product of these transistor current gains is usually larger than 1, this circuit will regenerate if the switch were closed; this is the SCR latching mechanism.

There are two problems with SCRs: they may draw large enough currents to destroy the IC or they will stay latched and stop the proper functioning of the IC until the latch is reset by sequencing the power supply voltages. (Turning the circuit OFF and then back ON again, assuming that the cause for triggering the SCR is no longer present.) In either case, the circuit is rendered nonfunctional, although in the first case, the IC must be replaced. (This is why triggering an SCR is considered as a serious fault mode.)

An example of how a parasitic SCR is possible in an IC layout is shown in Figure 6-11. The circled letters on this figure refer to the same regions that were shown for the SCR in the previous figure. Notice that a P-diffused resistor is serving as the emitter of a parasitic PNP transistor. This is a problem in an IC layout when these resistors are placed in an epi tub that is also the collector region of an NPN transistor. Notice that the P-diffused resistor is tied to the n^+ epi contact and the voltage drop that is caused by the flow of collector current through the resistance of the epi and the buried-layer regions causes the epi region under the resistor to drop in voltage and thereby turn ON the P-resistor to N-epi diode. This forward-biased P-type resistor injects holes (shown as ⊕) into the epi layer. The P-type base region of the NPN transistor then collects these holes: we have a PNPN (P-resistor, N-epi, P-base, N-emitter) SCR.

With this background of possible devices in mind, we will now take a look at some of the fault conditions that are possible with IC op amps.

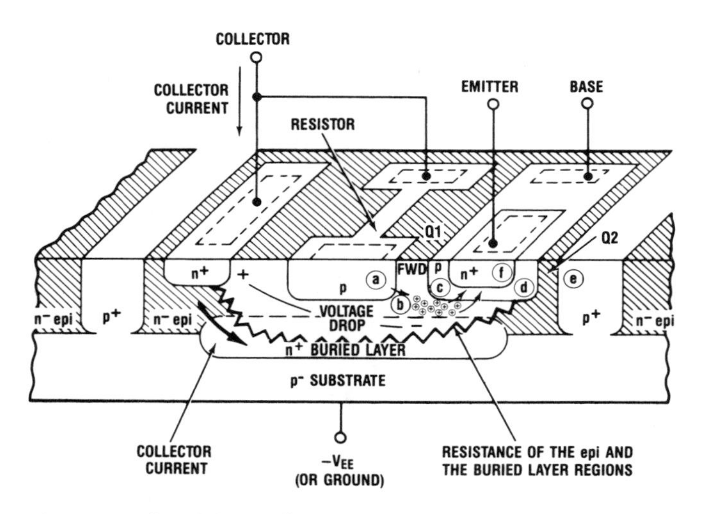

Fig. 6-11. One Way to Get a Parasitic SCR in an IC Layout

Limits on V_{IN} Differential

If both of the input leads to an op amp are kept within the bounds of the power supply voltages, there still may be a limit on the difference voltage that is allowed between these inputs. This is usually established by the base-emitter breakdown voltage of NPN transistors, which is typically 7 volts for most IC op amps. If PNP transistors are used at the input, then there is generally no problem because the emitter-base breakdown of these transistors is 30 to 60V. Bi-FET op amps are similar to PNP input op amps and also allow large differential input voltages without input-stage breakdown or other detrimental effects.

In some op amp designs (LM308, LM312, and LM318) there are internal diode clamps that are provided on the IC chip between the inputs and these will turn ON for a relatively small differential input voltage. With these op amps, it is necessary to externally limit the current to 10 mA or less that can be input to the IC when a large differential input voltage occurs.

Special problems can also result if large valued capacitors are used around these input diode-clamped op amps. Check the discharge path of these external capacitors when the supply is turned OFF to insure that the input of the op amp is not damaged. (Use external diode clamps and/or current limiting resistors.)

Plugging the Package in Backwards

Problems exist with dual and quad op amps if the dual-in-line packages are inadvertently plugged in backwards (180° from correct). A forward-biased diode results (the epi-substrate diode) directly across the power supply and there is no limit on the current flow, except what may be provided by the external circuitry. Although the power supplies may have current limiting, the large valued output capacitor that is used in a power supply is usually placed outside of this current limiting circuitry. The current surge that results from discharging this output capacitor can easily damage an IC.

Pulling V_{OUT} Above V_{CC} or Below $-V_{EE}$

In many IC op amps there is a two-diode path involving the collector-base junctions of two NPN transistors between V_{OUT} and V_{CC}, Figure 6-12. The first of these diodes is the collector-base junction of the short-circuit-current limit transistor, Q3. Its base-emitter junction is tied across the output current sampling resistor. The second diode is the collector-base junction of the output stage Darlington driver NPN, Q2. Additionally, a P-type diffused current-limit resistor can be forward-biased and this can create a one-diode drop path to V_{CC}. The feedback loop of the op amp will try to prevent an output over-voltage condition, but if the resulting short-circuit output current of the op amp is easily supplied by the external cause of this fault-mode, then these parasitic diodes will be forward biased to V_{CC}.

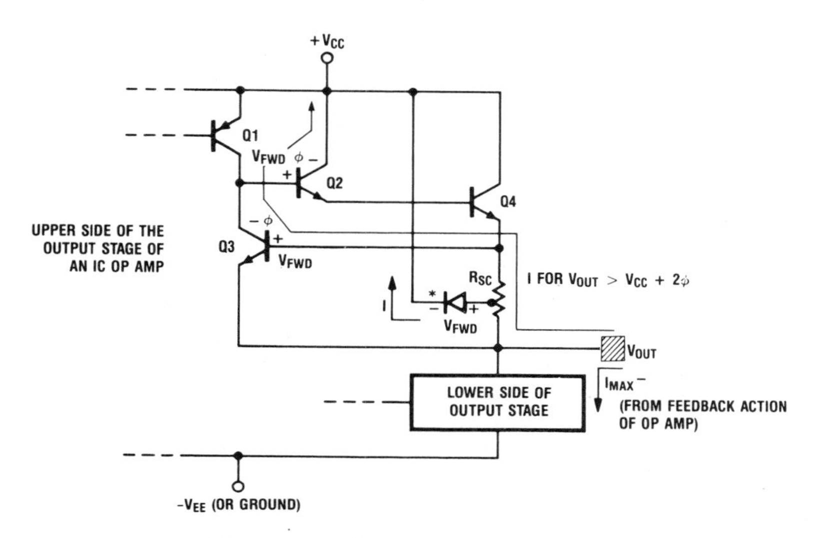

Fig. 6-12. Pulling V_{OUT} Above $+V_{CC}$

The output stage of the LM324 uses an NPN transistor current source to ground (or $-V_{EE}$) to insure that the upper Darlington transistors are conducting, even without an external load resistor. The collector of this current-source transistor, therefore, is available via the output pin of the op amp. If this output pin is pulled one forward-diode voltage below ground (or $-V_{EE}$), then an epi (the collector region)-substrate diode will be turned ON. If only a substrate PNP were used on the lower side of an output stage, then this transistor could tolerate a larger forced output voltage swing in the negative direction.

In all cases, the op amp feedback loop will generally be attempting to fix this output-voltage disturbance and therefore the full short-circuit-limited output current will be supplied and this can create a power dissipation problem. Again, for those outputs of op amps that must drive external lines to off-card circuitry, use protection diode clamps to the power supplies and series resistors, even if small values, to prevent large overvoltages directly at the IC, as shown in Figure 6-13, or use other external circuitry to limit the maximum fault-mode current to 10 mA or less.

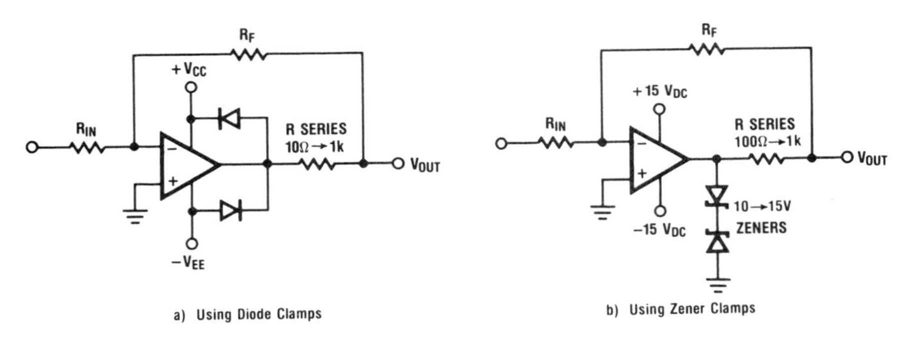

Fig. 6-13. Output Protection Circuits

Taking V_{IN} Above V_{CC} or Below $-V_{EE}$

When the input pins of a monolithic IC op amp are taken to voltages that are larger than the power supply voltages, many strange things can happen. The response to this overvoltage treatment is dependent upon the front-end design of the IC op amp and the details of the IC chip layout. A further consideration is, what is happening at the other input? This can add to the possible matrix of results.

There are many different responses to an input overvoltage. As a guide, look at the schematic diagram of the input stage of the op amp. If lateral or vertical PNPs are used, remember that these have very high emitter-base breakdown voltages. As an example, a schematic diagram of a 741 with an input taken above the power supply voltage is shown in Figure 6-14. Notice that the + 20V at the noninverting input will forward-bias the collector-base junction of the input NPN transistor, Q3. This allows access to additional internal circuitry. The bases of the PNP biasing transistors, Q1 and Q2, are now pulled above the V_{CC} level, but the large value of the E-B breakdown voltage of these transistors prevents current flow. (If the collector of Q3 tied directly to the $+V_{CC}$ line, only a single diode would exist from either input to V_{CC}.)

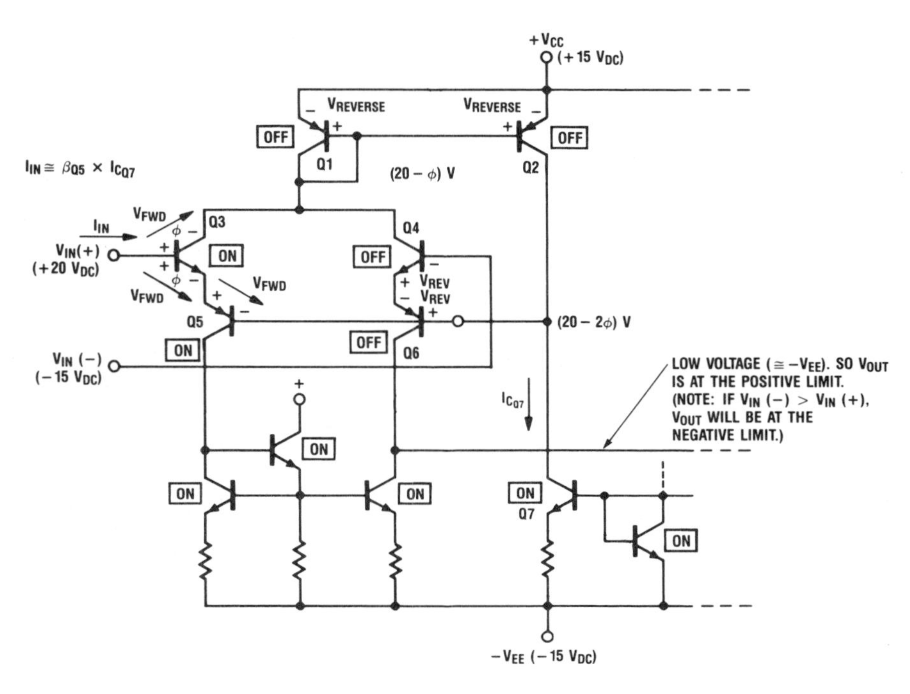

Fig. 6-14. A 741 With the Noninverting Input Taken Above the Supply Voltage

The base-emitter junction of Q3 is also forward biased. This causes the lateral PNP, Q5, to be turned ON. The input voltage to the second stage of the op amp is, therefore, held to approximately $-V_{EE}$, because Q6 is OFF and the output transistor of the current-mirror is ON, which will cause the output voltage of the op amp to be at its most-positive limit. (The signal inversion of the second stage of the op amp causes this to happen.)

Even though the internal-circuit node voltages are far from their normal values, no destructive high-current path is seen to exist for this input overvoltage condition, so this IC will survive if there are no additional parasitic-junctions associated with the particular IC layout.

It is all of these complexities that make the IC-supplier state: *don't operate the IC with any of the input pins outside the power supply range!* Some ICs can directly take this harsh treatment without degradation, but many need resistors in series with the input leads to place a limit on the current that can be forced into or pulled out of the IC. Again, this response to overvoltages depends on the particular IC that is selected and, in addition, remember that the particular IC layout that is used by each manufacturer can affect the action of the parasitic structures that are involved. (A further complication is that the performance of these parasitic devices is processing dependent.)

Turning ON a Parasitic Lateral NPN Transistor. The parasitic lateral NPN transistor that exists in all bipolar ICs can be demonstrated, Figure 6-15, by taking one input lead of an LM324 op amp below ground and limiting the resulting current flow to approximately 10 mA. The other input of the same op amp (to insure obtaining a closely-spaced parasitic collector) can then be raised to $+20\ V_{DC}$ while the op amp is powered from a single $+5\ V_{DC}$ supply. The strange thing is that approximately the full 10 mA of current can be measured in this second input-lead, the one held to $+20\ V_{DC}$ because of the parasitic lateral-NPN transistor that is activated. This is not destructive, but it is certainly curious. The additional power dissipation (20V × 10 mA or 200 mW) can usually be handled by the IC, but not if all four of the amplifiers are simultaneously placed in this fault mode.

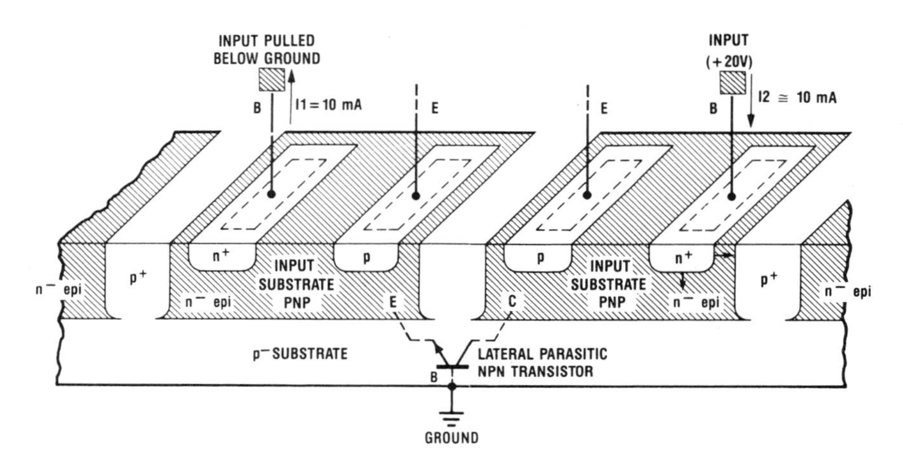

Fig. 6-15. Observing the Lateral NPN Between the Input Pins of the LM324

The collector-base junction (an N-epi to P-substrate diode) of a vertical PNP at the input or the gate (this is similarly tied to an epi tub) of a Bi-FET will be forward-biased if an input-lead is taken below the negative power-supply voltage. External current-limiting is required to prevent destroying these ICs. When an epi-substrate junction is forward-biased, a lateral-NPN parasitic transistor, shown previously in Figure 6-5, is turned ON. This can cause other amplifiers in the same package to be affected by this fault condition. (The feedback that is typically used in op amp application circuits tends to mask this interaction problem and therefore this internal coupling is more easily noticed with a multiple comparator IC such as the LM339.)

This excessive input voltage problem is generally limited to an op amp that must receive signals from an external source. It is always a good idea to add as much input overvoltage protection, using series resistors and clamp diodes to the power supplies, as is possible without degrading the signal because of the low-pass filters that are created or because of the diode leakage current at high temperatures on these input lines.

Current-Mode Inputs Provide Protection. Operation in the current-mode, that is made possible with the LM3900 current-differencing amp, has been used to obtain a large measure of protection in applications with analog signals or relatively low speed digital signals. This current-mode input circuitry can be designed to be rugged enough to be plugged directly into a 110V wall-outlet, or even worse, if desired.

The use of a conventional op amp as an inverting gain stage, with a relatively large value of input resistor, is also beneficial because this input resistor converts from high input voltages to relatively low input currents. If bounding circuits are also used on the amplifier, Figure 6-16, the op amp will not be driven into saturation (with the resulting slow recovery from overload and the loss of the protection benefits of the virtual ground) and relatively large input voltage-swings can be handled. Notice that large differential input voltages are prevented if the op amp stays active because the feedback action keeps the inverting input at a virtual ground. Additional diodes can be added directly at the inverting input (shown with dotted lines in the figure) to protect against the possibility of a large input voltage being applied when the power supplies for the op amp are OFF, for higher-frequency input voltages, or for high-speed voltage transients (but the effects of the leakage currents of all of these diodes may restrict the use of this approach).

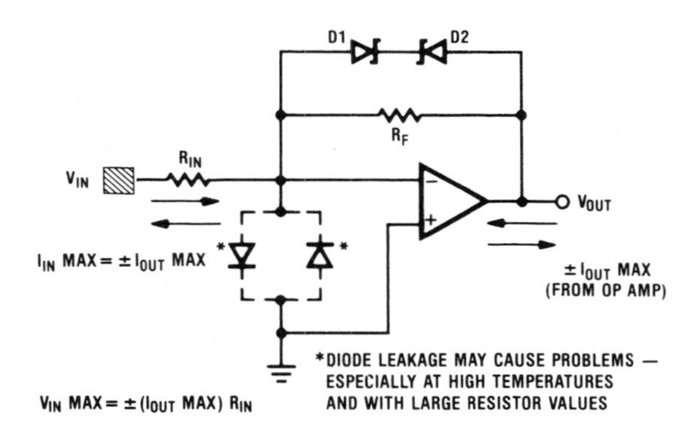

Fig. 6-16. An Input Voltage Protection Circuit

Protecting with Schottky Diodes. Schottky diodes are often used for IC protection because they will conduct at a smaller forward-voltage than the internal silicon diodes of the IC and they are also capable of operation at higher frequencies. In many instances, silicon 1N914 diodes can be used with an additional resistor placed in series with the input lead to the IC, as shown in Figure 6-17. Problems exist with this scheme because of the leakage currents of these added diodes, especially if wide operating temperatures are encountered and large resistor values are used.

Floating the $-V_{EE}$ Supply and Power Supply Sequencing

For those op amps with the input-pin connected to an epi tub, floating the $-V_{EE}$ supply is the same as pulling the input voltage below the $-V_{EE}$ supply because the op amp biasing current that would normally enter the $-V_{EE}$ supply will now forward-bias the substrate-to-epi diode. If an input of the op amp is at ground (as the noninverting input often is), this is the N-side of the PN junction that becomes forward-biased. This is easily prevented in split-supply op amp applications by *always using a diode at least across the $-V_{EE}$ supply-line to ground on each PC*

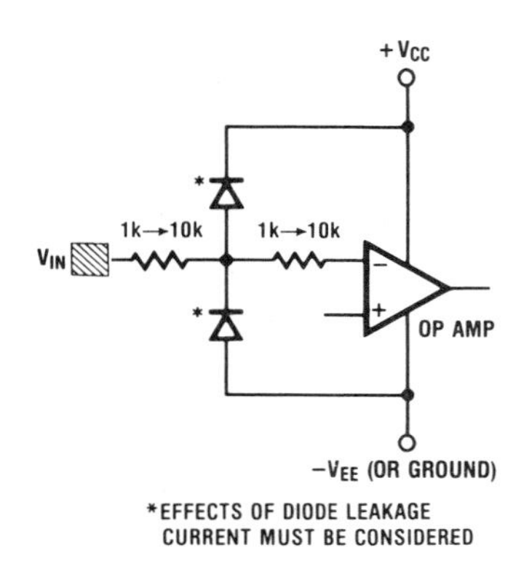

Fig. 6-17. An Input Overvoltage Protection Scheme

board. (It is also an advantage to have a diode across the + V_{CC} *supply line to also prevent power supply voltage reversal on this line under fault modes.)* In all split-supply applications, diodes across both supply lines are highly recommended. *When using Bi-FETs, at least the negative supply line diode should always be used.* With these diodes in place, there is never a concern for momentary polarity-reversal of the supply lines during power supply sequencing. (These diodes also prevent polarity reversals across electrolytic power supply bypass capacitors.)

Taking V_{CC} Above V_{CC} Maximum

Maximum power supply voltages that can be applied safely to an IC op amp are generally limited by internal collector-to-emitter breakdown mechanisms, usually of the NPN transistors. When these voltage limitations are exceeded, large values of supply current can flow. A problem exists when spikes of voltage are allowed on the power supply lines because these momentary high voltages can sometimes trigger a breakdown that folds back to a voltage value that is within the normal supply voltage range. When the power supply voltage returns to its normal value, the previous voltage spike has already triggered a breakdown mechanism so this normal power supply voltage can now supply a large current and destroy the IC. This makes voltage transients more dangerous than simple dissipation considerations would indicate.

Electrostatic Discharges Kill ICs

If you've ever walked across a carpet during the dry summer months and then been surprised by the 1- to 2-inch spark that jumped from your hand to a door knob, you have seen static electricity; the electrostatic discharge, ESD, at its best: 38 kV worth! At this voltage level, you know about, and feel the effects of ESDs. Relatively low voltages, say only 4 kV, can be discharged without you even realizing it, because discharge currents are less than approximately 1 mA. Does this bother ICs? No, *it destroys them! Even linear ICs.* (The comp cap is the most fragile component, but the base-emitter junctions of bipolar transistors at the inputs of an IC can also be permanently shorted by ESDs.) Bipolar ICs and the Bi-FETs are still both tougher than MOS ICs, but advances are constantly being made in MOS layouts to make them tougher.

Electrically, you are a mobile 100 pF (50 to 450 pF) capacitor in series with a 1.5 kΩ (100Ω to 100 kΩ) resistor. You typically have 1 to 5 kV across "your capacitor" that you acquired from inductive or capacitive pick-up, or from a triboelectric voltage generating mechanism. [This is the reason for the wrist- or heel-straps that ground (through 1 MΩ) electronic assembly workers.]

This triboelectric phenomenon exists when many common materials (one or both may even be insulators) that were once in contact are suddenly separated. *One of the materials steals electrons from the other*, and this charge transfer establishes a large electrostatic voltage. (Most plastics are good generators and should be used with care around ICs.)

Even ICs soldered on a PC board can be damaged by ESD. The problem is more severe with the input and output leads. These should be protected with voltage limiting networks. (Even an NE-2 neon lamp will help.) Any electronic equipment that people come in contact with also creates problems. When someone walks across a carpet on a dry day and then reaches out for a keyboard, 1- and 2-inch sparks can reach inside and zap the electronics.

Whenever you are experiencing IC failures, think about this ESD possibility and also look for inductors (relay coils, for example) that may be generating lethal voltages for any ICs that are nearby or even connected to different wires that are in the same cable bundle.

ESD damage is real. It is responsible for a large number of mysterious IC failures, and therefore a policy of protective and preventive measures should be established to minimize this problem. (For more information on ESD, see References 13 and 14.)

6.3 SPECIAL LOW CURRENT PROBLEMS

The world of low currents starts at approximately 10 nA. Today, the low cost Bi-FET op amps offer input currents of 50 pA and sometimes it is fair to ask; What is a pA? It takes a resistance of 15 TΩ ($15 \times 10^{12}\Omega$) to allow only 1 pA to flow from a 15V power supply. That is, if this pA is not actually flowing across the insulating surface of the resistor! When working with currents this small, materials that used to be considered insulators become almost short circuits. We will consider some of these low current problems to acquaint you with the practical aspects of this very low level of current flow.

Leakage Paths on the IC Package and PC Board

To say that a PC board is not a good enough insulator for your circuit, at first, may sound absurd. Measurements have been made on clean, high-quality, epoxy-glass PC board using two conductor stripes 0.05 inches apart that ran parallel for only 1 inch. At 125°C, the leakage resistance between these conductors measured $10^{11}\Omega$. This means, that with 15V across these copper runs, a current of 150 pA would flow, and much more if the boards were not cleaned. Even if first cleaned, using trichloroethylene or alcohol, they must be dried and than coated with epoxy, silicone rubber, a humidity sealer, or some protective coating to prevent further contamination.

Solder fluxes are a big problem and any moisture can really get things going, especially on dirty boards. (Moisture condensation is a problem whenever a cold board enters a warm room.) The effects on circuit performance owing to PC board leakage currents are very strange and erratic behavior is typical. It may take up to several hours for a sensitive circuit to stabilize following a major change in voltage. Any time such erratic behavior is seen, try cleaning the circuit board; even if you have to use hand soap and a toothbrush. A dishwasher is often found to be useful. Dry the board in an oven before retesting.

Using Guard Rings on the PC Board

When working with low level currents it is also advisable to make extensive use of guard rings. This is achieved by encircling sensitive nodes on a PC board with a copper run that is then tied to ground or to a voltage source that will reduce the voltage difference between the guard ring and the interior guarded node. For example, the summing junction of an inverting amplifier can be guarded by simply grounding the guard ring. A noninverting input can be guarded by connecting the guard ring to the low impedance feedback tap point that returns the feedback signal to the inverting input. (To reduce loading, a separate β network can be used, or a voltage follower can serve to isolate the loading.) These guard rings are easily made part of the PC board layout and are recommended for all low-current applications.

Plastic is NOT as Good as Teflon

Another surprise is that the *plastic* that is used for the *insulators* on alligator clips, for example, *is a conductor* when you are in the pA world and can cause current shunting. Also, the input current of a Bi-FET op amp can't be properly measured unless both a teflon socket and teflon wiring are used. (We will have more to say about this later in this chapter, section 6.6 Basic Op Amp Testing.) When the currents get small, trust only air and teflon insulation. (Below 0.01 pA, sapphire must be used.)

6.4 PASSIVE COMPONENTS CAN DEGRADE PERFORMANCE

The goal of using feedback is to have the transfer function of the op amp application circuit depend only on the external passive components of the β network. Having achieved this goal, it then makes sense to give considerable thought to the selection of these components, because *they now control performance.*

Selecting Resistors

Resistors are available in a wide range of values and initial tolerances and can be arranged in decreasing *quality* as: wire wound, metal film, carbon film, and carbon composition. The tolerances of resistors and other components can offer some surprises. Values tend to group near (but within) the tolerance limits. This is called a "double-humped" or "bimodal" distribution. A nice Gaussian distribution is often distorted in this way because the more accurate components were previously sorted out for the tighter tolerance spec components.

Resistors have a stray capacitance from end-to-end across the terminals (approximately 1.5 pF for 1/2W composition) that limits the high frequency response. This effect is less for physically small sized resistors, but is important to consider for all *large valued resistors.*

Wire-wound resistors are the most accurate (and the most costly) resistors available. Values larger than 200 kΩ are especially high priced. They are generally only used in the most demanding applications where the major concern is for accuracy, stability, and low temperature change. Wire-wound resistors are available that can maintain 0.01% (or better) stability with time and over moderate temperature change. Large size and poor ac performance, due to the inductance of the winding, are disadvantages of this type of resistor construction, although low inductance can be achieved by a special technique that uses two windings that are wound in opposite directions and then connected in series.

Metal-film resistors (available from 10Ω to 10 MΩ) are commonly used, except for the very low performance op amp application circuits, where carbon-composition resistors are often used. Metal-film resistors with tolerances of 1% are inexpensive and are therefore widely used. Popular temperature coefficients (often abbreviated as Tempco's or TC's) are ± 100 ppm/°C (designated as "TO") and ± 50 ppm/°C (designated as "T2"), although ± 10 ppm/°C units are also available. Special types are guaranteed to have only a positive or a negative change with temperature, if desired.

Metal-film resistors are also supplied on substrates and can be obtained with very close matching (ratio) and excellent Tempco tracking. The longterm drift characteristics and temperature drift are not as good as can be obtained with the best wire-wound units, but the small size provides excellent ac characteristics, and the tracking (ratio) characteristics can approach that of good wire-wound networks.

Carbon film, deposited on ceramic and glass substrates, provides increased stability, low noise, and improved high frequency characteristics when compared to carbon-composition resistors. Temperature coefficients vary from about -250 ppm/°C for low resistance values (that use 10Ω/square films) to -1000 ppm/°C for high resistance values (10 MΩ or higher). Good carbon-film resistors can be expected to change by less than 3% over five years if used conservatively.

Carbon-composition resistors are the common, low cost resistors that are used for breadboard checking of circuit ideas and noncritical final circuit applications. A *carbon noise*, that is proportional to current flow, exists with these resistors that increases the noise when compared with film and wire-wound resistors.

It has also been reported that two equal valued, carbon-composition resistors have been found to have temperature coefficients that differed by as much as 1000 ppm/°C! Water absorption from humid air can increase the resistance value by as much as several percent.

Soldering a resistor that is not completely dry can also cause shifts. These low cost resistors should be used with care (if at all) around precise op amp circuits; but, when obtained from a reputable manufacturer, they are inexpensive and very reliable when precision is not required.

Variable Resistors or Potentiometers

Potentiometers (pots) of any variety are usually undesirable, although sometimes necessary, and can be a problem in fast circuitry. There are many parasitic effects associated with the construction of a pot (for example, the capacitance from the resistor element to the case can be 17 pF) and, therefore, pots should only be used where the signal frequency, f, times the R value is not in excess of 10^6 Hz-Ω.

Pots also have a fixed amount of resistance at each end of the resistance element, the *end resistance*, that prevents wiper access to the ends of the resistor element. This restricts the minimum and maximum range that can be obtained with the pot. The wiper can also have a resistance of 10 to 20 ohms associated with its contact or construction. These factors, plus the initial tolerance of the resistor element have to be considered to insure obtaining a necessary adjustment range. In the most reliable designs, resistor networks are also placed around the pot (over to the wiper terminal if used as a potentiometer) so that system operation will still be obtained with the wiper set at either extreme position and even if the wiper is not in contact with the resistive element.

Multiturn pots are often claimed to have "infinite" resolution; but, mechanical backlash and shock-induced jumps may cut the usable resolution to values in the range of 0.02% to 0.5%. Good single-turn pots can be set as well, 0.05% to 0.5%, as good multiturn pots and are less likely to shift when rapped or vibrated, but for settings closer than 0.5% resolution, a lot of fiddling time is needed.

An alternative to the use of pots is to, instead, use a digital-to-analog converter that is programmed by mechanically cut links or by a computer. An additional low cost trimming approach is to make use of extra resistors on a PC board that can be selectively cut out to provide an adjustment capability at the time the board is electrically tested. (This trim technique is also more immune to arbitrary changes by the maintenance technicians and also works well under vibration and mechanical shock.)

Selecting Capacitors

Obtaining a good capacitor is more difficult, and, therefore, op amp circuits using capacitors are usually less accurate than if only resistors were used. Capacitors are not readily available in as wide a range of values, nor with as low initial tolerances, and they generally have poorer stability with time. In addition, they are not nearly as ideal.

One of the largest of the capacitor defects is the *memory effect* or dielectric absorption characteristic. Unfortunately, electric dipoles within the dielectric material will eventually respond to the electrification of the capacitor during both charging and discharging. These internal dipole alignments raise the capacitance value and, therefore, generally cause the largest memory problems with the larger dielectric constants: the high-K ceramic capacitors, for example.

The problem is that a capacitor, once charged, cannot be rapidly discharged. If insufficient time is taken for the slow acting internal electric dipoles to respond during this discharge cycle, the voltage across the capacitor will jump back to a certain percentage of the previous voltage. This can be from a fraction of a percent to as much as 25%, depending on the time duration of the discharge.

The parameter *dielectric absorption coefficient* DA is used to indicate the amount of this memory effect. It is determined by charging the capacitor for five minutes and then discharging it for ten seconds through a 5Ω resistor. The voltage that appears across the capacitor following this short discharge cycle is then measured. The ratio of the magnitude of this resulting voltage to the magnitude of the charging voltage is defined to be the dielectric absorption coefficient. A model of a capacitor is shown in Figure 6-18 that is used to account for this memory effect.

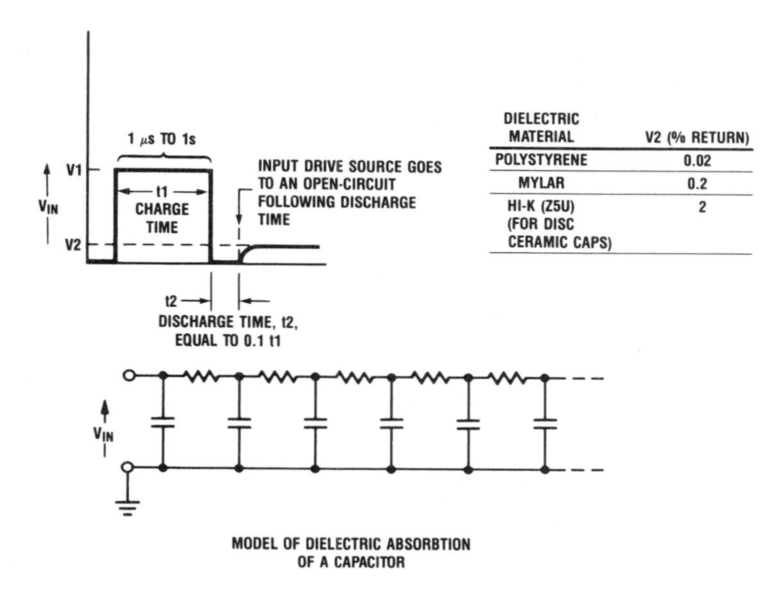

Fig. 6-18. Modeling the Dielectric Absorbtion of a Capacitor

Teflon (available from 100 pF to 1 μF with tolerances of 20% to 0.05% and TCs of −100 to −200 ppm/°C) and polystyrene (available from 100 pF to 10 μF with tolerances of 20% to 0.05% and tempcos of −110 ± 30 ppm/°C) are the best dielectric materials for minimizing this memory effect. The NPO or COG ceramic capacitors, available up to 0.05 μF, are also good in this respect. These capacitors are relatively large in size and high in cost. This is especially true for teflon. The maximum allowed temperature for polystyrene is +85°C, and for higher temperatures, teflon must be used. Low cost polystyrene caps can have approximately 0.05% or worse hysteresis in their capacitance values when temperature cycled, but good polystyrenes can do better than this. Polypropylene is the successor to polystyrene. Its dielectric absorption is as good as polystyrene, and it is cheaper, smaller, and more available.

Mica and glass capacitors are available to 0.01 μF and are also useful for the smaller capacitor values. Depending on the formulation, these dielectrics can exhibit dielectric absorption characteristics and tempcos nearly as good as polypropylene, but at times can be much worse. Polycarbonate capacitors are somewhat smaller in physical size than teflon or polystyrene and the dielectric absorption problem is somewhat greater, but should be considered for less demanding applications.

Mylar capacitors (available from 10 pF to 10 μF with tolerances of 20% to 1% and +500 ppm/°C) are inexpensive, have a large memory effect, and should be used in less critical applications.

The high dielectric-constant ("high-K") ceramic capacitors have many bad characteristics. The capacitance can change many percent owing to temperature, frequency, applied voltage, and time. Therefore, these ceramic capacitors should only be used for noncritical decoupling applications. These capacitors can have very low values of dc leakage current (with leakage resistance of 80,000 to 100,000 MΩ) and also have low values of inductance (limited by the inductance of the leads).

Tantalum capacitors are often recommended for power supply bypass applications because of a useful value of series resistance. These capacitors will often stop high frequency parasitic oscillations that involve resonant circuits in the power supply lines because the series resistance of the capacitor reduces the Q of the high frequency parasitic resonant circuits.

Soldering Disturbs the Circuit

Another problem to consider in high accuracy analog circuitry is that after soldering to change a component, a circuit may have to be baked or temperature cycled for a few hours to stabilize all components and to relieve the strains which may have been created by this localized high temperature soldering operation.

6.5 A COMMON TRANSISTOR CURRENT SOURCE BIASING ERROR

In many application circuits that use an external transistor as a current source, an error is often made in properly accounting for the base-emitter voltage drop of this transistor in what is hoped to be a temperature compensated design. This problem is shown in Figure 6-19. If you are not careful, the diode in series with the two resistors seems to compensate for the base-emitter drop of the transistor and therefore seems to cancel the drift in the voltage across the emitter resistor, R_E.

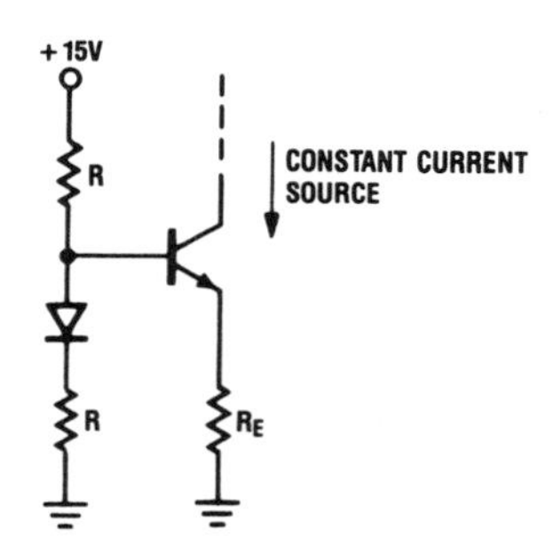

Fig. 6-19. A Common Biasing Error

A proper analysis of this circuit shows that, just as the equal resistors divide the upper 15V in half, they also divide the diode voltage in half, in providing the total voltage at the base of the transistor. Therefore, to temperature compensate this circuit will require two diodes, which can even be placed on the ground side of the lower resistor, where their effects at the tap point are more clearly seen by visual superposition.

In an attempt to temperature compensate the general resistive divider, we can place m-diodes in series at the top and n-diodes in series at the bottom and then write the equation for the voltage at the base of the transistor in terms of the resistor values and m and n. We will then look for an integer solution for m and n that best approximates the dc voltage and provides temperature compensation. This technique will work, but is limited to only certain resistor ratios or dc tap voltages and consumes a large number of diodes.

An alternative is shown in Figure 6-20, where a temperature coefficient, TC, is purposely put on the upper voltage source that is equal to the TC of one forward-diode drop. Now, with a diode also at the bottom of the resistor string, any desired voltage along this resistor string (that is for any values of R1 or R2) will provide the base bias voltage for a temperature compensated, transistor current source. (A simpler possibility may be to use an op amp current sink as discussed in Chapter 5 or a special current source IC, such as the LM334.)

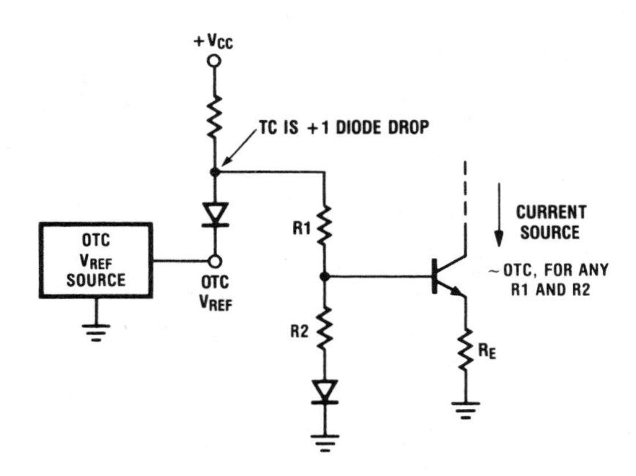

Fig. 6-20. Temperature Compensated Current Source Biasing

6.6 BASIC OP AMP TESTING

Op amp testing can be kept relatively straightforward or it can become quite complex, depending on the accuracy desired, the time allowed for testing, and the parameters that are to be measured.

It is important to always insure that the op amp is not saturated, that it is not oscillating, and that there is not an excess of 60 Hz or other noise present in the output voltage waveform.

Many timing problems associated with the test setup and measurement system are created with automatic test equipment that complicate the testing of linear ICs. Also, to avoid damaging the op amp, the signal generators that may be used in the test should be first turned OFF and then the power supplies turned OFF before removing the op amp from the test socket. After the insertion of the next op amp into the unpowered test socket, the power supplies should be turned ON first and then the signal generators can be turned ON. It is also a good idea to have clamp diodes from each supply line to ground to prevent reversal of the voltage polarity during power supply sequencing. (This will also prevent either power supply line from floating during a turn ON or a turn OFF sequence.)

A common testing mistake is made if the circuit is oversimplified in an attempt to save resistors. This trap is shown in Figure 6-21. Notice that as part of a test circuit there may be a separate load resistor as shown in Figure 6-21a; and further, because this load element is a resistor, the load current will be in phase with the output voltage. We have a real load, as opposed to a reactive load.

The simplification shown in Figure 6-21b saves a resistor by placing R_L from the output to the inverting input of the op amp. At first glance, this appears to be an equivalent circuit, but on closer examination we find that it is quite different. For example, as the test frequency approaches f_u of the op amp, the relatively large signal (at $f_{IN} = f_u$, $A_{VOL} = 1$ so $V_O = V_{IN}$) that exists at the inverting input affects the current flow through R_L. The phase shift between the voltages at each end of this resistor causes a reactive load current to flow. Therefore, the op amp is no longer driving simply a resistive load, and the phase margin can be degraded. Avoid this temptation to save a resistor in your test setup. Also notice that the input impedance falls from 1 KΩ (Figure 6-21a) to 20 Ω (Figure 6-21b).

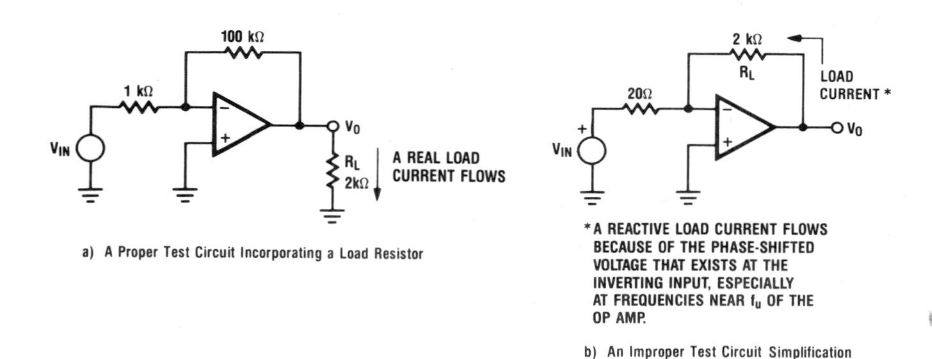

Fig. 6-21. Place the Load Resistor from Output to Ground

Determining the Offset Voltage, V_{OS}

A simple V_{OS} test circuit that uses the gain of the op amp to amplify its own offset voltage is shown in Figure 6-22. It is important to keep the value of the resistor at the input small enough (51Ω in this circuit) so that the input current of the op amp will not cause measurement errors.

This circuit does introduce an error of a few percent because it does not properly make the offset voltage determination with the output voltage of the op amp at zero volts. This shortcoming can be eliminated with the addition of an auxiliary op amp and an overall control loop, as shown in Figure 6-22b. To insure an overall negative feedback loop, the connections to the op amps should be as shown, or both changed, if desired. This testing loop is also useful to measure other op amp parameters, and can be made part of a basic op amp tester.

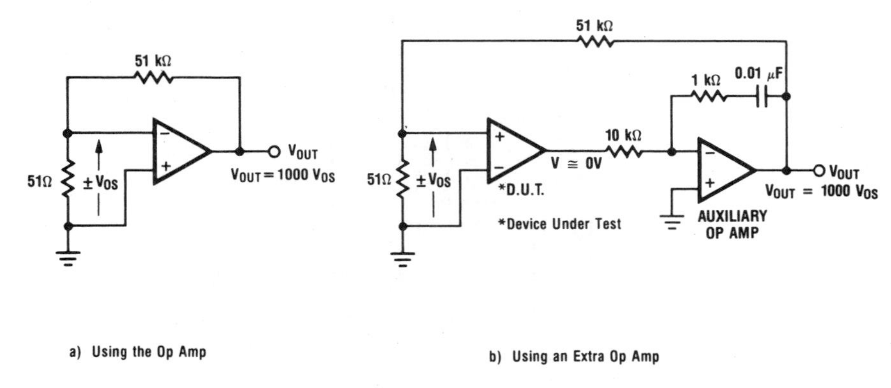

Fig. 6-22. Measuring the Offset Voltage of an Op Amp

A Way to Measure Open-Loop Gain

The auxiliary op amp loop technique is used to measure the large signal average voltage gain, as shown in Figure 6-23.

For an op amp operating with $\pm 15\ V_{DC}$ power supplies, the standard output voltage change is $\pm 10V$ as shown on the figure. For a single supply op amp operating with a $+15\ V_{DC}$ power supply, the output voltage change typically used is $+1V$ to $+11V$.

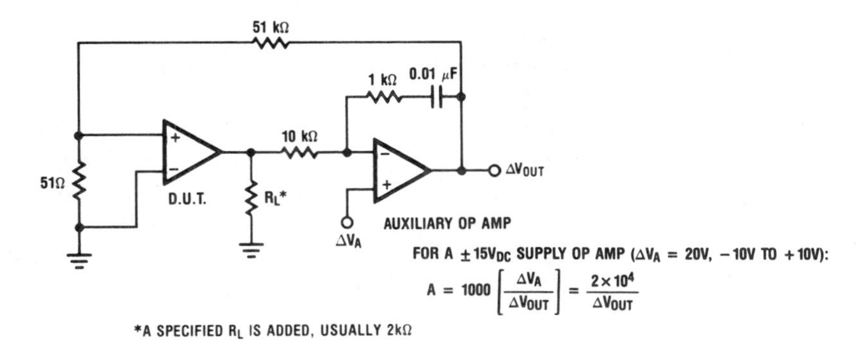

Fig. 6-23. Measuring the Large Signal Voltage Gain

Measuring I_B and I_{OS}

An op amp can be used to measure the effects of its own input current as shown in Figure 6-24. (The capacitors indicated on this circuit are used to reduce the noise contamination of the output voltage.)

It is important to keep the two resistors (labeled as R) well matched to prevent introducing an error in the measurement of I_{OS}. The value to be used for R is determined by the voltage drop that is produced by the input current. The idea is to keep this voltage drop large when

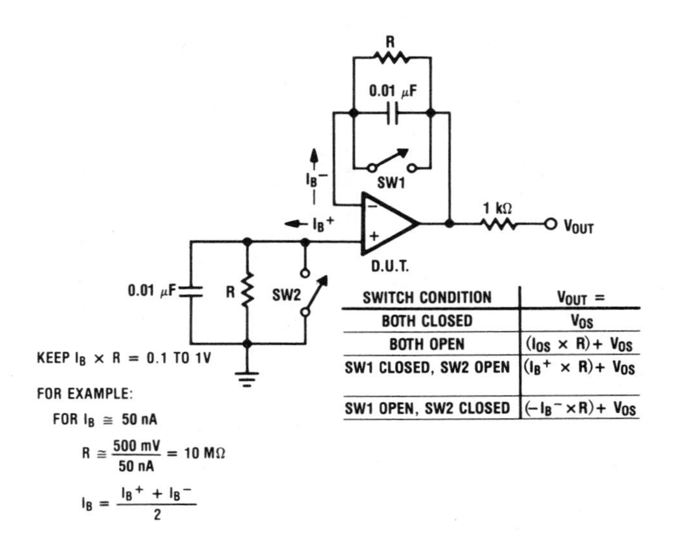

SWITCH CONDITION	V_{OUT} =
BOTH CLOSED	V_{OS}
BOTH OPEN	$(I_{OS} \times R) + V_{OS}$
SW1 CLOSED, SW2 OPEN	$(I_B^+ \times R) + V_{OS}$
SW1 OPEN, SW2 CLOSED	$(-I_B^- \times R) + V_{OS}$

Fig. 6-24. Measuring the Input Currents of an Op Amp

compared with the offset voltage of the op amp. It is very important to subtract the offset voltage from the I_{OS} measurement, because, for many op amps, I_{OS} can be a relatively small current.

The directions of the I_B currents shown on the figure are for an op amp with a PNP input stage. An op amp with an NPN input stage will have input currents that enter the op amp and the signs for the I_B voltage drops will correspondingly change from what is shown on the figure (in the table of V_{OUT} responses).

To measure I_B^+ of a Bi-FET op amp requires a much more complex circuit, as shown in Figure 6-25. The switch, SW1, is initially shorted and then is opened in synchronism with a zero crossing of the ac power line frequency. After one cycle of the power line frequency (16.67 ms), the output voltage, V_{OUT}, is measured. The input current, I_B^+, of the op amp is integrated by the 100 pF capacitor for this 16.67 ms time interval to provide a voltage across this capacitor that is given by

$$V_c = \frac{1}{C} \int I_B^+ \, dt = \frac{(I_B^+)\, \Delta t}{C}$$

or

$$V_c = \frac{16.67 \times 10^{-3}}{100 \times 10^{-12}} (I_B^+)$$

(The dc input current of the LH0052 is 1 to 2 pA and is therefore neglected.)

The resistive attenuation (R1 and R2) increases this voltage so V_{OUT} becomes

$$V_{OUT} = \left(\frac{R1 + R2}{R2}\right) V_c = \left(\frac{23{,}200 + 100}{100}\right) V_c = 233\ V_c$$

or

$$V_{OUT} = \frac{233\ (16.67 \times 10^{-3})}{100 \times 10^{-12}}\ (I_B^{\ +}) = 3.88 \times 10^{10}\ (I_B^{\ +})$$

For example, if $I_B^{\ +} = 50pA$, V_{OUT} would become

$$V_{OUT}\Big|_{I_B^{\ +} = 50\ pA} = (3.88 \times 10^{10})\ (50\ pA) = 1.94\ \text{Volts}$$

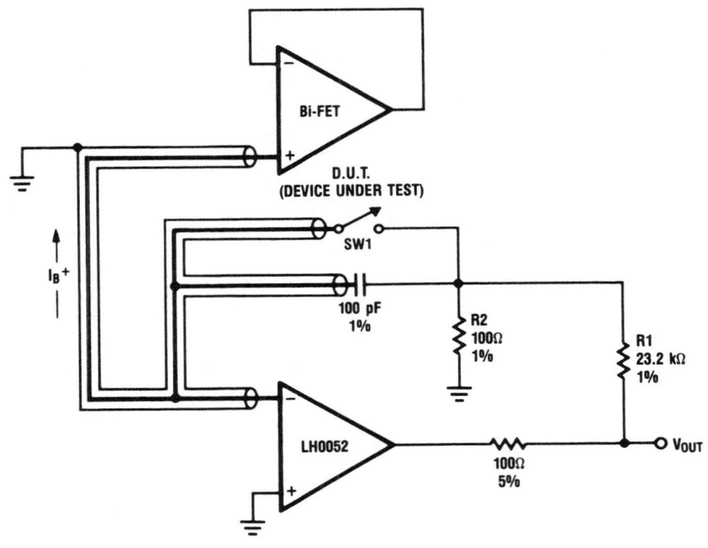

Fig. 6-25. Measuring the Noninverting Input Current of a Bi-FET Op Amp

A similar circuit, Figure 6-26, is used to measure $I_B^{\ -}$. An extra op amp, A_2, is added in a feedback loop to keep the Bi-FET op amp that is under test operating within its linear range.

For both of these circuits, all insulators that touch the input leads of the Bi-FET, *including that of the shielded cable, the insulated standoffs, and the test socket*, must be teflon.

The input currents for the Bi-FET op amps are specified for a junction temperature, T_j, of +25°C. Self-heating of the op amp can cause T_j to rise 10 to 30° above an ambient of +25°C. (The input current doubles for approximately every 10°C rise in T_j, so the actual input currents that are measured depend on the test time, the package used, the power dissipated, and whether a heat sink is used.)

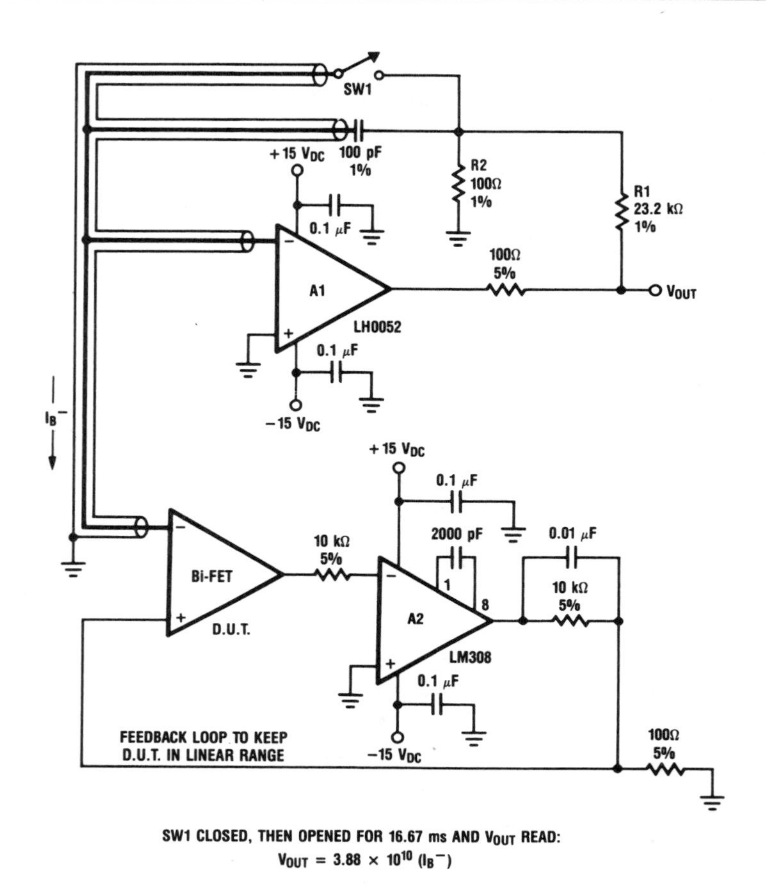

Fig. 6-26. Measuring the Inverting Input Current of a Bi-FET Op Amp

Extrapolating to Find the Unity-Gain Frequency

The unity-gain frequency is determined by making a gain measurement somewhere on the slope of the open-loop gain characteristic and then extrapolating this gain-bandwidth product at -6 dB/octave to an open-loop gain of unity. An estimation of f_u can also be easily made by measuring the small-signal bandwidth of a closed-loop amplifier and then similarly extrapolating this gain-bandwidth product to unity gain. In the example circuits of Figure 6-27, the test frequency is selected as 100 kHz (approximately 0.1 f_u).

In both of these circuits, the loading of the ac voltmeters can cause errors and therefore unity-gain op amp buffers may be needed to isolate the input capacitance of these meters.

The output signal amplitude should be kept small enough to avoid errors because of the slew rate limitations of the op amp. Check using an oscilloscope to insure that the output waveform does not show slew distortion.

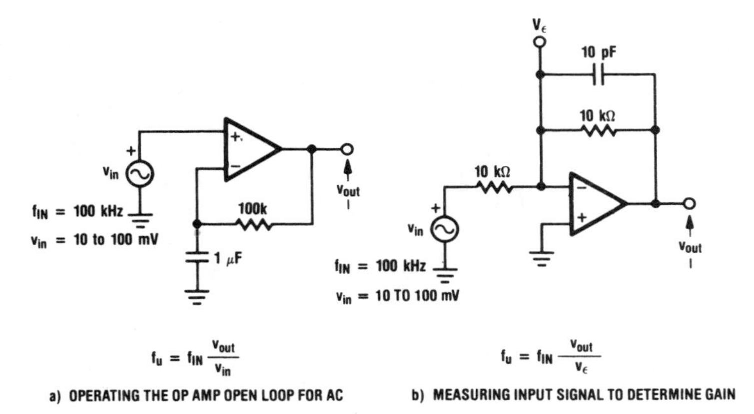

Fig. 6-27. Measuring the Unity-Gain Frequency

Use a Large Input Signal to Test the Slew-Rate Limit

The slew-rate limit of an op amp is defined to be the smaller of either the positive slope or the negative slope of the maximum rate of change of the output voltage waveform (neglecting any initial waveform disturbances). The circuit of Figure 6-28 will display both slopes, but unfortunately, the measurement of slew rate usually involves reading data from the screen of an oscilloscope.

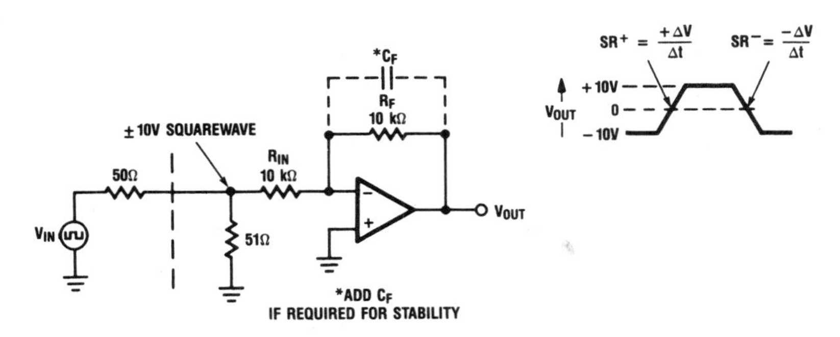

Fig. 6-28. Measuring the Slew Rate of an Op Amp

Measuring the DC Common-Mode Rejection Ratio

The measurement of the dc common-mode rejection ratio often presents difficulties. It is important to remember that the value obtained in this measurement is approximately the same as the value of the open-loop voltage gain. For this reason, the output voltage of the op amp has to be held at zero volts for the complete testing sequence.

An ideal CMRR test circuit is shown in Figure 6-29. The second op amp, #2, is holding the output voltage of the op amp under test at zero volts. (This second op amp has an ac coupled feedback network to insure stability of the overall loop of the two op amps.)

The measured parameter in this test circuit is the differential voltage that results from the common input terminal to the output of the second op amp as the common-mode voltage at the input is switched from + 10V to -10V or vice versa. This voltage can be as low as 2 mV if the CMRR is 120 dB or 60 mV for the more typical 90 dB CMRR performance of the high-volume op amps.

In high speed automatic testing, the CMRR test is often implemented in an equivalent, but different looking circuit, Figure 6-30. The supply voltages are generally under control of the test computer, which makes this circuit easier to implement. The equal valued resistors that are connected between the power supply voltage lines provide the reference voltage for the output of the op amp under test. This circuit is therefore equivalent to that of the previous figure as long as the voltage values shown for the power supplies are accurate to prevent introducing a power supply dependency in the test results.

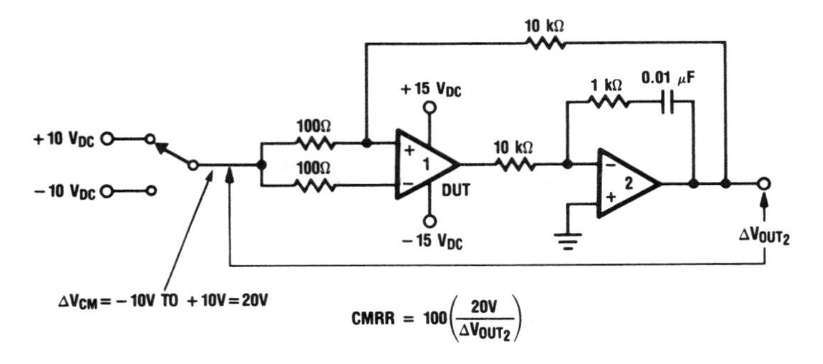

Fig. 6-29. Measuring the DC Common-Mode Rejection Ratio

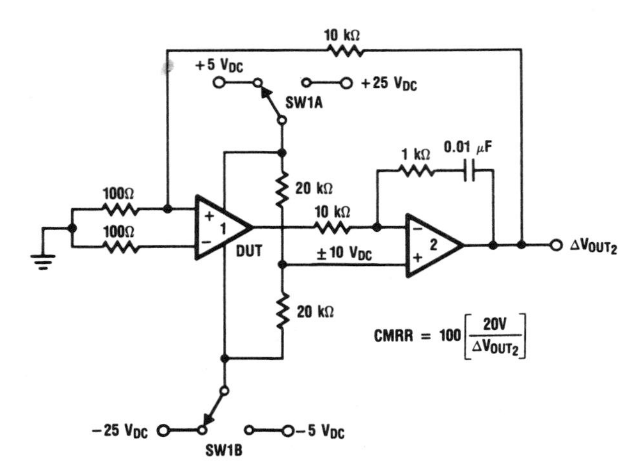

Fig. 6-30. An Alternate DC-CMRR Testing Technique

The ± 10V common-mode change at the input to the second op amp will introduce only a small change in the offset voltage of this op amp. This will cause a negligible error in the control of the output voltage of the op amp under test because this disturbance is within the overall feedback loop.

6.7 OVEN TESTING PROBLEMS

Making the proper tests on op amps while in an oven or temperature chamber creates special problems. Even if the internal temperature is kept uniform by air circulation within the chamber, the chamber is momentarily turned OFF so the high power cycling of the heater will not disturb readings, and the temperature readings have been verified; there are still more problems to consider. For example, thermal gradients across electrical connectors used in the oven door can cause thermocouple induced voltages. (Better results can be obtained with the connector located in the oven with the wires brought out. This is especially a problem when looking for microvolt levels during oven testing.)

Components in the Oven

A decision has to be made whether only the op amp or the complete application circuit should be subjected to temperature. Accurate results are usually only obtained if the resistors are included. (For example, to establish a fixed gain of 100 or 1000 to measure V_{OS}.) Any capacitors that may be involved are left outside, if possible. (Therefore, a dc evaluation is done without the capacitors to verify that the rest of the circuit is performing properly.) If the complete application circuit is placed in the oven, the temperature performance of the resistors and capacitors that are used with the op amp should be separately measured to isolate their temperature drifts from those that are caused by the op amp.

Getting Leads In and Out

The next problem concerns the leads that must be passed outside for measurement access. In general, 1 kΩ isolating resistors should be used, local to the test nodes, in series with the leads that carry the test signals. This will isolate the stray capacitive loading that is caused by the lead. (One #22 gauge insulated wire in a bundle can have approximately 40 pF/ft of stray capacitance.) This will also help prevent circuit response to the sampling *glitches* that propagate from some of the digital voltmeters.

The effects of capacitive loading and the possibilities for oscillations owing to long test leads must be considered and all points that are measured should also be observed with an oscilloscope to insure that the circuit remains stable. For example, disc-ceramic capacitors can severely change value as a result of changes in temperature. If these are part of a frequency compensation network that is included within the chamber, then temperature dependent oscillation problems can result. Better caps should, instead, be used.

Moisture Condensation

The most difficult temperature range for moisture condensation is around 0°C. Moisture will condense on the circuitry unless a dry air ambient is maintained within the chamber. (Moisture condensation will then occur if the cold circuitry is exposed outside the chamber.) The presence of this moisture will aggravate surface conduction problems on the PC board and will interfere with proper measurements. (If the oven is cooled by CO_2, carbonic acid can result, with disastrous consequences.)

6.8 HOW TO READ AN OP AMP DATA SHEET

To the uninitiated, an op amp data sheet can sometimes be confusing. A data sheet means different things to the different people who consider it. The circuit designer wants to use the data sheet to show the world how clever he is and how superior the specs of his part are. Marketing people also want to dress the part up to make it irresistible to customers. The testing engineers are under pressure to make increasingly difficult measurements on increasingly difficult products with no "escapes", those parts that slip through the test screens. And they want to do this with automatic equipment within a very short, total test time budget. Finally, the wafer production people want to make a profit and therefore would like to be able to ship as much of each wafer as is possible. A data sheet, therefore, is simply the document that results from all of these conflicting interests and it is not simple.

The Captivating First Page

The purpose of the first page of a data sheet is to quickly indicate what the part is, to list the outstanding qualities of the product, to show the areas of usefulness, and to provide a short description of what is unique. It is designed to catch your interest and to compel you to turn to the second sheet. If a circuit schematic is shown, it will generally be a simplified one so as to rapidly indicate the essential features of the design.

The Life Sustaining Absolute Maximum Ratings

At the top of the second page you will generally find a listing of the Absolute Maximum Ratings. There is some confusion with these specifications. These are not to be interpreted as Operating Limits. Exceeding these Absolute Maximum Ratings can cause permanent degradation, instant IC death, or degraded reliability of the product. On some data sheets you will find a separate listing of Operating Limits. This listing then shows the range of permitted parameter variations under operating conditions for such things as power supply voltages, for example.

Electrical Characteristics: The Guarantees

The Min and Max columns of the Electrical Characteristics provide the guarantees on the electrical performance of the op amp. Many of these are "one-sided" guarantees because the other side of the spec is not the limiting case. (Such as guaranteeing that the Min value of Input Offset Voltage is 0 mV.) Those specs that are listed with only a Typ value are used to provide an indication of the performance to expect from the *typical* part. No guarantee is implied.

The electrical testing of ICs is rapidly becoming one of the highest cost items associated with the production of a part. Therefore, attention is usually paid to the dominate performance specifications of an op amp such as Input Offset Voltage, Input Offset Current, Input Bias Current, Supply Current, Large Signal Voltage Gain, Common-Mode Rejection Ratio, Supply

Voltage Rejection Ratio, and Output Current. Custom testing of an IC adds to the cost and is not economic for low volume users. (A slightly higher cost op amp with better guarantees or customer testing is recommended.)

Many defining Notes are appended at the end of the Electrical Characteristics section. These notes are numbered and a parenthetical reference to a particular note is indicated in the body of the specs listing that shows where each note applies. For example, a note is usually referenced to the main heading that states that "unless otherwise stated, these test conditions apply over a (specified) range of supply voltage and a (specified) range of temperature (such as 0°C to 70°C for the lowest-cost products)." You will find specs specifically restricted to room temperature (25°C) and the same spec repeated with no temperature restrictions applied to indicate that this second spec now applies over the complete temperature ranges as listed in the note.

The general trend is toward tighter specs and more testing. Special high volume op amps have also become available that offer tighter specs at only a slight increase in cost. These are recommended for the more critical applications.

The statistical nature of the specs of components makes circuit performance also statistical. In many application circuits there is a large tolerance allowed for the specs of an op amp. It is also true that there are a few key circuits in every system that almost uniquely determine the overall accuracy of the system. An important part of system design is to realize where money can be saved by specifying low cost components with wide tolerances and to increase costs only in those areas that have a significant effect on the performance of the system.

Component suppliers can often be found that provide a low cost product that outperforms its data sheet. Use of these components in critical circuits can save money and make the designer appear very clever. If the components should, at some future time, *only* meet the data sheet limits, the component vendor is often cited as the bad guy. (Some form of incoming testing should be used to guard against this fateful event.) In general, automatic test systems will allow a part to pass if it is within the prescribed spec limits, even if it is only a relatively rare event. Therefore it is not wise to always count on receiving better-than-guaranteed parts. A test screen or an op amp with better specs should be used for the critical applications.

Typical Performance Characteristics

To give an indication of how the op amp will respond to a wide variety of conditions, a Typical Performance Characteristics section is included on the data sheet. This data is generally presented as a sequence of plotted multiparameter curves. The performance indicated in these curves is not guaranteed. It is supplied to show trends.

We will now consider some of the newer developments in op amps and speculate on the future of this favorite building block of the linear system designers.

CHAPTER VII

New Developments and the Future of Op Amps

It is natural to wonder what will happen to the future of the op amp: the popular SSI (small-scale integrated circuit) component of linear ICs. We can gain some insight into this speculation by looking at the history of digital circuits. Certainly, new applications for the digital SSI components are decreasing in favor of complete functions on LSI (large-scale integrated) and VLSI (very-large-scale integrated) circuit chips or even custom IC chips. This evolution to a more complex chip has also been slowly happening in the linear area and can be expected to accelerate in the years ahead.

The present ability of computers to design, layout, and test the next generation of computers is replacing the digital circuit designer and is a factor that favors using only digital circuits in the systems of the future. The layout of linear circuits involves many special geometries and is not of the highly regular, easily automated structures found in digital layouts. It is these (and other) complexities of linear circuit design that make automatic layout difficult and, unfortunately, will tend to cause traditional analog system solutions to continually give way to digital approaches as we pass further into the VLSI era.

The day is rapidly coming for complete linear or digital (or both) systems on a single chip.

This drive, to combine all of the electronics on one chip, causes the IC designers to look for those technologies that can combine both analog and digital circuits and not unreasonably compromise either. The performance benefits of the hybrid computer (mentioned in Chapter 1) have not yet been exploited on a silicon chip. Many modern system designers even go to great lengths to completely eliminate analog circuits. The move to I^2L (Integrated Injection Logic) was a bipolar process that combined digital circuitry with a linear chip and has allowed many useful products.

Today, the emphasis is on the MOS technologies. The concern, now, is how to add linear circuitry to these basic digital processes.

7-1 NEW CMOS LINEAR CIRCUITS

When a linear circuit designer looks at a CMOS process he finds excellent low cost analog switches, high-density circuitry (as a result of the small feature-size of the modern CMOS technologies, this has not yet appeared in the bipolar linear IC fabrication lines), excellent capacitors, parasitic bipolar-transistors, Zener diodes, and only relatively *mediocre complementary amplifying devices.*

The ground rules for analog design are therefore changing to the utilization of a new circuit component: *the analog switch*. This has already caused a shift toward *sampled-data systems*, where input signals are periodically sampled as opposed to the traditional *continuous systems* that have been the foundation of linear design. *The challenge today is to find new design approaches that make use of switches to accomplish linear functions. New VLSI chips are becoming available that combine linear sampled-data circuits with digital circuits on the same chip.*

Using Analog Switches

Problems arise when switches are used in the design of an op amp. For example, the presence of a local high-frequency clock that controls the sampling adds to the noise environment. Disturbances also tend to appear at the output of the op amp as a result of internal switching within the op amp. The designer of the traditional linear systems does not like to see all of this switching noise and it is too early to expect a significant shift to analog systems that operate in the sampled-data mode.

Once the application of sampled-data op amps is better understood, they will be found to offer advantages over the traditional linear op amps for some specialized circuits. Some of these sampled-data advantages are the ability to cancel offset voltage and also to eliminate drift in offset voltage. This is achieved with on-chip auto-zeroing or chopper-stabilized switching circuitry much like the chopper-stabilized op amp that was discussed in Chapter 1. (History is once again repeating itself.)

This sampling can also be used to cancel the relatively high flicker (1/f) noise of an MOS transistor. If the noise voltage is of low enough frequency that it is essentially constant between successive samples, it can be cancelled in the same manner as is used for dc drift reduction. (This is achieved by the use of a clock frequency that is at least two-times higher than the highest-frequency components of the noise that is to be cancelled.)

Improving Analog-to-Digital Converters

The design of one analog subsystem, the analog-to-digital converter, has been greatly simplified by the incorporation of a sampled-data approach. For this type of circuit, there can be periods of high background noise because the system can be designed to wait for a quiet spot in which to do the critical voltage comparisons. (This is not the case for a digital-to-analog converter, because these circuits are expected to provide a noise-free analog output voltage at all times.)

7.2 LINEAR MOS OP AMPS ON LSI CHIPS

The largest usage of the MOS op amp is expected to be as part of an MOS-LSI chip. Here, the design of the op amp can favor the important performance specs that are needed. The loading on the op amp is usually very light, often only a small valued capacitor has to be driven by these op amps.

ICs are available for the telecom marketplace and also general purpose, switched-capacitor filter circuits, both of which make use of multiple CMOS op amps on relatively large chips. The performance of these op amps, while adequate for the intended function, is not competitive, when all the specs are considered, with the bipolar op amps that are available today.

One of the design considerations for the high-performance CMOS op amps that were discussed in Chapter 1 was that they should make use of a standard digital CMOS process and not require any process changes just to support the op amps. Thus they are available to be used as standard cells to provide any required linear functions on LSI chips.

7.3 NEW POSSIBILITIES WITH BIPOLAR OP AMPS

The bipolar op amp is entering a relatively mature phase of development. We still find new circuits appearing to solve specialized application problems. For example, the LM11 bipolar op amp offers excellent dc input characteristics with input current even lower than is achieved with Bi-FETs and special low noise op amps are also available.

The appearance of the LF400 fast settling op amp came as a surprise to many observers. This has been considered an impossible design problem. It is hard to imagine improving this settling time by another order-of-magnitude, unless some new high-frequency active devices become available to the linear IC designers.

Two possibilities appear on the horizon that can improve the bipolar op amp. The first is to incorporate the small feature size (the ability of the modern high-resolution photolithography to pattern very small geometries) of the MOS processes into the bipolar linear fabrication lines. This will reduce the stray capacitances, reduce power drain, reduce input currents, and provide a higher-frequency op amp. It may also require low power supply voltages.

A second possibility is to move away from silicon and introduce gallium-arsenide (GaAs) op amps. This semiconductor material strangely allows only higher mobilities for the electrons (by a factor of 5) and therefore, N-channel GaAs JFETs (the MESFETs) and some newer ion-implanted MOSFETs can greatly increase the high-frequency capabilities of an op amp. [Gallium arsenide also makes low-leakage diodes (i_F or $i_R < 0.1$ pA @ ± 0.5V for an LED in the dark).]

There is much current research with gallium-arsenide. The relatively slow mobility of holes with this material reduces interest in bipolar transistors, except where high ambient operating temperatures are involved.

With the current trend toward more complex processes, a mixed technology (CMOS and bipolar) can also be expected in the future that will allow the design and performance benefits of both technologies on a single chip.

A more general replacement for linear circuits is possible with a Digital Signal Processing (DSP) approach where an analog voltage is periodically sampled, converted to a digital number, and then processed mathematically. The higher processing speeds being achieved in modern digital microprocessors and the continuing cost reductions for complex digital chips that include high-speed hardware multipliers are factors that will bring in this new DSP era.

Bibliography

1. Rogers, A. E., and T. W. Connolly: *Analog Computation in Engineering Design*, McGraw-Hill, New York, 1960.
   A good reference that details the, then, state-of-the-art in analog computation.

2. Meyer, R. G.: *Integrated-Circuit Operational Amplifiers*, IEEE Press, Wiley, New York, 1978.
   A collection of all of the basic papers on the IC op amp. Especially see: Solomon, J. E.: The Monolithic Op Amp: A Tutorial Study, pp. 12–30.

3. Thornton, R. D., C. L. Searle, D. O. Pederson, R. B. Adler, and E. J. Angelo, Jr.: *Multistage Transistor Circuits*, Vol. 5, Semiconductor Electronics Education Committee, Wiley, New York, 1965.
   One of a seven-volume book series. This collection is recommended as a good reference on basic semiconductor mechanisms, bipolar transistor action and circuit design.

4. Dostal, J.: *Operational Amplifiers*, Elsevier/North-Holland, Inc., New York, 1981.
   This book is highly recommended for all users of op amps. It is a valuable addition to the reference library.

5. Van Valkenburg, M. E.: *Introduction to Modern Network Synthesis*, Wiley, New York, 1960.
   A good discussion of the complex-frequency variable, poles and zeros and network functions. The text is mainly concerned with the synthesis of passive networks using s-plane techniques.

6. Smith, J. I.: *Modern Operational Circuit Design*, Wiley-Interscience, New York, 1971.
   A good book that discusses many useful op amp application circuits.

7. *Applications Manual for Computing Amplifiers*, Philbrick Researchers, Inc., Second Edition, June 1966.
   A good assortment of op amp application-circuits. May be hard to locate a copy of this today as it's become a collector's item.

8. Graeme, J. G.: *Applications of Operational Amplifiers*, McGraw-Hill, New York, 1973.
   Useful for additional application circuits.

9. Graeme, J. G., G. E. Tobey and L. P. Huelsman: *Operational Amplifiers, Design and Applications*, McGraw-Hill, New York, 1971.
   Contains application circuits and a good section on RC active-filters.

10. Johnson, D. E., J. R. Johnson, and H. P. Moore: *A Handbook of Active Filters*, Prentice-Hall, New Jersey, 1980.
A recommended text for RC active-filters. Especially useful for the design of many types of higher-order filters.

11. Roberge, J. K.: *Operational Amplifiers: Theory and Practice*, Wiley, New York, 1975.
A recommended text for applications of op amps.

12. Clayton, G. B.: *Linear Integrated Circuit Applications*, Lab Books, Blue Ridge Summit, Pennsylvania, 1977.
A useful collection of op amp application-circuits.

13. *ESD Training Manual*, Sept. 1980, NAVSEA SE 003-AA-TRN-010, Naval Publications and Forms Ctr., 5801 Tabor Ave., Philadelphia, PA 19120.
Recommended for increased information on electrostatic discharges.

14. *DOD-HDBK-263* and *DOD-STD-1686*, 2 May 1980, Department of Defense, Washington, DC 20360.
Recommended for increased information on electrostatic discharges.

15. Morrison, R.: *Grounding and Shielding Techniques in Instrumentation*, Second edition, Wiley, New York, 1977.
One of the rare books that addresses the practical problems of properly handling the many coupling, grounding, and shielding problems in building electronic hardware. Recommended for all who work with low-level analog signals.

Index

About the Author

Thomas M. Frederiksen, engineer, author, and seminar leader, founded his own company, Intuitive IC Seminars, to provide instruction to electronic design engineers and the many nonelectronically educated people who either work for, or with, a high-technology electronic company.

Upon earning his BSEE degree from California Polytechnic State University at San Luis Obispo, Frederiksen started his professional career as a development engineer with the Motorola Systems Development Laboratory. Subsequently he worked with the Microelectronics Group at Hughes Semiconductor Division and later became a senior project engineer at Motorola Semiconductor Products Division. Mr. Frederiksen then joined National Semiconductor Corporation, Santa Clara, California, where he developed custom ICs and standard single-supply building-block circuits for automotive and industrial applications. He designed the popular Quads: LM3900, LM324, and LM339. The LM324 is today's most popular operational amplifier. He has also been involved with analog-to-digital converters that will interface to microprocessors and other data acquisition circuits.

Mr. Frederiksen is the author of five books: *Intuitive IC Electronics, Intuitive CMOS Electronics,* and *Intuitive Op Amps* for engineers and technicians and *Intuitive Digital Computer Basics* and *Intuitive Analog Electronics,* which introduce the basics of electronics and also digital and analog circuits and systems, for nonelectronic professionals.

Mr. Frederiksen holds more than 40 patents on linear ICs and devices, has been a frequent contributor to the professional literature, and has given many major seminars on linear ICs for both Motorola and National Semiconductor in the United States and Canada and abroad. In 1977 he received the International Solid State Circuits Conference Best Paper Award.